New National Framework

9★

MATHEMATICS

M. J. Tipler   J. Douglas

Published in 2004 by:
Nelson Thornes Ltd
Delta Place
27 Bath Road
CHELTENHAM
GL53 7TH
United Kingdom

04 05 06 07 08 / 10 9 8 7 6 5 4 3 2 1

A catalogue record for this book is available from the British Library.

ISBN 0 7487 8614 7

Illustrations by Harry Venning
Page make-up by Mathematical Composition Setters Ltd

Printed and bound in Scotland by Scotprint

Acknowledgements

The publishers thank the following for permission to reproduce copyright material.

Digital Vision 7 (NT): 10; Corel 319 (NT): 35; Digital Stock 6 (NT): 35; Peter Adams/ Digital Vision BP (NT): 35; Jeremy Woodhouse/Digital Vision EP (NT): 35; Corel 447 (NT): 35; Casio: 41, 54; Corel 600 (NT): 119; Corel 74 (NT): 177; M.C. Escher's "Belvedere" © 2004 The M.C. Escher Company – Baarn – Holland. All rights reserved: 266; Corel 415 (NT): 307; Corel 62 (NT): 356; Corel 248 (NT): 356; Photodisc 66 (NT): 356; Photodisc 19 (NT): 357; Peter Adams/Digital Vision BP (NT): 362

Peter Adams/Digital Vision BP (NT): 11; Photodisc 54 (NT): 15; Digital Vision 6 (NT): 18; Digital Vision 15 (NT): 20; Corel 613 (NT): 23; Corel 667 (NT): 27; Casio: 65; Photodisc 19 (NT): 74; Corel 778 (NT): 79; Corel 600 (NT): 90; Stockbyte 29 (NT): 91; Corel 665 (NT): 154; Photodisc 72 (NT): 188; Corel 89 (NT): 257; M.C. Escher's "Belvedere" © 2004 The M.C. Escher Company – Baarn – Holland. All rights reserved: 269; Corel 1 (NT): 317; Corel 233 (NT): 368; Corel 62 (NT): 371; Corel 248 (NT): 371; Photodisc 66 (NT): 371; Digital Vision 12 (NT): 378; Corel 725 (NT): 394; Corel 495 (NT): 403

The publishers would like to thank QCA for permission to reproduce extracts from SATs papers.

The publishers have made every effort to contact copyright holders but apologise if any have been overlooked.

Contents

Contents

iv

Introduction

We hope that you enjoy using this book. There are some characters you will see in the chapters that are designed to help you work through the materials.

These are

 This is used when you are working with information.

 This is used where there are hints and tips for particular exercises.

 This is used where there are cross references.

 This is used where it is useful for you to remember something.

 These are blue in the section on number.

 These are green in the section on algebra.

 These are red in the section on shape, space and measures.

 These are yellow in the section on handling data.

Number Support

Place value and reading and writing numbers

Millions	Hundreds of thousands	Tens of thousands	Thousands	Hundreds	Tens	Units	tenths	hundredths	thousandths
4	0	0	6	5	0	2 •	3	0	8

This **place value chart** shows the number four million, six thousand, five hundred and two point three zero eight.

In 4 006 502·308 the 4 is 4 millions
                 the 6 is 6 thousands
                 the 5 is 5 hundreds
                 the 2 is 2 units
                 the 3 is 3 tenths
                 the 8 is 8 thousandths.

We use zeros not spaces as place holders.

17·384 is said as 'seventeen point three eight four'.
Sixteen and thirty-five hundredths is written as 16·35.

To **add or subtract 0·1**, add or subtract 1 to or from the tenths.

Examples   8·63 + 0·1 = 8·73   9·25 − 0·1 = 9·15

To **add or subtract 0·01**, add or subtract 1 to or from the hundredths.

Examples   5·837 + 0·01 = 5·847   6·357 − 0·01 = 6·347

Multiplying and dividing by 10, 100 and 1000

We use **place value to multiply and divide by 10, 100 and 1000**.
When we **multiply by 10, 100, 1000**, the digits move to the **left** by one place for 10, two places for 100 and three places for 1000.

Examples

5·7 × 10 = 57

7·23 × 100 = 723

When we **divide by 10, 100, 1000**, the digits move to the **right** by one place for 10, two places for 100 and three places for 1000.

Examples

96 ÷ 100 = 0·96

42·3 ÷ 100 = 0·423

Practice Questions **1, 2, 13, 25, 56**

Number

Putting numbers in order

To put **numbers in order** look at the digits with the same place value.
Start at the left.

Example   7546 > 7364
The thousands digits are the same.
The hundreds digits are 5 and 3.
    5 > 3
so 7546 > 7364

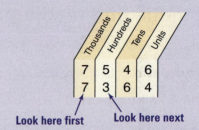

Look here first   Look here next

Example   8·59 < 8·63
The units digits are the same.
The tenths digits are 5 and 6.
5 < 6
so 8·59 < 8·63

Practice Questions 6, 8b, 10, 18, 55

Rounding

Rounding to the nearest 10, 100 or 1000

Example   6870 to the nearest 1000 is 7000.

6000   7000
                          6870

Numbers exactly halfway between are rounded up.

Example   465 to the nearest 10 is 470.

Rounding to the nearest whole number

To round to the **nearest whole number** we look at the digit in the tenths place.

Example   24·73 to the nearest whole number is 25.

The tenths digit is 7
so 24 becomes 25.

If the tenths digit is 5 or more, we
increase the units digit by one.

Rounding to one decimal place

To **round to one decimal place** we look at the digit in the hundredths place.

Example   4·53 to 1 d.p. is 4·5.
4·57 to 1 d.p. is 4·6.
4·55 to 1 d.p. is 4·6.

If the hundredths digit is
5 or more we increase
the tenths digit by one.

Practice Questions 8a, 33, 37

Integers

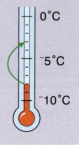

The numbers   ... , ⁻4, ⁻3, ⁻2, ⁻1, 0, 1, 2, 3, 4, ...   are called **integers**.

Example   The temperature is ⁻8 °C.
If it rises 5 °C, the temperature will then be ⁻3 °C.
    ⁻8 + 5 = ⁻3

We can **order integers** using a number line.

Example ⁻1, ⁻4, 2, 0, ⁻3 are shown on a number line.

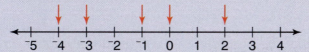

The further to the left a number is the smaller it is.

We can **add and subtract integers** using a number line.

Examples ⁻3 + 5 = 2

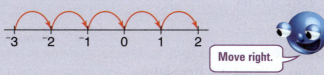

Move right.

3 + ⁻5 = ⁻2

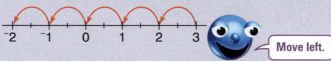

Move left.

3 − 5 = ⁻2

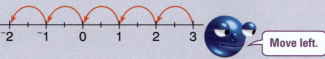

Move left.

3 − ⁻5 = 8

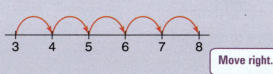

Move right.

Practice Questions 3, 11, 15, 22, 65

Divisibility

A number is **divisible by 2** if it is an even number.
A number is **divisible by 3** if the sum of its digits is divisible by 3.

Example 135 is divisible by 3 since 1 + 3 + 5 = 9 and 9 is divisible by 3.

A number is **divisible by 4** if the last two digits are divisible by 4.

Example 216 is divisible by 4 since 16 is divisible by 4.

A number is **divisible by 5** if its last digit is 0 or 5.
A number is **divisible by 6** if it is divisible by both 2 and 3.
A number is **divisible by 8** if half of it is divisible by 4.

Example Half of 96 = 48.   48 is divisible by 4 so 96 is divisible by 8.

A number is **divisible by 10** if the last digit is 0.

Practice Question 32

Multiples, factors, primes and squares

The **multiples** of a number are found by multiplying the number by 1, 2, 3, 4, 5, ...

Example   The multiples of 7 are 7, 14, 21, 28, 35, ...

The **factors** of a number are all of the numbers that will divide into that number leaving no remainder. We usually list them as pairs.

Example   The factor pairs of 36 are 1 and 36, 2 and 18, 3 and 12, 4 and 9, 6 and 6.

A **prime number** has exactly two factors, itself and 1.
1 is not a prime number.

A prime number can only be drawn as a rectangle that is a straight line of dots.

Example   7 can only be drawn as ●●●●●●● or ●
7 is a prime number.

4 can be drawn as ●●●● or ● or ●●
4 is not a prime number.

The prime numbers less than 30 are 2, 3, 5, 7, 11, 13, 17, 19, 23, 29.

Square numbers

5^2 is read as 'five squared' and means 5×5.

We get a square number when we multiply a whole number by itself.
The first 12 square numbers are:
$1^2 = 1,$ $\quad 2^2 = 4,$ $\quad 3^2 = 9,$ $\quad 4^2 = 16,$ $\quad 5^2 = 25,$ $\quad 6^2 = 36,$
$7^2 = 49,$ $\quad 8^2 = 64,$ $\quad 9^2 = 81,$ $\quad 10^2 = 100,$ $\quad 11^2 = 121,$ $\quad 12^2 = 144$

Practice Questions 4, 5, 7, 21, 31, 44, 57, 64

Mental calculation

We can use these ways to help us **add and subtract mentally**.

1 **Complements** in 1, 10 or 100

Examples   7 and 3 are complements in 10. $7 + 3 = 10$
55 and 45 are complements in 100. $55 + 45 = 100$
0·6 and 0·4 are complements in 1. $0·6 + 0·4 = 1$

> When two numbers add to 10 they are complements in 10.

2 **Partitioning**

Examples
$$23 + 54 = 20 + 3 + 50 + 4$$
$$= 20 + 50 + 3 + 4$$
$$= 70 + 7$$
$$= \mathbf{77}$$

$$49 - 26 = 49 - 20 - 6$$
$$= 29 - 6$$
$$= \mathbf{23}$$

3 Counting up

Example   460 – 280

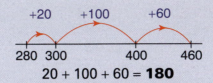

$$+20 \qquad +100 \qquad +60$$

$$280 \quad 300 \qquad\qquad 400 \qquad 460$$

$$20 + 100 + 60 = \mathbf{180}$$

4 Nearly numbers

Example $48 + 29 = (50 - 2) + (30 - 1)$
$\qquad\qquad\qquad = 50 + 30 - 2 - 1$
$\qquad\qquad\qquad = 80 - 2 - 1$
$\qquad\qquad\qquad = 78 - 1$
$\qquad\qquad\qquad = \mathbf{77}$

5 Adding or subtracting too much then compensating

Examples $127 + 38 = 127 + 40 - 2$
$\qquad\qquad\qquad = 167 - 2$
$\qquad\qquad\qquad = \mathbf{165}$

We can use these ways to **multiply and divide mentally**.

1 Partitioning

Example $12 \times 25 = (10 \times 25) + (2 \times 25)$ $\qquad 12 = 10 + 2$
$\qquad\qquad\qquad = 250 + 50$
$\qquad\qquad\qquad = \mathbf{300}$

2 Using factors

Example $32 \times 8 = 32 \times 2 \times 2 \times 2$ $\qquad 8 = 2 \times 2 \times 2$
$\qquad\qquad\qquad = 64 \times 2 \times 2$
$\qquad\qquad\qquad = 128 \times 2$
$\qquad\qquad\qquad = \mathbf{256}$

3 Doubling and halving

Example $26 \times 5 = 13 \times 10$
$\qquad\qquad\qquad = \mathbf{130}$

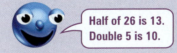

Half of 26 is 13.
Double 5 is 10.

Practice Questions 12, 14, 16, 17, 23, 26, 34, 35, 36, 41, 53, 58

Order of operations

We do operations in this order.

Brackets

↓

Squares (**I**ndices)

↓

Division and **M**ultiplication

↓

Addition and **S**ubtraction

Remember BIDMAS.

Examples

$$5 + 3 \times 2 = 5 + 6$$
$$= \mathbf{11}$$

Do first

$$3 + 5^2 - 6 = 3 + 25 - 6$$
$$= \mathbf{22}$$

Do first

$$3 \times (4 + 3) = 3 \times 7$$
$$= \mathbf{21}$$

Do first

Practice Questions **29, 40**

Estimating

An **estimate** is an approximate answer.
Always **estimate** the answer to a calculation first.

Example $8{\cdot}1 \times 3{\cdot}9$ is approximately $8 \times 4 = 32$

Practice Questions **9, 20**

Written calculation

We **add and subtract** by lining up digits with the same place value.
For decimals we line up the decimal points.
Always estimate the answer first.

Examples

```
   843
 + 791
  1634
    1
```

```
   5·68
  -3·49
   2·19
```

We can **multiply** like this.

Examples 186×8 is about $200 \times 8 = 1600$.

```
    186
 ×    8
   1488
    6 4
```

38×14 is about $40 \times 10 = 400$.

```
     38
 ×   14
    380      38 × 10
    152      38 × 4
    532
```

We can **divide** like this.

Example   208 ÷ 8 is about 200 ÷ 10 = 20.

8) 208
−160   8 × **20**
48
−48   8 × **6**
0

Answer **26**

Example   322 ÷ 14 is about 300 ÷ 15 = 20.

14) 322
−280   14 × **20**
42
−42   14 × **3**
0

Answer **23**

Practice Questions   30, 54

Using a calculator

When **using a calculator**, always check your answer.

Example   527 × 41   Key ⑤ ② ⑦ ⊗ ④ ① ⊜ to get 21 607
Check by:
- estimating   527 × 41 ≈ 500 × 40 = 20 000   ≈ means 'is approximately equal to'
- using inverse operations   21 607 ÷ 41 = 527

The inverse of multiplying is dividing.

Practice Question   39

Fractions, decimals and percentages

$\frac{5}{6}$ means 5 out of 6 or 5 ÷ 6 $\frac{5}{6}$ ← numerator
$\frac{5}{6}$ ← denominator

$\frac{5}{6}$ is called a **proper fraction**.

$\frac{27}{4}$ is called an **improper** fraction. The numerator is larger than the denominator.

$5\frac{1}{4}$ is called a **mixed** number.

We find **equivalent fractions** by multiplying or dividing both the numerator and denominator by the same number.

$\frac{2}{3}, \frac{4}{6}, \frac{6}{9}, \frac{8}{12}, \frac{10}{15}, \ldots$ are **equivalent fractions**.

Example

$\frac{4}{5}, \frac{8}{10}, \frac{20}{25}, \frac{24}{30}$ are equivalent fractions.

$\frac{4}{5}$ is a fraction in its simplest form.

4 and 5 have no common factor other than 1.

Number

When **writing one number as a fraction of another** make sure the numerator and denominator have the same units.

Example   19 hours as a fraction of a day is $\frac{19}{24}$.   **1 day is 24 hours.**

To **convert a decimal to a fraction**, write it with a denominator of 10 or 100.

Examples $0 \cdot 6 = \frac{6}{10}$

$= \frac{3}{5}$

Always simplify fractions.

$0 \cdot 45 = \frac{45}{100}$

$= \frac{9}{20}$

$$\frac{45}{100} \xrightarrow[\div 5]{\div 5} = \frac{9}{20}$$

To **convert a fraction to a decimal** write the fraction with a denominator of 10 or 100.

Examples

$$\frac{4}{5} \xrightarrow[\times 2]{\times 2} = \frac{8}{10}$$

$= 0 \cdot 8$

$$\frac{7}{20} \xrightarrow[\times 5]{\times 5} = \frac{35}{100}$$

$= 0 \cdot 35$

To **convert a fraction or a decimal to a percentage**, write it with a denominator of 100.

Examples

$$\frac{4}{25} \xrightarrow[\times 4]{\times 4} = \frac{16}{100}$$

$= 16\%$

$0 \cdot 37 = \frac{37}{100}$

$= 37\%$

To **convert a percentage to a decimal or fraction** write it as a number of parts per hundred.

Example $23\% = \frac{23}{100}$

$= 0 \cdot 23$

We can **add and subtract** simple **fractions** mentally.

Example $\frac{1}{4} + \frac{1}{2} = \frac{3}{4}$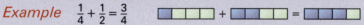

Example $\frac{7}{12} + \frac{1}{12} = \frac{7 + 1}{12}$

denominators the same

$= \frac{8}{12}$ **divide numerator and denominator by 4 to simplify.**

$= \frac{2}{3}$

We can find a **fraction of a quantity**.
In maths, 'of' means 'multiply'.

Examples $\frac{1}{5}$ of $35 = \frac{1}{5} \times 35$

$= 35 \div 5$

$= 7$

$\frac{2}{3}$ of $27 = 2 \times \frac{1}{3} \times 27$

$= 2 \times 9$

$= 18$

To find a **percentage of a quantity** remember these.

$10\% = \frac{1}{10}$ $5\% = $ half of 10% $1\% = \frac{1}{100}$ $25\% = \frac{1}{4}$

Examples 5% of $60 = 3$

15% of $80 = 12$

10% of 60 is 6
so 5% of 60 is 3
10% of 80 is 8
5% of 80 is 4
so 15% of 80 is 10% + 5%

Practice Questions 12a, 19, 24, 27, 28, 38, 42, 45, 46, 48, 49, 50, 51, 52, 59, 60, 61

Ratio and proportion

Proportion compares part to whole.
Ratio compares part to part.

Example   The ratio of red to yellow in this diagram is 2 : 3.

The proportion of red on the whole diagram is $\frac{2}{5}$ ⟵ red part
⟵ total number of parts

We write this as $\frac{2}{5}$ or 40% or 0·4.

We can **solve problems involving ratio and proportion**.

Example   7 packets cost £5.
So 14 packets cost twice as much, which is £10.

We can **divide in a given ratio**.

Example   There are 42 pupils to be divided into groups A and B in the ratio 4 : 3.
Out of every 7 pupils, 4 go into group A and 3 into group B.

$\frac{4}{7}$ of 42 = $4 \times \frac{1}{7} \times 42$ $\frac{3}{7}$ of 42 = $3 \times \frac{1}{7} \times 42$

$= 4 \times 6$ $= 3 \times 6$
$= 24$ $= 18$

24 go into group A.   18 go into group B.

Practice Questions 43, 47, 62, 63

Practice Questions   **Except for question 39.**

1   What is the value of the digit 6 in these?
   a   8642   **b**   6·53   **c**   7·163   **d**   5004·64

2   Write these as decimal numbers.
   a   five thousand and six point seven
   b   seventeen thousand, point five three
   c   six and nine hundredths
   d   two hundred and seven and eighty-four hundredths

T   **3**

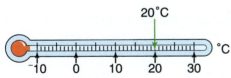

Use a copy of this.
   a   Draw an arrow by the thermometer to show 8 °C.
       Label your arrow 8 °C.
   b   Draw an arrow by the thermometer to show ⁻5 °C.
       Label your arrow ⁻5 °C.
   c   In York the temperature was ⁻3 °C.
       In London the temperature was 8 °C warmer.
       What was the temperature in London?
   d   Put   20 °C,   8 °C,   ⁻5 °C,   ⁻3 °C   in order, coldest first.

Number

4 Write down the factors of these.
    a 15   **b** 24   **c** 30

5 Write down the **factor pairs** of 24.

6 Which is bigger?
    a 3650 or 3560   **b** 23·05 or 23·5   **c** 46·207 or 46·027

7 I am a prime number between 0 and 10.
    I am a factor of 42.
    I am not a multiple of 3.
    What numbers could I be?

8 This table gives the area of some of the world's largest islands.
    a Round each area to the nearest 1000.
    b Put the islands in order from largest to smallest.

Island	Area in miles^2
Baffin Island	195 916
Great Britain	84 195
Greenland	839 852
Honshu	87 799
Sumatra	104 990

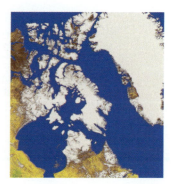

9 Which is the best approximation for these?
    a 20·8 − 7·2   **A** 20 − 7   **B** 21 − 8   **C** 208 − 72   **D** 21 − 7
    b 7·3 × 5·9   **A** 7 × 5   **B** 8 × 5   **C** 7 × 6   **D** 8 × 6

10 Here are some number cards:

    [1]   [7]   [3]   [5]

    You can use each card once to make the number 1735, like this:

    [1][7][3][5]

    a What is the **biggest** number you can make with the four cards?
    b Explain why you cannot make an even number with the four cards.
    c
    [1]   [7]   [3]   [5]

    Use some of these four number cards to make numbers that are as close as possible to these.
    i 50   **ii** 60   **iii** 4000

    Example   is the number you can make closest to 80.

11 a Ben noted these midnight temperatures.
Put them in order, coldest first.
⁻7 °C   ⁻3 °C   1 °C   4 °C   ⁻9 °C   0 °C

b Fran got these scores in six short quizzes.
Put them in order, lowest first.
⁻5, 4, ⁻2, ⁻3, 3, 0

12 a Find the missing numbers.
18 + ___ = 27   100 − ___ = 29
50% of ___ = 16   a quarter of ___ = 20

b What numbers could go in each space to make the calculations correct?
___ × ___ = 25
___ ÷ ___ = 25

13 Calculate these.
a 69×10 **b** 12×100 **c** $390 \div 10$ **d** $8300 \div 100$ **e** $1{\cdot}2 \times 10$
f $1{\cdot}9 \times 100$ **g** $36 \div 10$ **h** $5{\cdot}8 \div 10$ **i** $3{\cdot}6 \times 1000$ **j** $472 \div 100$

14 Use complements to find the missing numbers.
a $4 + \square = 10$ **b** $37 + \square = 100$ **c** $100 - \square = 56$
d $0{\cdot}3 + \square = 1$ **e** $1 - \square = 0{\cdot}8$

15 The temperature at midnight was ⁻15 °C.
How much has it risen if it is now
a 0 °C   **b** ⁻8 °C   **c** 10 °C?

16 Find the answers mentally.
a $5 + 8 + 16$ **b** $18 - 9 + 3$ **c** $80 - 20 - 10$ **d** $90 + 40 - 40$
e $40 + 30 + 12$ **f** $80 + 30 - 7$ **g** $42 + 30$ **h** $94 - 40$
i $83 - 29$ **j** $67 + 19$ **k** $47 + 17$ **l** $46 + 4 + 31 + 9$
m $41 - 19$ **n** $175 + 26$ **o** $196 - 84$ **p** $295 - 132$

17 $5{\cdot}7 + 3{\cdot}8 = 9{\cdot}5$
What is the answer to $9{\cdot}5 - 3{\cdot}8$?

18 A class was divided into five groups to do an
experiment.
Each group weighed some powder.
These are the masses.
   1·03 kg,   1·2 kg,   1·8 kg,   1·08 kg,   1·3 kg
Put the masses in order from heaviest to lightest.

Number

19 Find the missing numbers that go in the boxes.

a $\dfrac{3}{5} \xrightarrow{\times 4} = \xrightarrow[\times 4]{} \dfrac{12}{\Box}$ b $\dfrac{2}{3} \xrightarrow{\times 4} = \xrightarrow[\times 4]{} \dfrac{\Box}{12}$ c $\dfrac{24}{30} \xrightarrow{\div 6} = \xrightarrow[\div 6]{} \dfrac{\Box}{5}$

d $\dfrac{15}{50} = \dfrac{3}{\Box}$ e $\dfrac{2}{3} = \dfrac{\Box}{30}$ f $\dfrac{42}{36} = \dfrac{\Box}{6}$ g $1 = \dfrac{\Box}{9}$

20 Estimate the answers to these.

a $91 + 89$ b $572 - 302$ c 98×22 d $92 \div 31$
e $4 \cdot 8 + 3 \cdot 02$ f $19 \cdot 4 - 5 \cdot 8$ g $4 \cdot 7 \times 21$

21 Write down the first six square numbers.

22 London is 30 m *above* sea level.
The Dead Sea is 400 m *below* sea level.
What is the difference in height between London and the Dead Sea?

23 Here is the 65 times table.
Use the 65 times table to help you fill in the
missing numbers.

$65 \times 5 = \underline{\quad}$
$390 \div 65 = \underline{\quad}$
$12 \times 65 = \underline{\quad}$
$20 \times 65 = \underline{\quad}$

$1 \times 65 = 65$	$6 \times 65 = 390$
$2 \times 65 = 130$	$7 \times 65 = 455$
$3 \times 65 = 195$	$8 \times 65 = 520$
$4 \times 65 = 260$	$9 \times 65 = 585$
$5 \times 65 = 325$	$10 \times 65 = 650$

24 What fraction of

a   1 metre is 57 centimetres   b   1 hour is 24 minutes?

25 a   Adam filled 10 test tubes with 16·5 mℓ of solution.
How much solution did he need altogether?
b   Mao measured the thickness of 10 sheets of card as
56 mm.
How thick is each sheet?

26 Rachel filled each of 6 tubes with 1·2 mℓ of acid.
How much acid did she need altogether?

27 Change these to mixed numbers.

a $\dfrac{7}{4}$ b $\dfrac{8}{5}$ c $\dfrac{14}{8}$ d $\dfrac{32}{7}$ e $\dfrac{45}{8}$

28 Change these to improper fractions.

a $1\dfrac{1}{2}$ b $1\dfrac{2}{3}$ c $3\dfrac{4}{5}$ d $6\dfrac{5}{6}$ e $10\dfrac{5}{8}$

29 Calculate these mentally.
  a $4 \times 7 - 3$ **b** $8 + 3 \times 4$ **c** $16 - 2 \times 7$ **d** $\frac{9+7}{8}$ **e** $\frac{36}{3 \times 2}$
  f $3(16 - 12)$ **g** $5(25 - 15)$ **h** $\frac{35 - 10}{5}$ **i** 2×3^2 **j** $8^2 - 10$
  k $4^2 + 9$ **l** $7^2 - 3^2$

'Mentally' means 'in your head'.

30 Calculate these.
  a $76 + 93$ **b** $283 - 51$ **c** $817 + 564$ **d** $588 - 97$ **e** $982 + 67$
  f $179 + 87$ **g** $402 - 53$ **h** $8325 - 1214$ **i** $8280 + 469$
  j $6217 - 364$ **k** $4 \cdot 36 + 2 \cdot 79$ **l** $7 \cdot 96 - 3 \cdot 42$ **m** $5 \cdot 87 - 1 \cdot 34$

[T]

31 Use a copy of this.
  Shade
  a the prime numbers
  b the numbers that have 3 as a factor
  c the factors of 28.
  d the multiples of 7.
  What shape does the shading make?

65	100	76	82	92	44
87	29	15	2	42	99
35	6	28	60	11	19
300	21	17	7	30	33
77	4	1	63	72	14
110	55	74	40	200	76

[T]

32 Use a copy of this.
  Shade triangles with numbers that are
  a divisible by 10
  b divisible by 5
  c divisible by 4
  d divisible by 3.
  What shape does the shading make?

70	273 / 470	840
65	747 / 830	56
864	295 / 836	312
587	31 / 869	277

33 The population of Guildford to the nearest thousand is 127 000.
  What is the population to the nearest 10 000?

34 Bevan bought three sausages, two scoops of chips
  and a burger.
  a How much did this cost?
  b How much change did he get from £10?
  Paula bought 4 burgers.
  She was asked to pay £10·40.
  c How can you tell this is wrong without doing the calculation?

Chips 70p
Sausages £1·10
Burgers £1·40

35 Find the answers to these mentally.
  a $0 \cdot 8 + 0 \cdot 6$ **b** $0 \cdot 9 + 0 \cdot 7$ **c** $0 \cdot 5 + 0 \cdot 9$ **d** $1 \cdot 2 - 0 \cdot 4$ **e** $7 \cdot 8 - 1 \cdot 2$

36 What must be added or subtracted to or from 19·7 g of salt to get
  a 19·9 g   **b** 18·5 g   **c** 19·6 g   **d** 20·7 g?

Number

37 Round these numbers.
   a 369 to the nearest 100
   b 4195 to the nearest 1000
   c 56·4 to the nearest whole number
   d 43·56 to 1 d.p.

38 Write these fractions in their lowest terms.
   The first two are started for you.

   a
   $$\frac{4}{8} \overset{\div 4}{\underset{\div 4}{=}} —$$
   b
   $$\frac{7}{21} \overset{\div 7}{\underset{\div 7}{=}} —$$
   c $\frac{2}{8}$ **d** $\frac{10}{15}$ **e** $\frac{80}{100}$ **f** $\frac{18}{24}$

39 Use your calculator to find these.
   a 426×72 **b** $836 + 794$ **c** $5937 - 1698$
   Check your answers.
   Explain how you checked them.

40 **a** Write the answers. $(4 + 2) \times 3 =$ ___
                          $4 + (2 \times 3) =$ ___
   b Work out the answer to $(2 + 4) \times (6 + 3)$.
   c Put brackets in the calculation to make the answer 50.
       $4 + 5 + 1 \times 5$
   d Now put brackets in the calculation to make the answer 34.
       $4 + 5 + 1 \times 5$

41 Find the answers to these mentally.
   a $39 \div 3$ **b** $950 \div 2$ **c** $144 \div 8$ **d** $5·6 \div 8$ **e** $3·6 \div 6$

42 Find the answers to these.
   a $\frac{1}{2} + \frac{1}{2}$ **b** $\frac{1}{2} - \frac{1}{4}$ **c** $\frac{3}{4} + \frac{3}{4}$ **d** $\frac{2}{5} + \frac{2}{5}$ **e** $\frac{11}{12} - \frac{5}{12}$

43 **a** Five apples cost £2.
       How much do fifteen apples cost?
   b Three oranges cost £1·10.
       How much do nine oranges cost?

44 What goes in the gaps?
   a 3^2 is ___   **b** 16 is ___ squared   **c** six squared is ___
   d 100 is ___ squared **e** 5^2 is ___   **f** seven squared is __

45 Write these as decimals.
   Some have been started.
   a $\frac{1}{2} = \frac{\square}{10} =$ **b** $\frac{1}{4} = \frac{\square}{100} =$ **c** $\frac{1}{10}$ **d** $\frac{1}{100}$
   e $\frac{3}{10}$ **f** $\frac{39}{100}$ **g** $\frac{6}{100}$ **h** $60\% = \frac{60}{100} =$
   i 80% **j** 75% **k** 35%

You want the denominator of the fraction to be 10 or 100.

46 Write these as fractions in their simplest form.

   a $0.3 = \frac{3}{\square}$ **b** $0.7 = \frac{\square}{10}$ **c** $0.8 = \frac{\square}{10} = \frac{\square}{5}$ **d** 0.5

   e 0.6 **f** $0.41 = \frac{41}{\square}$ **g** $0.77 = \frac{\square}{100}$ **h** 0.83

   i 0.25 **j** 0.14

> Write it first as a fraction with denominator 10 or 100.

47 On a class trip there were 5 pupils for each adult.
   a What is the *ratio* of pupils to adults on the trip?
   b What is the *proportion* of pupils on the trip?

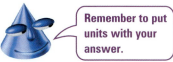

> Remember to put units with your answer.

48 Do these mentally.
   a $\frac{1}{10}$ of £50   **b** $\frac{1}{5}$ of 25 m   **c** $\frac{1}{7}$ of 14 cm   **d** $\frac{3}{4}$ of 16 ℓ

   e $\frac{2}{3}$ of £24   **f** $\frac{5}{6}$ of 36 cm   **g** $\frac{3}{8}$ of 32 g

49 Bess got 17 questions out of 25 correct in her science test.
   a What fraction did she get correct?
   b What percentage did she get in her test?

50 **a** I am equivalent to $\frac{1}{2}$.

      My denominator is 3 more than my numerator.
      What fraction am I?

   b I am equivalent to $\frac{12}{32}$.

      My numerator is a prime number.
      What fraction am I?

51 Find a percentage from the box for each sentence.
   a 7 out of 25 students surveyed liked comedy better than dramas
      on TV.
   b 6 out of 20 students surveyed liked maths better than science.

                                                                                             25% 28%
                                                                                                 40% 30%

52 Find the answers to these mentally.
   a 10% of 60   **b** 50% of 110   **c** 25% of 160   **d** 50% of 4·8
  ∗e 15% of 50

53 Find the answers to these mentally.
   a 72 ÷ 4   **b** 95 ÷ 5   **c** 246 ÷ 6   **d** 486 ÷ 9
   e 714 ÷ 7   **f** 708 ÷ 3   **g** 2008 ÷ 8

54 Find the answers to these.

   a   321   **b**   864   **c**   52   **d**   82
        ×   9   ×   7   ×12   ×14
        ———   ———   ———   ———

   e   327   **f** 387 ÷ 9   **g** 384 ÷ 16
        × 24
        ———

55

   a Make the largest number you can with these cards.
There must be at least one digit after the decimal point.
   b Make the second largest number with at least one digit after the decimal point.

56 Cam was building a set for a play.
He wrote down a measurement as 4·83 m.
What should it have been if the measurement he wrote down is
   a 0·1 m too long
   b 0·01 m too short
   c 0·1 m too short
   d 0·01 m too long?

57 **a** The numbers 846, 864 and 486 all have the digits 4, 6 and 8.
What other numbers can be made with these three digits?
   b What is the largest multiple of 4 that you can make from the digits 4, 6 and 8?

58 Calculate these mentally.
   a $2 \times 5 \times 6$ **b** $3 \times 4 \times 5$ **c** 120×3 **d** 30×20 **e** 48×2
   f 32×5 **g** 82×6 **h** $1 \cdot 2 \times 3$ **i** $4 \cdot 6 \times 4$ **j** $9 \times 0 \cdot 7$

59 Write these as percentages.
   a $\frac{33}{100}$ **b** $\frac{7}{10} = \frac{\square}{100} =$ **c** $\frac{1}{4} = \frac{\square}{100} =$ **d** $\frac{7}{20}$
   e $0 \cdot 73 = \frac{\square}{100}$ **f** $0 \cdot 04$ **g** $0 \cdot 19$

60 There are 40 pupils on a geography field trip.
60% have paid.
How many have paid?

61 Write the equivalent fraction that has a denominator of 12.
   a $\frac{1}{2}$ **b** $\frac{3}{4}$ **c** $\frac{2}{3}$ **d** $\frac{5}{6}$

62 **a** What is the ratio of fertiliser to water in the mix?
   b What proportion of the mix is fertiliser?
Give your answer as a percentage.
   c What proportion of the mix is water?
Give your answer as a fraction.
  *d If you made up 10 litres of Plant Food Mix,
how much of it would be fertiliser?

> **Plant Food Mix**
> 1 litre of fertiliser
> 4 litres of water

63 Matt and Olly share £45 in the ratio 2 : 3.
  a What fraction of the money does Matt get?
  b How much money does Matt get?

64 Two prime numbers are subtracted.
The answer is 4.
What could the numbers be?
Find as many answers as possible.

Only use the prime numbers less than 50.

65 Pia has these number cards.

  a She has to choose a card to give the answer 4.
    What does she choose?
  b She has to choose two cards to give the lowest possible total.
    What is this total?

66 Jamie and Simon were having a race to put the numbers from ⁻3 to 5 in the circles so that each line has the total given.
Where could they put the numbers?
  a 3

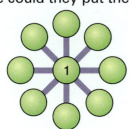

  b 6

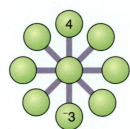

1 Place Value, Ordering and Rounding

You need to know

✓ place value   page 1
✓ multiplying and dividing by 10, 100 and 1000   page 1
✓ putting whole numbers in order   page 2
✓ rounding   page 2

Key vocabulary

ascending, billion, descending, index, power

 Planet X

On Planet X there are only five number symbols.

When they get to 5 they have no more symbols.

They have to write 5 as 1 lot of 5 and 0 ones.

This is how some numbers are written.

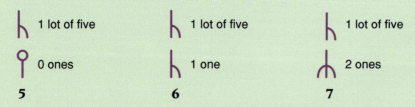

Write these numbers in the Planet X number system.

8   9   10

Powers of ten

$10 = 10^1$ ← This is called an index.

$100 = 10 \times 10 = 10^2$

$1000 = 10 \times 10 \times 10 = 10^3$

$10\,000 = 10 \times 10 \times 10 \times 10 = 10^4$

$10^1, 10^2, 10^3, 10^4, \ldots$ are called **powers of ten**

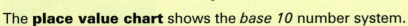

Notice that 1000 has **3** zeros and we write it as 10^3.

The **place value chart** shows the *base 10* number system.
Each place to the left in a number is one power of ten bigger.

	Millions			Thousands			Hundreds		
(Thousands of millions) Billions	Hundreds of millions	Tens of millions	Millions	Hundreds of thousands	Tens of thousands	Thousands	Hundreds	Tens	Units
10^9	10^8	10^7	10^6	10^5	10^4	10^3	10^2	10^1	1

Exercise 1

1 Write these without the index.
The first two have been started.

a $10^2 = 10 \times 10$ **b** $10^4 = \underline{} \times \underline{} \times \underline{} \times \underline{}$

     $= \underline{}$ $= \underline{}$

c 10^1 **d** 10^3 **e** 10^8 **f** 10^9 **g** 10^{11}

2 Write these as powers of ten.
The first two have been started.

a one thousand $= 1000$ **b** one million $= 1\,000\,000$

               $= 10^?$ $= 10^?$

c one hundred   **d** one hundred thousand   **e** ten million

*f one billion   ***g** ten   ***h** ten billion

3 Write the numbers in red in words.
The first one has been done.

a A heart beats about **37 000 000** times each
year.   **thirty-seven million**

b Most people blink about **5 625 000** times each
year.

c One megabyte is **1 048 576** bytes.

d Africa has an area of **30 271 000** km^2.

e The galaxy NGC2207 is **114 100 000** light years
from the Earth.

19

Number

T

*4 Use a copy of this.
What number goes in the gap?
Shade the answer in the grid.
What letter does the shading make?

1000	8	10
10^3	100 000	10^5
100	7	10^6
10 000 000	4	10 000

a 10 000 = 10—   b 10^3 = _____
c one hundred million = 10—
d 10^5 = _____   e 10 000 000 = 10—
*f ten billion = 10—

The Earth is one hundred and forty-nine million, six hundred thousand kilometres from the Sun.

We write this as 149 600 000 km or as $1·496 \times 10^8$ km.

We say $1·496 \times 10^8$ as 'one point four nine six times ten to the power of eight'.

Worked Example
Write, in figures, the number that is 4 more than a half million.

Answer
A half million is   500 000.
4 more than this is   **500 004**.

Exercise 2

1 Write the numbers in red in words.
The first one is done.
   a Saturn's rings are **$2·7 \times 10^5$** km in diameter.
      two point seven times ten to the power of five
   b Jupiter's largest moon is **$5·262 \times 10^3$** km in diameter.
   c Jupiter's largest moon is **$1·07 \times 10^6$** km from Jupiter.
   d Saturn is **$1·427 \times 10^9$** km from the Sun.

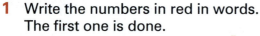
This is linked to science.

*2 Write in figures the number that is
   a 10 more than three hundred and fifty
   b 3 more than twelve thousand
   c 1 less than four thousand
   d 10 more than two and a half thousand
   e 12 more than half a million.

 Puzzle

1 Jackie chose three of these number cards.
Using her three cards she made the biggest number she could. Then she made the smallest number she could.
She added these together and got 848.
What were the three cards she chose?
Is there more than one answer?

`1` `2` `6` `7` `9`

Choose three cards and check if the biggest and smallest numbers add to 848. If not choose three more.

*2 I am a 6-digit number between 400 and 500.
All of my digits are different.
The difference between my units digit and my
thousandths digit is the same as the difference
between my tens digit and my hundredths digit.
My tenths digit is three times my hundredths digit.
My units digit is twice my tenths digit.
The sum of my digits is 21.
What number am I?

My first digit
must be 4.

Multiplying and dividing by 10, 100 and 1000

Remember

To **multiply by 10**, move each digit one
place to the left.
To **multiply by 100**, move each digit
two places to the left.
To **multiply by 1000**, move each digit
three places to the left.

4·63 × 1000 = 4630

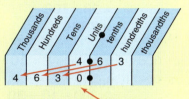

We use a zero rather than a space to
show there are no units. The zero is
called a place holder.

To **divide by 10**, move each digit one
place to the right.
To **divide by 100**, move each digit two
places to the right.
To **divide by 1000**, move each digit three
places to the right.

17 ÷ 100 = 0·7

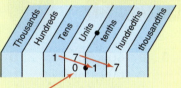

This zero stops the decimal
point from getting lost.

Exercise 3

1 Find the answers to these.
  a 42×10
  b 42×100
  c $42 \div 10$
  d $42 \div 1000$
  e $1·4 \times 100$
  f $19·6 \div 10$
  g $432 \div 1000$
  h $4·02 \times 100$
  i $5 \div 100$
  j $78·6 \times 1000$
  k $93·421 \div 100$
  l $0·007 \times 100$

2 Which of 10, 100 or 1000 goes in the box?
  a $54 \times \square = 540$
  b $3120 \div \square = 312$
  c $9 \div \square = 0·9$
  d $72 \div \square = 7·2$
  e $3·5 \times \square = 350$
  f $642 \div \square = 6·42$
  g $9 \div \square = 0·009$
  h $0·03 \times \square = 30$
  i $4·5 \div \square = 0·0045$

Number

3 Jon used these digit cards.
He made the number 43·2.

   a Make the number 10 times bigger than Jon's.
   b Make the number 10 times smaller than Jon's.
   c Make the number 1000 times bigger than Jon's.
   d Make the number 1000 times smaller than Jon's.

We can use place value to multiply by **multiples of 10, 100 and 1000**.

Worked Example
Find these.   **a** 40×70 **b** $2·4 \times 20$ **c** $600 \div 20$

Answer

a
$$40 \times 70 = 4 \times 10 \times 7 \times 10$$
$$= 4 \times 7 \times 10 \times 10$$
$$= 28 \times 100$$
$$= \mathbf{2800}$$

b
$$2·4 \times 20 = 2·4 \times 2 \times 10$$
$$= 4·8 \times 10$$
$$= \mathbf{480}$$

or

	2	0·4
2	4	0·8

Answer = **4·8**

$4·8 \times 10 = 48$

c
$600 \div 20 = 600 \div 2 \div 10 = 30$
because $600 \div 2 = 300$
$300 \div 10 = \mathbf{30}$

or

$600 \div 2 = 300$
$600 \div 20 = \mathbf{30}$

Exercise 4		**This exercise is to be done mentally.** **You may use jottings.**

1 **a** 40×30 **b** 50×60 **c** 200×40 **d** $900 \div 30$
    e $800 \div 400$ **f** 300×600 **g** $60\,000 \div 300$ ✱**h** $2000 \div 50$
  ✱**i** $20 \times 70\,000$ ✱**j** $600 \div 2000$

2 **a** 41×20 **b** 34×20 **c** $3·3 \times 30$ **d** $5·4 \times 20$
    e $840 \div 20$ ✱**f** $46 \div 20$ ✱**g** $5·3 \times 30$ ✱**h** $9·6 \div 30$

3 Jamie can type 20 words in one minute.
    How many words can Jamie type in
    a 30 minutes   ✱**b** 1 hour 10 minutes?

4 Angela trained on a 400 m track.
    Each week she ran a total of 8000 m.
    How many laps of the track did Angela run each week?

✱**5** What numbers could go in the gaps?
    ___ × ___ = 800   ___ × ___ = 800   ___ × ___ = 800

✱**6** What numbers could go in these gaps?
    ___ ÷ ___ = 20   ___ ÷ ___ = 20

Multiplying and dividing by 0·1 and 0·01

Investigation

× and ÷ by 0·1 and 0·01

You will need a calculator or spreadsheet package.

Fill in a copy of the table.
Use a calculator or spreadsheet to help.

Number	× 10	÷ 0·1	× 100	÷ 0·01	÷ 10	× 0·1	÷ 100	× 0·01
4·7								
0·8								
0·92								
164								

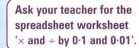

Ask your teacher for the spreadsheet worksheet '× and ÷ by 0·1 and 0·01'.

What do you notice about multiplying by 10 and dividing by 0·1?
What about multiplying by 100 and dividing by 0·01?
What about dividing by 10 and multiplying by 0·1?
What about dividing by 100 and multiplying by 0·01?

Discussion

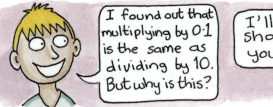

I found out that multiplying by 0·1 is the same as dividing by 10. But why is this?

I'll show you.

Robert

Nishi

Nishi wrote:

$$5·6 \times 0·1 = 5·6 \times \frac{1}{10}$$
$$= 5·6 \div 10$$
so 5·6 × 0·1 is the same as 5·6 ÷ 10.

Remember:
$0·1 = \frac{1}{10}$.

Use Nishi's way to explain why multiplying by 0·01 is the same as dividing by 100.

Multiplying by 0·1 is the same as dividing by 10.
Multiplying by 0·01 is the same as dividing by 100.
Dividing by 0·1 is the same as multiplying by 10.
Dividing by 0·01 is the same as multiplying by 100.

Number

Exercise 5   **This exercise is to be done mentally.**

1 **a** Which of these gives the same answer as $8 \div 0 \cdot 1$?
          A 8×1 **B** $8 \times 0 \cdot 1$ **C** 8×10 **D** 8×100
    b Which of these gives the same answer as $14 \times 0 \cdot 1$?
          A 14×100 **B** $14 \div 10$ **C** $14 \div 100$ **D** 14×10
    c Which of these gives the same answer as $832 \div 0 \cdot 01$?
          A 832×100 **B** $832 \div 10$ **C** 832×10 **D** $832 \div 100$
    d Which of these gives the same answer as $4 \cdot 8 \div 0 \cdot 1$?
          A $4 \cdot 8 \times 10$ **B** $4 \cdot 8 \div 10$ **C** $4 \cdot 8 \div 100$ **D** $4 \cdot 8 \times 100$
    e Which of these gives the same answer as $1 \cdot 6 \times 0 \cdot 1$?
          A $1 \cdot 6 \times 10$ **B** $1 \cdot 6 \div 10$ **C** $1 \cdot 6 \times 100$ **D** $1 \cdot 6 \div 0 \cdot 1$
    f Which of these gives the same answer as $0 \cdot 024 \times 0 \cdot 1$?
          A $0 \cdot 024 \times 10$ **B** $0 \cdot 024 \div 100$ **C** $0 \cdot 024 \div 10$ **D** $0 \cdot 024 \div 0 \cdot 1$

2 **a** $8 \times 0 \cdot 1$ **b** $32 \times 0 \cdot 1$ **c** $50 \div 0 \cdot 1$ **d** $17 \times 0 \cdot 1$
    e $832 \times 0 \cdot 01$ **f** $1 \cdot 7 \times 0 \cdot 01$ **g** $0 \cdot 86 \div 0 \cdot 1$ **h** $4 \cdot 8 \div 0 \cdot 01$

3 In a science experiment, Lyle filled six beakers with $0 \cdot 01$ ℓ of a salty solution.
    How much salty solution did he use altogether?

*4 Write down the next 3 lines of these patterns.
    a $3 \times 1 = 3$ **b** $7 \times 1 = 7$
            $3 \times 0 \cdot 1 = 0 \cdot 3$ $7 \times 0 \cdot 1 = 0 \cdot 7$
            $3 \times 0 \cdot 01 = 0 \cdot 03$ $7 \times 0 \cdot 01 = 0 \cdot 07$

    c $5 \times 4 = 20$ **d** $7 \times 3 = 21$
            $0 \cdot 5 \times 4 = 2$ $7 \times 0 \cdot 3 = 2 \cdot 1$
            $0 \cdot 05 \times 4 = 0 \cdot 2$ $7 \times 0 \cdot 03 = 0 \cdot 21$

Putting decimals in order

The two longest jumps at sports day were 2·034 m and 2·043 m.
Martin wants to know which is the longest jump.

Start at the left.
The units and tenths digits are the same.
Look at the hundredths digits.
    2·043 > 2·034.   4 > 3
2·043 m was the longest jump.

Worked Example
Which of < or > goes between these?
a 0·267   0·289   b 8·106 kg   8109 g

Answer
a Starting at the left the units and tenths digits are the same.
   Look at the hundredths digits.
   8 > 6
   So **0·267** < 0·289.

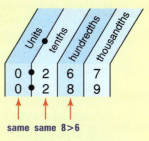

b To compare measurements the units must be the same.
       1 kg = 1000 g   so   8·106 kg = 8106 g.
   Starting at the left, the first digits which are different are the
   thousandths.
   9 > 6
   So **8109 g** > **8·106 kg**.

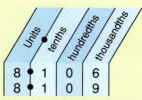

Exercise 6

1 Which is smaller?
   a 0·7 or 0·9   b 0·35 or 0·42   c 0·47 or 0·42
   d 1·65 or 1·56   e 4·27 or 4·279   f 16·405 or 16·045
   g 0·037 or 0·073   h 10·012 or 10·102

2

					C		
1·247	0·512	1·247	0·512	0·511	**0·83**	1·247	0·512
1·240	10·64	10·61	0·511	**50**	0·511	10·64	0·083	6·53	6·35
10·64	0·511	6·35	0·083	1·247	6·35	6·35

Use a copy of the box.
Write the letter beside each question above its answer in the box.
Which is bigger?
C 0·81 or 0·83 = **0·83**   I 10·64 or 10·61   E 6·53 or 6·35
S 6·35 or 3·65   M 0·038 or 0·083   F 10·06 or 10·61
N 0·511 or 0·512   T 0·501 or 0·511   A 1·247 or 1·240
L 1·240 or 1·204

Number

3 Are these true or false?

    a 0·86 > 0·76   **b** 0·1051 < 0·0991   **c** 0·6854 < 0·6872   **d** 26·31 < 26·301

4 **a** This list gives the masses, in kilograms, of mice in a 'cutest mouse' contest. Put the masses in ascending order.

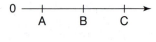
Ascending means from smallest to largest.

        0·081 kg   0·128 kg   0·11 kg   0·082 kg   0·08 kg

    b This list gives the distance in metres, that 6 boys threw a slipper in a contest.

'Descending' means from largest to smallest.

        Put the distances in descending order.

        20·45 m   21·017 m   24·592 m   21·07 m   20·543 m

∗5 Which of these numbers goes at A, at B and at C on the number line?

    7·234   7·259   7·22

0   A   B   C

To compare measurements the units must be the same.

1 m = 1000 mm
1 km = 1000 m
1 cm = 10 mm
1 m = 100 cm
1 ℓ = 1000 mℓ
1 kg = 1000 g

∗6 Which of < or > goes in the box?

    a 7·2 m ☐ 7204 mm   **b** 4·68 km ☐ 497 m

    c 7·63 cm ☐ 768 mm   **d** 9·04 ℓ ☐ 909 mℓ

    e 5·029 m ☐ 52·79 cm   **f** 5869 g ☐ 0·5896 kg

∗7 **a** Use all of the cards to make the largest decimal number.
There must be at least one digit, which is not 0, after the decimal point.

    b Use all of the cards to make the smallest decimal number.
There must be at least one digit, which is not 0, before the decimal point.

5   4   9   0   ·

5   2   0   0   ·

∗8   2·7 < 2·8 **but** ⁻2·7 > ⁻2·8

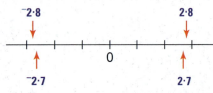
⁻2·8   2·8

    0

⁻2·7   2·7

Remember: The further left a number is on the number line the smaller it is.

Does < or > go in these boxes?

    a 2·2 ☐ ⁻2·1   **b** ⁻3·6 ☐ 3·6   **c** 0 ☐ ⁻4·5

    d ⁻1·2 ☐ ⁻9   **e** ⁻1·2 ☐ ⁻2·1   **f** ⁻10·3 ☐ ⁻13·1

*9 Which number is halfway between these?

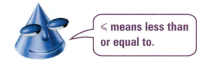

a 7·4 and 7·6   b 6·2 and 6·8   c 2·8 and 3·4
d 3·25 and 3·27   e 7·2 and 7·3   f 0·8 and 0·9

Use a number line to help.

*10   1·52 ⩽ x ⩽ 1·62
Give 10 values that x could have.

⩽ means less than or equal to.

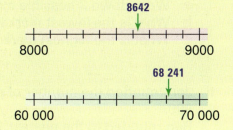

Guess my number – a game for a group

To play

- Choose a leader.
- The leader writes down a four-digit decimal number so no one else can see it.
- The others in the group try to guess it. They take turns to ask the leader questions which begin 'Is it bigger/smaller than ...?' The leader can only answer Yes or No.

  Example Sam was a leader.
  He wrote 98·64.
  He was asked these questions.

- The person who guesses the number is the new leader.

Sam wrote down 98·64	
Questions	Answers
Is it bigger than 100?	No
Is it smaller than 50?	No
Is it bigger than 90?	Yes
Is it bigger than 95?	Yes
Is it smaller than 97?	No
Is it bigger than 98?	Yes
Is it bigger than 99?	No
Is it bigger than 98·5?	Yes
Is it bigger than 98·7?	No
Is it bigger than 98·65?	No
Is it 98·64?	Yes

Rounding to powers of ten

The population of York is 124 609.
A magazine article about York rounded this to the nearest 10 000.
124 609 to the nearest 10 000 is 120 000.

120 000   130 000
124 609 is nearer to   124 609
120 000 than to 130 000.   York

Examples 8642 to the nearest thousand is 9000.

8642
8000   9000

68 241 to the nearest ten thousand is 70 000.

68 241
60 000   70 000

When a number is halfway between two numbers we round up.
8500 to the nearest thousand is 9000.
25 000 to the nearest ten thousand is 30 000.

27

Number

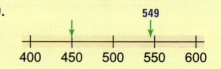

Worked Example

To the nearest 100 the number of cars in a town is 500.
What is the largest and smallest number of cars that
could be in the town?

549

400 450 500 550 600

Answer

The **largest number of cars is 549**.
One more car would make 550 and this would round up to 600.
The **smallest number of cars is 450**.

Exercise 7

1 Round
  a 324 to the nearest 10   **b** 5862 to the nearest 100
  c 5682 to the nearest 1000   **d** 1825 to the nearest 10
  e 28 562 to the nearest 10 000   **f** 89 500 to the nearest 1000
  g 725 000 to the nearest 100 000   **h** 69 782 to the nearest 1000
  i 138 987 to the nearest 100 000   **j** 5 864 321 to the nearest million.

2 Nick's school raised £946 for charity.
  a The charity said that nearly £1000 was raised.
     Was this given to the nearest 10, 100 or 1000?
  b The school newsletter said nearly £950 was raised.
     Was this given to the nearest 10, 100 or 1000?

£950 raised
for charity

3 Peter's grandmother told him she was 80, to the nearest 10 years.
  Write down three possible ages she could be, other than 80.

4 The River Thames is 350 km long, to the nearest 10 km.
  Write down 3 possible lengths that this river could be.

Give your answers
to the nearest km.

5 Mary's end of year mark in maths was 70, to the nearest 10.
  a Write down 3 possible marks Mary could have got.
  b What is the lowest mark Mary could have got?
  c What is the highest mark Mary could have got?

6 Jan rounded the number of pupils at Rosebank School to the nearest 100 and got
  800.
  Ben rounded to the nearest 10 and got 850.
  Write down three possible numbers of pupils at Rosebank.

Rounding to decimal places

Discussion

- 12 staff in an office won £6719 in a raffle.
  How do you think they should share the prize?

- Three families picked 12·4 kg of raspberries altogether.
  They wanted to share them equally between the three families.

    12·4 ÷ 3 = 4·133333.

  How do you think the families should share the raspberries?

Sometimes we round to the **nearest whole number**.

Examples   £12·70 to the nearest pound is £13.
8·43 cm to the nearest cm is 8 cm.

To **round to a given number of decimal places**, for example, to round 8·74862 to 2 decimal places (d.p.)

1   keep the number of digits you want

2   look at the next decimal place.
    If this is 5 or greater, increase the last digit you are keeping by 1.

3   8·74862 to 2 d.p. is 8·75.

8·74 | 862

↑
keep

8·74 | 862
↑
look at this digit.
It is >5 so add 1 to 4

Examples   10·306 to 2 d.p. is 10·31
7·03 to 1 d.p. is 7·0
4·635 to 2 d.p. is 4·64

To give a sensible answer to a calculation we often need to **round** the answer.

Example   Seven friends share a 250 g bag of popcorn.
To find how many grams each gets we key (250) (÷) (7) (=) to get
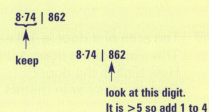
35.714285711.
We could give the answer as 35·7 g to 1 d.p.
Or we could give it as 36 g to the nearest whole number.

Exercise 8   **Except for questions 5, 6 and 7.**

1   Round these to the nearest whole number.
    a   2·8237 m   b   6·827 ℓ   c   0·7348 ℓ   d   1·292 m   e   2·896 kg
    f   12·995 kg   g   4·0038 km   h   22·881 cm   i   0·859 mm

2 Round these to 1 d.p.

    a 3·4215   **b** 24·21   **c** 7·68   **d** 0·06   **e** 21·409

    f 21·05   **g** 5·03   ∗**h** 13·99

3 Round these to 2 d.p.

    a 1·638   **b** 22·424   **c** 13·6452   **d** 60·135   **e** 2·309

    f 15·104   ∗**g** 13·995   ∗**h** 0·096

4 Use some or all of these cards.

    a Make a number that rounds to 25 to the nearest whole number.

    b Make a number that rounds to 23·6 to the nearest 1 d.p.

  ∗**c** Make a number that rounds to 6·53 to the nearest 2 d.p.

5 Do these on the calculator.

    Round the answers to 1 d.p.

    a 7·59 ÷ 6   **b** 0·57 ÷ 4   **c** 8 ÷ 3   **d** 43 ÷ 9

    e 2·69 ÷ 11   **f** $\frac{22·9}{3}$ **g** $\frac{86·47}{5}$ **h** $\frac{72·791}{11}$

6 Do these on the calculator.

    Round the answers to 2 d.p.

    a 18 ÷ 7   **b** 53 ÷ 9   **c** 9·41 ÷ 6   **d** $\frac{0·78}{11}$ **e** $\frac{22·9}{3}$

←0·85 m→

∗**7** The area of a door is 1·8 m².

    The width of the door is 0·85 m.

    How high is the door?

    Give the answer in metres to 2 d.p.

1·8 m²

Remember:
Area = height × width.

8 Round the answers to these sensibly.

    Say what you have rounded them to.

    a It costs £4·95 to run a heater for 10 hours.

       How much does it cost each hour?

    b Lee drank 13·25 ℓ of water in a week (7 days).

       How many litres did he drink each day? He drank the same amount each day.

  ∗**c** A tap that drips once each second wastes 1250 mℓ each hour.

       How much water is in each drip?

Summary of key points

A This chart shows the **powers of ten**.

In 10^3, 3 is called an index.

(Thousands of millions) Billions	Hundreds of millions	Tens of millions	Millions	Hundreds of thousands	Tens of thousands	Thousands	Hundreds	Tens	Units
10^9	10^8	10^7	10^6	10^5	10^4	10^3	10^2	10^1	1

$6·3 \times 10^4$ is read as 'six point three times ten to the power of four'.

To **multiply by 10, 100, 1000**, ... each digit moves 1 place to the left for each zero in 10, 100, 1000, ...

To **divide by 10, 100, 1000**, ... each digit moves 1 place to the right for each zero in 10, 100, 1000, ...

Examples   3·5 × 1000 = 3500   4250 ÷ 100 = 42·5

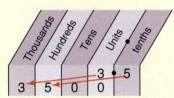   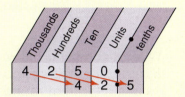

 We can use place value to **multiply and divide by multiples of 10, 100, 1000**, ...

Examples   1·3 × 20 = 1·3 × 2 × 10   360 ÷ 90 = 360 ÷ 9 ÷ 10
                    = 2·6 × 10   = 40 ÷ 10
                    = 26   = 4

 Multiplying by 0·1 is the same as dividing by 10.
Multiplying by 0·01 is the same as dividing by 100.
Dividing by 0·1 is the same as multiplying by 10.
Dividing by 0·01 is the same as multiplying by 100.

Examples   36 × 0·1 = 36 ÷ 10   5·2 ÷ 0·01 = 5·2 × 100
                    = 3·6   = 520

 To put **decimals in order**, compare digits with the same place value.

Work from left to right to find the first digits that are different.

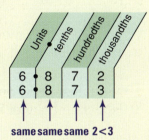

Examples

6·872 < 6·873   starting at the left, the digits are the same until we get to the 2 and 3.
2 is less than 3 so 6·872 < 6·873

⁻3·5 > ⁻6·5   ⁻6·5 is further to the left on the number line so it is smaller.

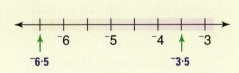

Number

F **Rounding to powers of 10**

487 248   to the nearest thousand is

487 000

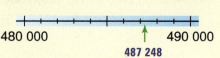

487 248

487 000   488 000

to the nearest ten thousand is

490 000

480 000   490 000

487 248

When a number is halfway between two numbers we round up.

Example   25 500 to the nearest thousand is 26 000.

G To round to a given number of **decimal places**:

1 keep the number of digits asked for after the decimal point.

2 delete any following digits.

If the first digit to be deleted is 5 or more, increase the last digit kept by 1.

Examples $8 \cdot 6832 = 8 \cdot 7$ to 1 d.p.

$53 \cdot 0535 = 53 \cdot 05$ to 2 d.p.

Rounding to the nearest whole number is rounding to 0 d.p.

Example $81 \cdot 235$ to the nearest whole number is 81.

Test yourself   **Except for questions 15 and 16.**

1 Write these without the index.   **A**
   a 10^3 **b** 10^6 **c** 10^1

2 Write these as a power of ten.   **A**
   a one hundred   **b** one hundred thousand   **c** one million   **d** one billion

3 The Sun is about $1 \cdot 49 \times 10^8$ km from Earth.   **A**
   Write $1 \cdot 49 \times 10^8$ in words.

4 Write, in figures, the number that is 4 more than three and a half thousand.   **A**

5 Find the answers to these.   **B**
   a 36×10 **b** 21×1000 **c** $57 \div 10$ **d** $360 \div 100$
   e $1 \cdot 2 \times 100$ **f** $475 \div 1000$ **g** $0 \cdot 026 \times 100$ **h** $1 \cdot 024 \div 100$

6 Find the answers to these.   **C**
   a 20×70 **b** 700×30 **c** $800 \div 20$ **d** $12\,000 \div 600$ ***e** $200 \div 4000$

7 Find the answers to these.   **D**
   a $57 \times 0 \cdot 1$ **b** $473 \div 0 \cdot 1$ **c** $8 \cdot 3 \times 0 \cdot 01$ **d** $0 \cdot 68 \div 0 \cdot 01$

8 Which of < or > goes between these?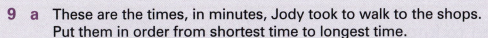
  a 5·68   5·67   **b** 4·378   4·387   *c ⁻1·8   ⁻2·8

9 **a** These are the times, in minutes, Jody took to walk to the shops.
  Put them in order from shortest time to longest time.
      4·27,   4·72,   4·7,   4·07
  b Tom practised using some new laboratory scales.
  He got these masses.
  Put them in order from lightest to heaviest.
      0·869 kg,   0·689 kg,   0·06 kg,   0·007 kg,   0·798 kg

10 Enid did a survey on the masses of cats.
  This shows part of her frequency table.

Mass (kg)	Tally	Frequency
$1·0 < x \leqslant 1·5$		
$1·5 < x \leqslant 2·0$		
$2·0 < x \leqslant 2·5$		

  a Explain what $1·0 < x \leqslant 1·5$ means.
  b Enid's kitten weighs 1·5 kg.
  In which colour box should Enid put the tally mark?

11 Round these to   **i** the nearest thousand   **ii** the nearest hundred thousand.
  a 623 471   **b** 832 314   **c** 7 083 741   **d** 985 500

12 The number of people on a ski slope, to the nearest 10, was 320.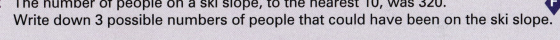
  Write down 3 possible numbers of people that could have been on the ski slope.

13 April chose a 4-digit number.
  She rounded it to the nearest 1000 and got 8000.
  a What is the smallest number April could have chosen?
  b What is the biggest number April could have chosen?

14 Approximate these to the given number of decimal places.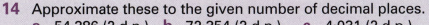
  a 54·386 (2 d.p.)   **b** 72·354 (2 d.p.)   **c** 4·031 (2 d.p.)   **d** 16·257 (1 d.p.)
  e 4·605 (1 d.p.)   **f** 16·423 (0 d.p.)   *g 105·004 (2 d.p.)   *h 9·998 (2 d.p.)

 15 Round your answers to 1 d.p.
  a $9 \div 5$ **b** $20 \div 3$ **c** $6·45 \div 7$ **d** $\frac{32·6}{5}$

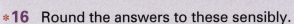 *16 Round the answers to these sensibly.
  a In design and technology, Julia weighed the 6 muffins she had made.
  The total mass was 287 g. About what was the mass of each muffin?
  b In chemistry, Olivia's group of five were given 0·46 kg of powder to share.
  How much powder did each get?

2 Integers, Powers and Roots

You need to know

Key vocabulary

cube, cubed, cube root, highest common factor (HCF),
indices, index, lowest common multiple (LCM), prime factor,
square, squared, square root

▶▶ Shaping Up!

1

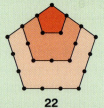

1   5   12   22

These diagrams show the first four pentagonal numbers.
Why are 1, 5, 12, 22 called pentagonal numbers?

Draw a diagram to find the fifth pentagonal number.
What is it?

2

1   6   15

These diagrams show the first three hexagonal numbers.
Why are 1, 6, 15 called hexagonal numbers?

Draw a diagram to show the fourth hexagonal number.
Write down the first four hexagonal numbers.

Working with integers

This number line shows the **integers**.

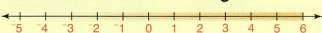

The arrow shows it continues in both directions.

The further to the left a number is, the smaller it is.
A temperature of ⁻7 °C is lower than a temperature of ⁻3 °C.

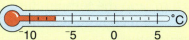

We can **add and subtract integers** using a number line.

Examples

⁻2 + 5

⁻2 is where you start.
+5 means move right 5 places.
The answer is 3.

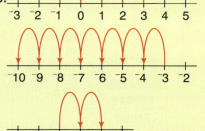

3 + (⁻4)

3 is where you start.
+(⁻4) means move left 4 places.
The answer is ⁻1.

⁻3 – 7

⁻3 is where you start.
–7 means move left 7 places.
The answer is ⁻10.

1 – (⁻2)

1 is where you start.
–(⁻2) move right 2 places.
The answer is 3.

When adding a negative or subtracting a positive move left.
When adding a positive or subtracting a negative move right.

Exercise 1

1 These were the temperatures in 5 cities on one night.

Calgary	Athens	London	Dublin	Warsaw

⁻7 °C	8 °C	0 °C	⁻3 °C	⁻1 °C

a Which city had the coldest night?
b Which city had the warmest night?
c Put the temperatures in order.
Put the coldest first.

Number

2 Naseem wrote down the temperature each hour.
This table shows her results.

Time	Temperature
9 a.m	−8°C
10 a.m	−7°C
11 a.m	−3°C
12 p.m	−1°C
1 p.m	2°C
2 p.m	3°C
3 p.m	4°C
4 p.m	0°C
5 p.m	−3°C

a How much did the temperature rise between 9 a.m. and 2 p.m.?

b At which two times was the temperature the same?

c Between which times did the temperature fall from 2 °C to ⁻3 °C?

d What is the difference between the temperature at 10 a.m. and the temperature at 3 p.m.?

e During which hour was there the greatest rise in temperature?

f During which hour was there the greatest fall in temperature?

T

3 Use a copy of this.
When the wind blows it feels colder.
The stronger the wind, the colder it feels.
Fill in the gaps in the table.
The first row is done for you.

[SATs Paper 1 Level 4]

Wind Strength	Temperature out of the wind (°C)	How much colder it feels in the wind (°C)	Temperature it feels in the wind (°C)
Moderate breeze	5	7 degrees colder	⁻2
Fresh Breeze	⁻8	11 degrees colder	.....
Strong Breeze	⁻4	..... degrees colder	⁻20
Gale	.....	23 degrees colder	⁻45

T

4 The arrow on this thermometer shows a temperature of 10 °C.
Use a copy of this.

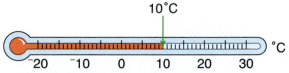

a Draw an arrow on the thermometer to show a temperature of 24 °C.
Label the arrow 24 °C.

b Draw an arrow to show these temperatures.
Label your arrows.   ⁻4 °C   ⁻16 °C   ⁻11 °C

c The temperature was ⁻10 °C.
It went up 15 °C.
What is the temperature now?

d The temperature was ⁻13 °C.
It went up 7 °C.
What is the temperature now?

e The temperature went from ⁻2 °C to 17 °C.
How much did it rise?

f The temperature at 3 p.m. was 7 °C.
At 10 p.m. it was ⁻5 °C.
What is the difference in these temperatures?

5 Which sign, > or <, should go in the box?
a ⁻3☐4 b ⁻3☐0 c 5☐⁻1
d ⁻17☐⁻4 e ⁻2☐⁻11 f ⁻6☐⁻7

6 | ⁻5 | ⁻4 | ⁻3 | ⁻2 | ⁻1 | 0 | 1 | 2 | 3 | 4 | 5 |

a Write down two numbers from the box which have a difference of 8.
b Write down two numbers from the box which have a sum of ⁻5.

7 ⁻5 0 1 ⁻9 5 9 ⁻4 ⁻1

a Choose a number card to give the answer ⁻5.

b Choose a number card to give the answer 5.

8 | ⁻3 | ⁻2 | ⁻1 | 0 | 1 | 2 | 3 |

What goes in the gap? Choose from the box.
a ⁻6 + ____ = ⁻7   b ⁻3 – ____ = ⁻1
c ⁻3 + 2 – ____ = ⁻2   d ⁻2 + 3 – ____ = ⁻2

Use a number line
to help.

☐ ***9** Use a copy of these.
In **a** and **b** add two numbers to get the number in the circle above.
What number goes in the top circle of each?
The second row of **a** is done for you.

a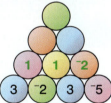

b

c Subtract the number on the right from the number on the left.
What number goes in the top circle?

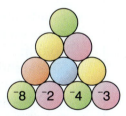

*10 a Two integers are added.
The answer is ⁻24.
What could the integers be?

b ☐ – ☐ = ⁻21
What integers might go in the boxes?

 Puzzle

Each line of circles, shown by the arrows, adds to
the number given in red.
Replace the letters with numbers to give the totals
shown.

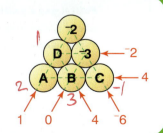

Integers — a game for a group or class

You will need to draw a diagram like this for each turn.

To play
- Choose a leader.
- The leader calls out 4 integers between ⁻10 and 10.
- As each one is called, put it in one of your blue boxes.
- Whoever gets the biggest total is the leader for the next turn.

Multiplying and dividing integers

Discussion

- What would the next 3 lines of each of these tables be? **Discuss.**

2 × 3 = 6	2 × ⁻3 = ⁻6
1 × 3 = 3	1 × ⁻3 = ⁻3
0 × 3 = 0	0 × ⁻3 = 0
⁻1 × 3 = ⁻3	⁻1 × ⁻3 = 3
⁻2 × 3 = ⁻6	⁻2 × ⁻3 = 6

Look for patterns.

T

- Use a copy of this table. **Discuss** how to fill it in. Shade the positive answers one colour. Shade the negative answers another colour. Shade the zeros another colour.

Look for patterns in the rows and columns.

+	⁻3	⁻2	⁻1	0	1	2	3
3	⁻9	⁻6	⁻3	0	3	6	9
2				0	2	4	6
1				0	1	2	3
0				0	0	0	0
⁻1							⁻3
⁻2							⁻6
⁻3							⁻9

Use the table to find the answers to these.

3 × ⁻2, 2 × ⁻1, ⁻3 × 2, ⁻2 × 3
⁻3 × ⁻2, ⁻2 × ⁻2, ⁻3 × ⁻1, ⁻2 × ⁻1

- Katie wrote this.

  Multiplying a positive number and a negative number gives a negative number.
  Multiplying two negative numbers gives a positive number.

You could check using a calculator. Use the (–) to key a negative number.

Is Katie correct? **Discuss**

When we **multiply a positive number and a negative number** together we get a **negative answer**.

Examples $5 \times {}^-2 = {}^-10$ ${}^-3 \times 8 = {}^-24$ $7 \times {}^-8 = {}^-56$ ${}^-5 \times 7 = {}^-35$

positive × negative = negative

When we **multiply two negative numbers** together we get a **positive answer**.

Examples ${}^-3 \times {}^-4 = 12$ ${}^-6 \times {}^-7 = 42$ ${}^-8 \times {}^-3 = 24$ ${}^-5 \times {}^-9 = 45$

negative × negative = positive

Multiplying and dividing are inverse operations.
Dividing by ${}^-2$ is the inverse of multiplying by ${}^-2$.

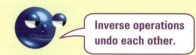

Inverse operations
undo each other.

${}^-4 \rightarrow$ | multiply by ${}^-2$ | $\rightarrow {}^-8$

${}^-4 \leftarrow$ | divide by ${}^-2$ | $\leftarrow {}^-8$

Worked Examples

a ${}^-7 \times 3$ **b** ${}^-5 \times {}^-7$ **c** ${}^-12 \div 3$ **d** $\frac{{}^-30}{5}$

Answer

a ${}^-7 \times 3 = {}^-\mathbf{21}$ negative × positive = negative
b ${}^-5 \times {}^-7 = \mathbf{35}$ negative × negative = positive
c ${}^-12 \div 3 = {}^-\mathbf{4}$ negative ÷ positive = negative
d $\frac{{}^-30}{{}^-5} = \mathbf{6}$ negative ÷ negative = positive

Exercise 2

1 Write down the next 3 lines of these.

a $2 \times 4 = 8$
$1 \times 4 = 4$
$0 \times 4 = 0$
${}^-1 \times 4 = {}^-4$
${}^-2 \times 4 = {}^-8$

b $2 \times {}^-4 = {}^-8$
$1 \times {}^-4 = {}^-4$
$0 \times {}^-4 = 0$
${}^-1 \times {}^-4 = 4$
${}^-2 \times {}^-4 = 8$

c ${}^-2 \times 4 = {}^-8$
${}^-1 \times 4 = {}^-4$
$0 \times 4 = 0$
$1 \times 4 = 4$
$2 \times 4 = 8$

2 Which ones have a negative sign in the box?

a $3 \times {}^-5 = \square\,15$ **b** $4 \times {}^-6 = \square\,24$ **c** ${}^-5 \times {}^-3 = \square\,15$ **d** ${}^-2 \times 8 = \square\,16$

e ${}^-4 \times {}^-7 = \square\,28$ **f** ${}^-3 \times 9 = \square\,27$ **g** ${}^-2 \times 9 = \square\,18$ **h** ${}^-3 \times {}^-4 = \square\,12$

i ${}^-4 \times {}^-5 = \square\,20$ **j** ${}^-4 \times 8 = \square\,32$ **k** ${}^-10 \times 3 = \square\,30$ **l** $10 \times {}^-3 = \square\,30$

m ${}^-10 \times {}^-3 = \square\,30$

Number

T **3** Use a copy of this.

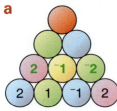 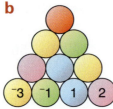 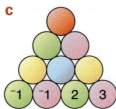 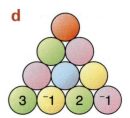

a   b   c   d

Each pair of numbers is multiplied to get the number above.
What numbers go in the red circles?

T **4** Use a copy of these multiplication squares.
Fill them in.

a

×	2	⁻5	⁻3	4
⁻1	⁻2	5		
3	6			
⁻2				
5				

b

×	⁻1	2	3	⁻4
3				
⁻1				
⁻4				
⁻6				

c

×	⁻6	2	⁻1	8
⁻4				
7				
⁻6				
⁻1				

d

×	⁻3	7	⁻2	⁻8
6				
⁻1				
⁻7				
2				

T **5** $3 × ⁻2 = ⁻6$ $⁻2 × 3 = ⁻6$ $⁻6 ÷ 3 = ⁻2$ $⁻6 ÷ ⁻2 = 3$

These four facts all use the numbers 3, ⁻2 and ⁻6.
Use a copy of this.
Fill in the gaps with more facts using the numbers given.

a $4 × ⁻1 = ⁻4$ **b** $⁻9 × 2 = ⁻18$ **c** $⁻7 × ⁻3 = 21$ ***d**   _____

                                        $⁻3 × ⁻7 = 21$

_____   _____   _____

   $⁻4 ÷ ⁻1 = 4$ $⁻60 ÷ 15 = ⁻4$

_____ $⁻18 ÷ 2 = ⁻9$ _____   _____

T **6** Use a copy of this.
Use the clues to find the numbers in the yellow squares.

Clues

Square 2. The answer to **1.** divided by 2.
Square 3. The answer to **2.** divided by ⁻2.
Square 4. The answer to **3.** multiplied by ⁻1.
Square 5. The answer to **4.** multiplied by ⁻5.
Square 6. The answer to **5.** divided by 3.
Square 7. The answer to **6.** multiplied by ⁻4.
Square 8. The answer to **1.** multiplied by ⁻2.
Square 9. The answer to **8.** divided by ⁻6.
Square 10. The answer to **9.** divided by 4.
Square 11. The answer to **10.** multiplied by 3.
Square 12. The answer to **11.** multiplied by ⁻4.
Square 13. The answer to **12.** divided by ⁻3.

*7 Make up some multiplications and some divisions with this answer.

I got −8 as the answer.

Integers on a calculator

The **sign change key** on the calculator is $(-)$.

Example   −5 is keyed as $(-)\ 5$.

Worked Example
Use the calculator to find these.   **a** 85 − −92   **b** 37 × −5

Answer
a Key $85\ -\ (-)\ 92\ =$ to get **177**.
b Key $37\ \times\ (-)\ 5\ =$ to get **−335**.

Exercise 3

1

										F
−140	−102	475		24·9	−140	8·25	25·62	−65	−26	**326**

24·9	−140	8·25	1620	−1827	25·62	−1827	−140	−65

−1827	24·9		3844	−24·9	338·4	338·4	475	25·62

−8·25	−26	182·7	−26	338·4	−26	−2·6	−65

Use a copy of this box. Write the letter beside each question above its answer in the box.

F −84 + −242 = **326**   **H** 247 + −349   **E** 348 − −127   **I** −29 × 63
C −31 × −124   **L** −94 × −3·6   **O** 1456 ÷ −56   **M** −792 ÷ 96
N −29 × −6·3   **G** 145·6 ÷ −56   **U** −792 ÷ 96   **Y** −416 ÷ 6·4
T −84 + −98 − −42   **P** 144 × −25 × −0·45 **S** −8·3 × 3·9 ÷ −1·3   **A** −8·3 ÷ 1·3 × 3·9
D −4·2 × −6·1

T 2 Use a copy of these addition and multiplication squares.
Fill them in.

a

+	⁻84	253	⁻461	514
⁻198				
⁻257				
346				
⁻98				

b

×	47	⁻14	⁻28	8·6
⁻6·4				
⁻72				
95				
⁻69				

Multiples, factors and common factors

The **factors** of a number are all those numbers which will divide into that number with no remainder.

Example   24 can be divided by 24, 12, 8, 6, 4, 3, 2, 1.
So the factors of 24 are 24, 12, 8, 6, 4, 3, 2, and 1.

The multiples of a number are found by multiplying the number by 1, 2, 3, 4, 5, ...

Example   Some of the multiples of 6 are:
$6 \times 1 = \textbf{6}$
$6 \times 2 = \textbf{12}$
$6 \times 3 = \textbf{18}$
$6 \times 5 = \textbf{30}$
$6 \times 10 = \textbf{60}$
$6 \times 15 = \textbf{90}$

6, 12, 18, 30, 60 and 90 are all multiples of 6.
What are some other multiples of 6?

Discussion

Lorna drew these diagrams to find the factors of 12.

1 row of 12

2 rows of 6

3 rows of 4

The factors of 12 are 1, 12, 2, 6, 3 and 4.

How could you draw diagrams to show how to find the factors of 24?
How could you draw diagrams to show how to find the factors of these numbers?

20   18   32   15   25   17

Can you always use diagrams to find the factors of a number? **Discuss.**

Exercise 4

1 Use a copy of this.
   Shade the
   a multiples of 3   b multiples of 8
   c factors of 24   d factors of 35.

2 a Write down the factors of 18.
   b Write down the factors of 32.
   c Write down all the common factors of 18 and 32.

3 a Write down the factors of 30.
   b Write down the factors of 42.
   c Write down all the common factors of 30 and 42.

4 Write down all the common factors of
   a 12 and 16   b 20 and 36   *c 48 and 112.

? **Puzzle**

Rachel, Rick and Ray all start climbing some stairs.
Rachel climbs them 1 at a time.
Rick climbs them 2 at a time.
Ray climbs them 3 at a time.
Which is the first step they will all step on?
Which is the second step they will all step on?

Prime numbers and prime factors

Remember
A **prime number** has just two factors, itself and 1.

1 is not a prime number.

Example   The only factors of 11 are 1 and 11.
             11 is a prime number.

A prime number can only be drawn as a rectangle that is a straight line of dots.

Example   11 can only be drawn as ○○○○○○○○○○○ or ○

             6 is not a prime number.
             It can be drawn as ○○○○○○ or ○

   or ○○○ or ○○

The prime numbers less than 30 are 2, 3, 5, 7, 11, 13, 17, 19, 23, 29.

Number

Exercise 5

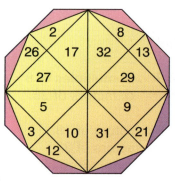

T **1** Use a copy of this.
Colour the prime numbers.

2 Write down the largest prime number
   a less than 20   **b** less than 30   **c** less than 40.

*3 Test to see which three numbers between 80 and 100 are prime.

A **prime factor** is a factor that is a prime number.

Example   The factors of 12 are 1, 2, 3, 4, 6 and 12.
Of these, 2 and 3 are prime numbers.
The prime factors of 12 are 2 and 3.

We can find the prime factors using a **factor tree**.

Example
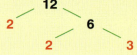
$12 = 2 \times 2 \times 3$
2 and 3 are the prime factors of 12.

$12 = 2 \times 2 \times 3$ is 12 written as a **product of prime factors**.

Fun with Factors – a game for two players (P and Q)

You will need a copy of this grid.

To play

- Player P crosses out any number.

- Player Q then crosses out as many factors as possible of this number and adds them together. This is Q's score.

- Player Q then crosses out a number that is still left.

- Player P then crosses out as many factors as possible of this number that are not crossed out already and adds them. This is P's score.

- Then P crosses out a number that is still left.

- The game continues until all the numbers have been crossed out.

- The winner is the player with the highest score.

1	2	3	4	5	6
7	8	9	10	11	12
13	14	15	16	17	18
19	20	21	22	23	24
25	26	27	28	29	30
31	32	33	34	35	36

continued

Example   If P crosses out 27 the game starts like this.

1	2	3	4	5	6
7	8	9	10	11	12
13	14	15	16	17	18
19	20	21	22	23	24
25	26	27	28	29	30
31	32	33	34	35	36

P crosses out 27.

X	2	3	4	5	6
7	8	9	10	11	12
13	14	15	16	17	18
19	20	21	22	23	24
25	26	27	28	29	30
31	32	33	34	35	36

Q crosses out 1, 3, 9.
Q's score = 1 + 3 + 9
= 13

X	2	3	4	5	6
7	8	9	10	11	12
13	14	15	16	17	18
19	20	21	22	23	24
25	26	27	28	29	30
31	32	33	34	35	36

Q crosses out 24.

X	2	3	4	5	6
7	8	9	10	11	12
13	14	15	16	17	18
19	20	21	22	23	24
25	26	27	28	29	30
31	32	33	34	35	36

P crosses out 2, 4, 6, 8, 12.
P's score = 2 + 4 + 6 + 8 + 12
= 32

X	2	3	4	5	6
7	8	9	10	11	12
13	14	15	16	17	18
19	20	21	22	23	24
25	26	27	28	29	30
31	32	33	34	35	36

P crosses out 13.

X	2	3	4	5	6
7	8	9	10	11	12
13	14	15	16	17	18
19	20	21	22	23	24
25	26	27	28	29	30
31	32	33	34	35	36

Q cannot cross out any
factors of 13.
Q crosses out 33.

Exercise 6

T

1 Use a copy of these factor trees.
Fill in the gaps.

a 24   **b** 20   **c** 36

2 Use your answers to question **1** to write down the prime factors of these.
   a 24   **b** 20   **c** 36

3 Milly used a table to find the prime factors of 12.

$12 = 2 \times 2 \times 3$

The prime factors of 12 are **2** and **3**.

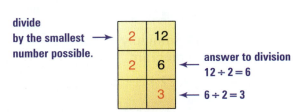

divide by the smallest → number possible.

answer to division
$12 \div 2 = 6$

$6 \div 2 = 3$

Number

Use a copy of these tables
Complete them to find the prime factors of the bold number.

a

2	16

b

	28

c

	45

d

2	72

4 Use your answers to question **3** to write each of these as a
product of prime factors: **a** is started for you.

  a $16 = 2 \times 2 \times \cdots$ **b** 28   **c** 45   **d** 72

Highest common factor and lowest common multiple

The **HCF (highest common factor)** of two numbers is the largest factor common to both.

Example   The factors of 18 are 1, 2, 3, **6**, 9, 18.
    The factors of 30 are 1, 2, 3, 5, **6**, 10, 15, 30.
    The HCF of 18 and 30 is **6**.

> We use HCF when cancelling fractions. See page 91.

The **LCM (lowest common multiple)** of two numbers is the smallest number that is a multiple of both.

Example   The multiples of 9 are 9, 18, 27, 36, **45**, 54, 63, ...
    The multiples of 15 are 15, 30, **45**, ...
    The LCM of 9 and 15 is **45**.

We can use prime factors to find the HCF and LCM.

Example

$20 = \textbf{2} \times \textbf{2} \times 5$
$36 = \textbf{2} \times \textbf{2} \times 3 \times 3$
HCF of 20 and 36 $= \textbf{2} \times \textbf{2}$
              $= \textbf{4}$

$20 = \textbf{2} \times \textbf{2} \times 5$
$36 = \textbf{2} \times \textbf{2} \times 3 \times 3$
LCM of 20 and 36 $= \textbf{2} \times \textbf{2} \times 3 \times 3 \times 5$
              $= \textbf{180}$

> We just write 2×2 once.

We could write 20 as $2^2 \times 5$ and 36 as $2^2 \times 3^2$.

We can use a diagram to find the HCF and LCM.

To find the HCF, multiply the numbers where the circles overlap.

    HCF $= 2 \times 2 = \textbf{4}$

To find the LCM, multiply all the numbers in the diagram.

    LCM $= 5 \times 2 \times 2 \times 3 \times 3$
          $= \textbf{180}$

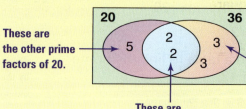

These are the other prime factors of 20.

These are the other prime factors of 36.

These are the prime factors of *both* 20 and 36.

Exercise 7

1 a Write down the factors of these.
   i 12   **ii** 28
 b Write down the HCF of 12 and 28.

2 a Write the first six multiples of 3.
 b Write the first six multiples of 5.
 c Write down the LCM of 3 and 5.

3 a Copy and complete the diagrams.
   i $36 = 2 \times 2 \times 3 \times 3$ **ii** $40 = 2^3 \times 5$
     $90 = 2 \times 3 \times 3 \times 5$ $140 = 2^2 \times 5 \times 7$

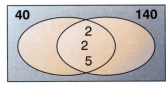

 b Use the diagrams to find the HCF and LCM of
   i 36 and 90   **ii** 40 and 140.

4 Find the HCF and LCM of these.
 a 16 and 12   **b** 24 and 56
 c 10 and 65   **d** 60 and 150

Squares and square roots

Remember
3^2 is read as 'three squared', and means 3×3.
3^2 could also be read as 'three to the power of 2'.

The first 12 square numbers are:

$1^2 = 1$ $2^2 = 4$ $3^2 = 9$ $4^2 = 16$ $5^2 = 25$ $6^2 = 36$
$7^2 = 49$ $8^2 = 64$ $9^2 = 81$ $10^2 = 100$ $11^2 = 121$ $12^2 = 144$

The first 12 square roots are:

$\sqrt{1} = 1$ $\sqrt{4} = 2$ $\sqrt{9} = 3$ $\sqrt{16} = 4$ $\sqrt{25} = 5$ $\sqrt{36} = 6$ $\sqrt{49} = 7$ $\sqrt{64} = 8$
$\sqrt{81} = 9$ $\sqrt{100} = 10$ $\sqrt{121} = 11$ $\sqrt{144} = 12$

Squaring and finding the square root are inverse operations.

$3 \rightarrow$ | square | $\rightarrow 9$

$3 \leftarrow$ | square root | $\leftarrow 9$

Number

$\sqrt{16}$ is read as 'the **square root** of 16'.
To find $\sqrt{16}$ we find what number squared equals 16.

$4 \times 4 = 16$ and also $^-4 \times {}^-4 = 16$.
4 and $^-4$ are both square roots of 16.

$\sqrt{16} = 4$ or $^-4$.

When we are asked for the square root of a number we usually just give the positive square root.

Examples $\sqrt{49} = 7$ but $^\pm\sqrt{49} = 7$ or $^-7$ and $\sqrt{144} = 12$ but $^\pm\sqrt{144} = 12$ or $^-12$

On the calculator $5 \cdot 6^2$ is keyed as (5·6) (x^2) (=) to get 31·36.

$\sqrt{196}$ is keyed as (√) (196) (=) to get 14.

Exercise 8

except questions 6–12

1 a 10 is the square root of _____.   **b** 9 is the square root of _____.
   c _____ is the square root of 25.   **d** _____ is the square root of 16.
   e 2 is the square root of _____.   **f** _____ is the square root of 64.
   g The square root of 1 is _____.

2 a $5^2 = 25$ and $\sqrt{25} = $ _____   **b** $4^2 = 16$ and _____ $= 4$
   c $11^2 = 121$ and $\sqrt{121} = $ _____   **d** _____ $^2 = 64$ and $\sqrt{64} = 8$
   e _____ $^2 = 4$ and $\sqrt{4} = 2$

3 Calculate mentally.
   a $\sqrt{8+1}$ **b** $\sqrt{21-5}$ **c** $\sqrt{50-1}$ **d** $\sqrt{30+6}$ **∗e** $\sqrt{73-3^2}$

4 Find the answers to these mentally.
   a $(2-1)^2$ **b** $(3+5)^2$ **c** $(16-9)^2$ **d** $(^-8)^2$ **e** $(^-6)^2$

5 Give the positive *and* negative square roots of these.
   a $^\pm\sqrt{9}$ **b** $^\pm\sqrt{25}$ **c** $^\pm\sqrt{49}$ **d** $^\pm\sqrt{144}$ **e** $^\pm\sqrt{121}$

6 Anna made some mistakes when doing her homework.
   Are her answers right (✓) or wrong (✗)?
   a $24^2 = 576$ **b** $35^2 = 1220$ **c** $\sqrt{324} = 16$
   d $\sqrt{96} = 9 \cdot 8$ (1 d.p.)   **e** $3 \cdot 7^2 = 13 \cdot 69$ **f** $\sqrt{14} = 3 \cdot 84$ (2 d.p.)
   g $14 \cdot 8^2 = 219 \cdot 04$ **h** $\sqrt{10 \cdot 5} = 3 \cdot 24$ (2 d.p.)

7 Use your calculator to find these.
Round **f** to 2 d.p.

    a 21^2 **b** $3 \cdot 5^2$ **c** $16 \cdot 2^2$ **d** $\sqrt{225}$
    e $\sqrt{484}$ ***f** $\sqrt{52}$

8 Use a copy of this.
Fill in the missing numbers in the chain.

    $1296 \xrightarrow{\ \surd\ } \square \xrightarrow{\ \surd\ } \square \xrightarrow{\text{square}} \square$

9

			A					**A**	
20·1	3·24	31	**169**	2·3	1·96		2·1	**169**	0·3

								A	
21·3	1·5	2·3	1·3	8·6	6724		1·3	729	
								169	729

			A							**A**		
1·5	729	20·1	**169**	2·3	0·36	8·6	6724		0·16	2·3	**169**	196	8·6

Use a copy of the box.
Use your calculator to find these.
Put the letter that is beside each above its answer.

    A $13^2 = \textbf{169}$ **V** 14^2 **N** 27^2 **D** 82^2 **O** $1 \cdot 8^2$
    Z $\sqrt{961}$ **W** $\sqrt{4 \cdot 41}$ **S** $\sqrt{0 \cdot 09}$ **E** $\sqrt{73 \cdot 96}$ **K** $0 \cdot 6^2$
    G $0 \cdot 4^2$ **T** $1 \cdot 4^2$ **U** $\sqrt{2 \cdot 25}$ **I** $\sqrt{1 \cdot 69}$ **B** $\sqrt{453 \cdot 69}$
    R $\sqrt{5 \cdot 29}$ **M** $\sqrt{404 \cdot 01}$

10 Give the answers to these to 1 d.p.

    a $\sqrt{8}$ **b** $\sqrt{21}$ **c** $\sqrt{129}$ **d** $\sqrt{6 \cdot 8}$ **e** $\sqrt{17 \cdot 4}$

*11 Use your calculator to find these.
Give the answers to **e** onwards to 2 d.p.

    a $5 \cdot 3^2$ **b** $0 \cdot 06^2$ **c** $6 \cdot 7^2$ **d** $11 \cdot 3^2$ **e** $\sqrt{207}$
    f $\sqrt{425 - 113}$ **g** $(^-8 \cdot 41)^2$ **h** $\sqrt{118 - 47}$ **i** $\dfrac{(2+3)^2}{14}$ **j** $\sqrt{4^2 + 7^2}$

*12 What are the two smallest numbers that are both triangular and square?
The first few triangular numbers are   1   3   6   10...

*13 The difference of the squares of two consecutive even numbers is 20.
Find these two numbers.
You could start like this.

    Try $8^2 - 6^2 = 64 - 36$
                $= 28$ ✗

> Consecutive even numbers are even numbers which are next to each other, e.g. 6 and 8.

Discussion

The square root key on Kelly's calculator is broken.
She finds $\sqrt{11}$ like this.

3 x^2 = to get 9   **too small**

4 x^2 = to get 16   **too big**

3·5 x^2 = to get 12·25   **too big**

3·3 x^2 = to get 10·89   **too small**

3^2 9
4^2 16

How could Kelly find $\sqrt{11}$ to 1 d.p. **Discuss.**
What about $\sqrt{17}$
What about $\sqrt{5}$?

T

Investigation

You could use a
spreadsheet to help.

Happy Numbers

Is 13 a happy number?

Step 1   Square each digit and find the sum. $1^2 + 3^2 = 1 + 9 = 10$
Step 2   Keep squaring the digits and finding the sum until you get a single
       digit. $1^2 + 0^2 = 1 + 0 = 1$

If the single digit is 1, the number is a happy
number. 13 is a happy number.

Is 865 a happy number?

Step 1 $8^2 + 6^2 + 5^2 = 64 + 36 + 25 = 125$
Step 2 $1^2 + 2^2 + 5^2 = 1 + 4 + 25 = 30$
       $3^2 + 0^2 = 9 + 0 = 9$

865 is not a happy number.

Is your house number a happy number?
Is your birthdate (MM DD YY) a happy number?
Is your telephone number a happy number? **Investigate**

 Puzzle

1 Find the missing digits.
▲ can stand for different digits.

a $(▲2)^2 = 144$ b $▲▲^2 = 100$ c $(▲4)^2 = ▲9▲$

*d $(▲8)^2 = ▲8▲$ *e $(▲▲)^2 = ▲▲25$

2 Jody's teacher said that in 4 years time his age would be $\sqrt{2025}$ years. What is his age now?

Investigation

Last digits

1 $4^2 = 16$.

The last digit of 4^2 is 6.

     $3^2 = 9$.

The last digit of 3^2 is 9.

Copy and complete this list for the last digits of $1^2, 2^2, 3^3, ... 9^2$.

     1, 4, 9, 6, 5, ___, ___, ___, ___

Describe the pattern.

     $10^2 = 100$.

The last digit of 10^2 is 0.

     $11^2 = 121$.

The last digit of 11^2 is 1.
Find the last digits of $12^2, 13^2, 14^2, 15^2$.

What do you think the last digits of $16^2, 17^2, 18^2, 19^2$ and 20^2 will be?
Use your calculator to check your answers.

Explain why 73, 28, 122, 137 could not be square numbers.

*2   9604 is a square number.
▲■^{2} = 9604
■ could be 8.
What other number could ■ be?
What could ■ be for these square numbers?

     ▲■^{2} = 2209   ▲■^{2} = 3025   ▲■^{2} = 1681

Cubes and cube roots

Discussion

You might like to use small cubes of side 1 cm to help in this discussion.

How many cubes of side 1 cm are in the
cube of side 2 cm?
How many cubes of side 1 cm are in the
cube of side 3 cm?
What if the cube was of side 4 cm?
What if the cube was of side 5 cm?
What if the cube was of side 10 cm?

Cube of side 2 cm

Cube of side 3 cm

A cube is made from 27 small cubes. How many small cubes are along each side?
What if the cube was made from 8 small cubes?
What if the cube was made from 64 small cubes?
What if the cube was made from 125 small cubes?
What if the cube was made from 1000 small cubes?
What if the cube was made from just 1 small cube? **Discuss.**

8^3 is read as 'eight **cubed**' and means $8 \times 8 \times 8$.

$$1^3 = 1, \qquad 2^3 = 8, \qquad 3^3 = 27, \qquad 4^3 = 64, \qquad 5^3 = 125, \qquad 10^3 = 1000$$

The 3 is an **index**.

1, 8, 27, 64, 125 and 1000 are **cube numbers**.

$\sqrt[3]{27}$ is read as 'the cube root of 27'.

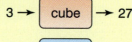

8^3 can be read as
8 to the power of 3.

The plural of
'index' is 'indices'.

To find $\sqrt[3]{27}$ we ask 'What number cubed equals 27?'
$$3 \times 3 \times 3 = 27 \text{ so } \sqrt[3]{27} = 3.$$
$$\sqrt[3]{1} = 1 \quad \sqrt[3]{8} = 2 \quad \sqrt[3]{27} = 3 \quad \sqrt[3]{64} = 4 \quad \sqrt[3]{125} = 5 \quad \sqrt[3]{1000} = 10$$

Exercise 9

1 Write these using indices.
  a 6×6　　**b** 4×4　　**c** $5 \times 5 \times 5$　　**d** $8 \times 8 \times 8$　　**e** $12 \times 12 \times 12$

2 Find the missing numbers mentally.
  a 3 cubed equals ___.　　**b** 64 is the cube of ___.　　**c** ___ is the cube of 3.
  d 1000 is the cube of ___.　　**e** 4 cubed equals ___.　　**f** 125 is the cube of ___.
  g The cube root of 27 is ___.　　　**h** The cube root of 64 is ___.
  i ___ is the cube root of 8.　　　　**j** ___ is the cube root of 27.
  k The cube root of 1 is ___.　　　　**l** The cube root of 64 is ___.
  m ___ is the cube root of 1000.

T

3 Use a copy of this.
Shade the factors of 20
the square numbers
the cube numbers
the multiples of 5.
What animal does your shading make?

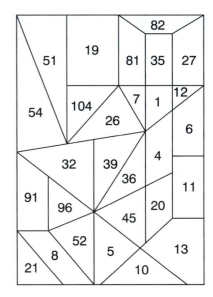

4 Find the answers to these mentally.

a $\sqrt[3]{8}$ b 2^3 c $\sqrt[3]{64}$ d 3^3 e $\sqrt[3]{1000}$

f 5^3 g 1^3 h $\sqrt[3]{125}$ i $\sqrt[3]{1}$ j 4^3

5 Belinda wrote this number pattern. The last
two lines are smudged. What are they?

$1^3 = 1$
$2^3 = 3 + 5$
$3^3 = 7 + 9 + 11$

*6 There is just one 2-digit number that is both a square number and a cube number.
Which 2-digit number is this?

Make it Special – a game for a group

You will need 2 dice, paper and a pen.

To play

- Take turns to roll the two dice.

- Make a 2-digit number with the numbers rolled.

- If you can make **a multiple of 3, take 1 point.**
If you can make **a prime number, take 2 points.**
If you can make **a square number, take 3 points.**
If you can make **a cube number, take 12 points.**

Example   Caroline rolled these.
She could make 63 or 36.
She made 36, which is a multiple of 3 and a square
number. She took 4 points.

- The winner is the person with the most points after 10 rounds.

Number

To **find cubes using the calculator** we use the x^3 key.

To find cube roots we use the $\sqrt[3]{}$ key.

Examples $8 \cdot 1^3$ is keyed as

$8 \cdot 1$ x^3 $=$ to get $531 \cdot 441$.

$\sqrt[3]{91}$ is keyed as

$\sqrt[3]{}$ 91 $=$ to get $4 \cdot 5$ (1 d.p.).

Cubing and finding the cube root are inverse operations.

4 → cubed → 64

4 ← cube root ← 64

Some calculators do not have a $\sqrt[3]{}$ key.

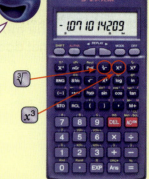

Exercise 10

1

A									
9261	40	3375	5·6	9		4·913	11	9	551·368

			A						A
3375	9	12	9261	6	6	551·368	⁻729		9261

5·6	7	20	4·913	⁻50·653	6	551·368

Use a copy of the box.
Find the answer to these.
Use your calculator.
Put the letter that is beside each above its answer.

A $21^3 = \mathbf{9261}$ **I** 15^3 **D** $(^-9)^3$ **N** $1 \cdot 7^3$ **E** $8 \cdot 2^3$

T $(^-3 \cdot 7)^3$ **R** $\sqrt[3]{343}$ **L** $\sqrt[3]{216}$ **S** $\sqrt[3]{729}$ **C** $\sqrt[3]{1728}$

P $\sqrt[3]{64\,000}$ **O** $\sqrt[3]{1331}$ **U** $\sqrt[3]{8000}$ **G** $\sqrt[3]{175 \cdot 616}$

2 Give the answers to 1 d.p.
Use your calculator.

a $\sqrt[3]{12}$ **b** $\sqrt[3]{18}$ **c** $\sqrt[3]{40}$ **d** $\sqrt[3]{60}$ **e** $\sqrt[3]{960}$ **f** $\sqrt[3]{124}$

? Puzzle

Find the missing digits.
▲ can stand for different digits.

a $(\blacktriangle\blacktriangle)^3 = 1000$ **b** $(1\blacktriangle)^3 = \blacktriangle\blacktriangle 28$

c $(\blacktriangle 7)^3 = \blacktriangle 91\blacktriangle$ **d** $(\blacktriangle\blacktriangle)^3 = \blacktriangle\blacktriangle\blacktriangle 7$

Summary of key points

 We can **add and subtract integers** using a number line.

Examples $^-3 + 2 = {}^-1$

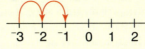

$4 + (^-3) = 1$

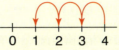

$^-2 + {}^-3 = {}^-5$

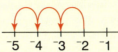

$2 - (^-1) = 3$

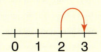

Multiplying (or dividing) two negative numbers gives a positive number.
Multiplying (or dividing) one negative and one positive number
gives a negative number.

Examples $^-3 \times 4 = {}^-12$ **negative × positive = negative**
            $^-2 \times {}^-5 = 10$ **negative × negative = positive**
            $^-12 \div 6 = {}^-2$ **negative ÷ positive = negative**
            $^-14 \div {}^-2 = 7$ **negative ÷ negative = positive**

We can use a calculator to do calculations with integers.

(–) changes the sign of a number.

Example   For $27 - {}^-76$ **key** to get 103.
            For $192 \div {}^-16$ **key** to get $^-12$.

 The **factors** of a number are all the numbers that will divide exactly into
that number.

Example   The factors of 36 are 1, 2, 3, 4, 6, 9, 12, 18, 36.

The **multiples** of a number are found by multiplying the number by 1, 2,
3, 4, 5, 6, ...

Example   The first six multiples of 7 are:
            $7 \times 1 = 7$
            $7 \times 2 = 14$
            $7 \times 3 = 21$
            $7 \times 4 = 28$
            $7 \times 5 = 35$
            $7 \times 6 = 42$ ...

C A **prime number** has exactly two factors, 1 and itself.

Example   2, 3, 5, 7, 11, 13 ... are the first six prime numbers.

Prime factors are the factors of a number which are prime numbers.

Example   The factors of 10 are 1, 2, 5, 10.
The prime factors of 10 are 2 and 5.

Prime factors can be found using a **factor tree**.

Example

$2 \times 9 = 18$
$3 \times 3 = 9$

D The **HCF (highest common factor)** of two numbers is the largest factor of both numbers.

Example   The factors of 12 are 1, 2, 3, 4, **6**, 12.
The factors of 18 are 1, 2, 3, **6**, 9, 18.
The HCF of 12 and 18 is **6**.

The **LCM (lowest common multiple)** of two numbers is the smallest number that is a multiple of both numbers.

Example   The multiples of 6 are 6, 12, **18**, 24, ...
The multiples of 9 are 9, **18**, 27, 36, ...
The LCM of 6 and 9 is **18**.

We can use prime factors or Venn diagrams to find HCFs and LCMs.

Example $12 = \mathbf{2} \times 2 \times \mathbf{3}$ $12 = 2 \times \mathbf{2} \times \mathbf{3}$
$18 = \mathbf{2} \times \mathbf{3} \times 3$ $18 = \mathbf{2} \times \mathbf{3} \times 3$
HCF of 12 and 18 $= \mathbf{2} \times \mathbf{3}$ LCM of 12 and 18 $= 2 \times \mathbf{2} \times \mathbf{3} \times 3$
$\qquad\qquad = \mathbf{6}$ $\qquad\qquad\qquad = \mathbf{36}$

We write the 2×3 once.

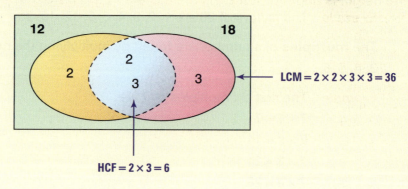

E 1, 4, 9, 16, ... are called **square numbers**.

$1^2 = 1$, $2^2 = 4$, $3^2 = 9$, $4^2 = 16$, $5^2 = 25$, $6^2 = 36$,
$7^2 = 49$, $8^2 = 64$, $9^2 = 81$, $10^2 = 100$, $11^2 = 121$, $12^2 = 144$

$\sqrt{1} = 1$, $\sqrt{4} = 2$, $\sqrt{9} = 3$, $\sqrt{16} = 4$, $\sqrt{25} = 5$, $\sqrt{36} = 6$, $\sqrt{49} = 7$,
$\sqrt{64} = 8$, $\sqrt{81} = 9$, $\sqrt{100} = 10$, $\sqrt{121} = 11$, $\sqrt{144} = 12$

$\sqrt{25}$ is read as 'the **square root** of 25'.
To find $\sqrt{25}$ we find what number squared gives 25.

$5 \times 5 = 25$ and $^-5 \times {}^-5 = 25$
$\pm\sqrt{25} = 5$ or $^-5$

> Usually we just give the positive square root.

We can **use a calculator** to find squares and square roots.

Example $3{\cdot}5^2$ is keyed as ⟨3·5⟩ ⟨x^2⟩ ⟨=⟩ to get 12·25.

$\sqrt{225}$ is keyed as ⟨$\sqrt{}$⟩ ⟨225⟩ ⟨=⟩ to get 15.

 F 1, 8, 27, 64, 125 and 1000 are all **cube numbers**.

$1^3 = 1$ $2^3 = 8$ $3^3 = 27$ $4^3 = 64$ $5^3 = 125$ $10^3 = 1000$
$\sqrt[3]{1} = 1$ $\sqrt[3]{8} = 2$ $\sqrt[3]{27} = 3$ $\sqrt[3]{64} = 4$ $\sqrt[3]{125} = 5$ $\sqrt[3]{1000} = 10$

Examples   To find $\sqrt[3]{729}$ key ⟨$\sqrt[3]{x}$⟩ ⟨729⟩ ⟨=⟩ to get 9.

To find $2{\cdot}4^3$ key ⟨2·4⟩ ⟨x^3⟩ ⟨=⟩ to get 13·824.

Test yourself

1 This table gives the height above sea level of some places.
If the height is negative it means the place is below sea level.

Place	Height above sea level
Mediterrenean sea	0 m
Jerusalem	800 m
Jericho	$^-260$ m
Dead Sea	$^-400$ m
London	30 m

  a Which place is furthest below sea level?
  b Which place is highest above sea level?
  c How much higher than London is Jerusalem?
  d How much higher than Jericho is London?
  e What is the difference in height between London and the Dead Sea?

2 Write these temperatures in order, coldest first.

    3 °C $^-7$ °C   1 °C $^-1$ °C   4 °C $^-10$ °C

Number

T **3** Use a copy of this.
What number goes in the red circles?
  a In the green circles, two numbers are added to get the number above.
  b In the blue circles, the number on the right is subtracted from the number on the left to get the number above.

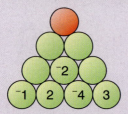
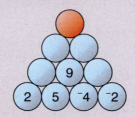

T **4** Use a copy of this. A
Fill it in.

×	2	⁻3	5	⁻5
2				
⁻1				
⁻4				
3				

T **5** There are four multiplication and division facts using 7, ⁻2 and ⁻14.

$7 \times {}^-2 = {}^-14$ ${}^-2 \times 7 = {}^-14$ ${}^-14 \div 7 = {}^-2$ ${}^-14 \div {}^-2 = 7$

Use a copy of this.
Fill in the gaps with more facts using the numbers given.

  a $3 \times {}^-5 = {}^-15$ **b** ${}^-6 \times {}^-4 = 24$ **c** _____

     _____   _____   _____

     ${}^-15 \div 3 = {}^-5$ $24 \div {}^-6 = {}^-4$ $35 \div {}^-7 = {}^-5$

     _____

T **6** Use a copy of these.
Fill them in.
Use your calculator.

  a

+	⁻29	79	⁻82
46			
⁻47			
⁻38			

  b

×	⁻5·5	42	⁻4·5
⁻34			
8·6			
⁻3·8			

7 Write down all the common factors of
   a 14 and 35   **b** 24 and 42

8 Write down all the prime numbers between 10 and 40.

T

9 Use a copy of these.
   Fill in the gaps.
   a

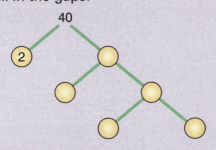

   b

10 a Use your answers to question **9** to write down the prime
     factors of these.
     i 40   **ii** 48
   b Now write each of these as a product of
     prime factors.
     i 40   **ii** 48

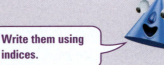

Write them using indices.

11 a Write down the factors of 18 and 42.
   b Write down the HCF of 18 and 42.
   c Write down the first six multiples of 4.
   d Write down the first six multiples of 6.
   e Write down the LCM of 4 and 6.

12 a $20 = 2 \times 2 \times 5$
     $48 = 2 \times 2 \times 2 \times 2 \times 3$
     Copy and complete the diagram for
     20 and 48.
   b Use the diagram to find the HCF and
     LCM of 20 and 48.

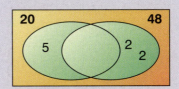

13 What goes in the gaps?
   a ___ squared is 64.   **b** 25 is the square of ___.
   c The square of ___ is 81.   **d** 4 is the square root of ___.
   e The square root of 9 is ___.   **f** The square root of ___ is 6.

14 Calculate these mentally.

    a $\sqrt{20+5}$ **b** $(6-4)^2$ **c** $(2+5)^2$ **d** $\sqrt{2^2+5}$

15 Use your calculator to find these.

    Give the answers to **c** and **d** to 1 d.p.

    a $6 \cdot 8^2$ **b** 29^2 **c** $\sqrt{33}$ **d** $\sqrt{109}$

T

16 Use a copy of this.

    Cross out all the factors of 12.

    Cross out all the multiples of 5.

    Cross out all the prime numbers
that are left.

    Cross out the square numbers
that are left.

    Cross out the cube numbers
that are left.

    What number is left?

36	9	29	19	35	2
59	4	100	1	10	8
200	28	13	23	53	3
5	49	12	64	17	150
7	81	11	15	6	27

17 **a** 2 cubed equals ___.   **b** 27 is the cube of ___.

    c ___ is the cube of 4.   **d** 125 is the cube of ___.

    e 10 cubed equals ___.   **f** The cube root of 8 is ___.

    g ___ is the cube root of 27.   **h** 4 is the cube root of ___.

18 Give the answers to **c** and **d** to 1 d.p.

    Use your calculator.

    a 12^3 **b** 17^3 **c** $\sqrt[3]{54}$ **d** $\sqrt[3]{1070}$

3 Calculation

You need to know

Key vocabulary

estimate, complements, order of operations, partitioning

 A multiple of mysteries!

1
```
    **
×   **
  ─────
  *2*
```
What digit does ∗ stand for?

2 a A number is a multiple of 4.
      It is between 0 and 40.
      When it is divided by 5 it has remainder 4.
      What could it be?

   b A number divided by 5 has no remainder.
      When it is divided by 6 it has remainder 5.
      It is between 20 and 40
      What could it be?

Mental calculations

Adding and subtracting

These strategies can be used to **add and subtract mentally**.

1 Complements in 1, 10, 50, 100, 1000

Look for complements when adding and subtracting.

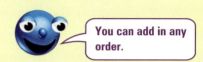
You can add in any order.

Examples
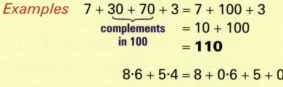

$7 + \underbrace{30 + 70}_{\substack{\text{complements} \\ \text{in } 100}} + 3 = 7 + 100 + 3$
$= 10 + 100$
$= \mathbf{110}$

0·6 and 0·4 are *complements* in 1.

$8·6 + 5·4 = 8 + 0·6 + 5 + 0·4$
$= 8 + 5 + 1$
$= \mathbf{14}$

2 Partitioning

Examples

$5·3 + 4·5 = 5 + 0·3 + 4 + 0·5$
$= 9 + 0·8$
$= \mathbf{9·8}$

$4·7 - 3·9 = 4·7 - 3 - 0·9$
$= 1·7 - 0·9$
$= 1·7 - 0·7 - 0·2$
$= 1 - 0·2$
$= \mathbf{0·8}$

3 Counting up

Example $3700 - 1800$

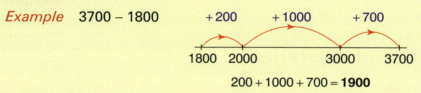

$200 + 1000 + 700 = \mathbf{1900}$

Example $17·8 + 4·5$

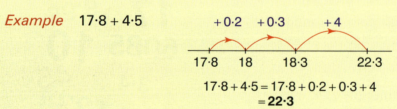

$17·8 + 4·5 = 17·8 + 0·2 + 0·3 + 4$
$= \mathbf{22·3}$

4 Nearly numbers

Example $5·8 + 3·9 = (6 - 0·2) + (4 - 0·1)$
$= 6 + 4 - 0·2 - 0·1$
$= 10 - 0·2 - 0·1$
$= 9·8 - 0·1$
$= \mathbf{9·7}$

5 Adding or subtracting too much then compensating

Examples

$1·83 + 2·9 = 1·83 + 3 - 0·1$
$= 4·83 - 0·1$
$= \mathbf{4·73}$

$6·25 - 2·9 = 6·25 - 3 + 0·1$
$= 3·25 + 0·1$
$= \mathbf{3·35}$

6 Using facts you already know

Example 0·28 + 0·43
We know that 28 + 43 = 71.
so 0·28 + 0·43 = 0·71.

Exercise 1   **This exercise is to be done mentally.
You may use jottings.**

1 a 6 + 3 + 10   **b** 12 + 6 − 3   **c** 7 + 10 − 2 − 4
  d 20 + 16 − 5 − 3   **e** 19 − 3 − 15 + 7 − 8   **f** 16 + 20 − 5 − 7 + 14

2 a 0·3 + 0·4 − 0·1   **b** 0·4 + 0·6 − 0·3   **c** 0·8 + 0·3 − 0·4 + 0·7
  d 1·2 − 0·6 + 0·4 − 0·7   **e** 1·4 − 0·8 + 1·1 − 0·6 − 0·8

3 Use complements to find x.
  a 100 = x + 27   **b** 50 = 28 + x
  c 10 = 4·71 + x **d** 100 − x = 71
  e 50 − x = 17   **f** 452 = 1000 − x **g** 5·36 = 10 − x

Remember:
Complements add
to a multiple of 10.

4 In this pyramid, to get each number we added the two
numbers above it.

Find the number that goes in the yellow square.

15	18	12
	33	30
		63

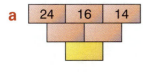   **a** 24 16 14   **b** 5 30 8 6   **c** 7 3 6 9 5

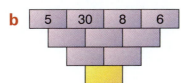

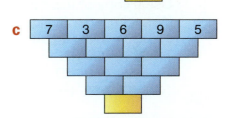

5 A German artist, Dürer, made a wood cut called 'Melancholia'.
It had this 4 × 4 magic square in it.

16	3	2	13
5	10	11	8
9	6	7	12
4	15	14	1

  a What does each row, column and diagonal add to?
  b If you add 2 to each of the numbers in the 'Melancholia'
    square, do you get a new magic square?
  c Use a copy of this magic square.
    Fill in the gaps.
    In what way is this magic square the same as the 'Melancholia'
    square?

1	14	15	
12			6
		11	10
13		3	16

Number

6 **a** 420 + 360 **b** 320 – 260 **c** 530 + 860 **d** 940 – 370
   e 5800 + 6300 **f** 3200 – 1800 **g** 5200 – 3500

7 Is it possible to get these totals by throwing four darts at this
   dart board? If so, how? Assume no darts miss the board.
   a 110 **b** 135 **c** 130 **d** 145
   e 140

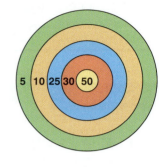

5 | 10 | 25 | 30 | 50

T **8**

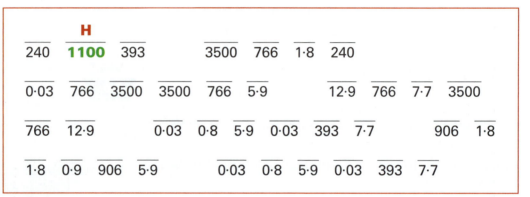

	H					
240	**1100**	393		3500	766	1·8	240

| 0·03 | 766 | 3500 | 3500 | 766 | 5·9 | | 12·9 | 766 | 7·7 | 3500 |

| 766 | 12·9 | | 0·03 | 0·8 | 5·9 | 0·03 | 393 | 7·7 | | 906 | 1·8 |

| 1·8 | 0·9 | 906 | 5·9 | | 0·03 | 0·8 | 5·9 | 0·03 | 393 | 7·7 |

Use a copy of this box.
Write the letter beside each calculation above its answer in the box.
H 730 + 370 = **1100**   **T** 730 – 490   **S** 0·2 + 1·6   **A** 1·7 – 0·9
I 526 + 380   **F** 8·3 + 4·6   **O** 468 + 298   **M** 5200 – 1700
E 690 – 240 – 57   **N** 8·4 – 2·5   **K** 4·6 – 3·7   **C** 0·08 – 0·05
R 8·6 + 3·2 – 4·1

9 Find the answers to these mentally.
   a Add 592 and 750.
   b What is the **sum** of 753 and 132?
   c Find the **difference** between 8·3 and 3·2.
   d Add 8·4, 8·6 and 8·2.
   e Decrease 19·5 by 12·6.
   f Increase 4860 by 3320.

*10 **a** Choose **five** digits from the box to make this true.
            □□ + □□ + □ = 100
   b Choose **six** digits from the box to make this true.
           □□ + □□ – □□ = 100

In each part use a
digit only once.

| 1 | 3 | 5 | 7 |
| | 4 | | 6 |

Puzzle

1 There are five different ways Paul can go from home to school.
He goes a different way each day and comes home the same way.

Distances are in miles

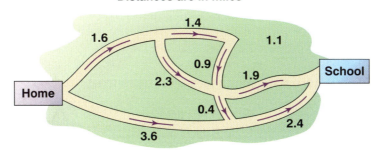

a How far is each way?
b How far does Paul travel in total each week?

2 The numbers on the lines are
the differences between the
numbers in the circles.
Find some ways to fill in the
circles.

a
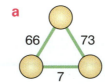

b
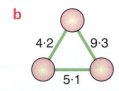

*Investigation

Magic Squares

1

12
1

4	15

16	3

7
14

2	13
11	8

5	10
9	6

Each of the six small blocks is to be placed on the large
grid to make a magic square.

Investigate ways of doing this if the [12/1] block is placed as shown.

What if the [12/1] block was put somewhere else?

continued

65

2 The numbers in this magic square have been mixed up. Change the numbers around so that the sum of each row, each column and each diagonal is 3·3.

Can this be done in more than one way? **Investigate.**

1·5	0·9	0·3
0·7	1·3	1·7
1·1	0·5	1·9

 Puzzle

1 Replace the letters with digits to make these true.

a
```
  N I N E
– _ T E N
  T W O
```

b
```
    S U N
+ B U R N
F E V E R
```

To help you, in a N = 1 in b N = 7

＊2
```
  A·B B
+ B·A A
D C·C D
```

Replace A, B, C and D with digits to make the addition correct.

Multiplying and dividing

These ways can be used to **multiply and divide mentally**.

Multiplying and dividing by multiples of 10 and 100

Examples

$800 \times 4 = 8 \times 100 \times 4$
$= 8 \times 4 \times 100$
$= 32 \times 100$
$= \mathbf{3200}$

$1600 \div 800 = \dfrac{1600}{8 \times 100}$
$= 1600 \div 8 \div 100$
$= \mathbf{2}$

$1600 \div 8 = 200$
$200 \div 100 = 2$

Partitioning

Examples

$12 \times 3{\cdot}5 = (10 \times 3{\cdot}5) + (2 \times 3{\cdot}5)$
$= 35 + 7$
$= \mathbf{42}$

$4{\cdot}2 \times 6 = (4 \times 6) + (0{\cdot}2 \times 6)$
$= 24 + 1{\cdot}2$
$= \mathbf{25{\cdot}2}$

$2 \times 6 = 12$
$0{\cdot}2 \times 6 = 1{\cdot}2$

Factors

Example

$3{\cdot}5 \times 8 = 3{\cdot}5 \times 2 \times 2 \times 2$
$= 7 \times 2 \times 2$
$= 14 \times 2$
$= \mathbf{28}$

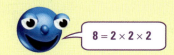
$8 = 2 \times 2 \times 2$

Near tens

Example

$43 \times 11 = 43 \times 10 + 43$
$= 430 + 43$
$= \mathbf{473}$

Known facts

We know that $5 = 10 \div 2$ and that $25 = 100 \div 4$.

Example $48 \times 25 = 48 \times 100 \div 4$
$= 4800 \div 2 \div 2$
$= 2400 \div 2$
$= \mathbf{1200}$

Doubling and halving

We can double one number and halve the other.

Examples $46 \times 5 = 23 \times 10$ $16 \times 2{\cdot}5 = 8 \times 5$
$= \mathbf{230}$ $= \mathbf{40}$

Half of 16 is 8.
Double 2·5 is 5.

Place value

We make the numbers into whole numbers by multiplying by 10, 100 or 1000.

Examples $83 \times 0{\cdot}01 = \mathbf{0{\cdot}83}$
$\downarrow \times 100 \quad \uparrow \div 100$
$83 \times 1 \quad = \mathbf{83}$

We multiplied by 100 so we must divide the answer by 100.

There is more about multiplying by 10, 100, 1000 on page 1.

$7 \times 0{\cdot}4 = 7 \times 4 \div 10$ $0{\cdot}8 \times 0{\cdot}6 = 8 \times 6 \div 10 \div 10$
$= 28 \div 10$ $= 48 \div 100$
$= \mathbf{2{\cdot}8}$ $= \mathbf{0{\cdot}48}$

Exercise 2   **This exercise is to be done mentally. You may use jottings.**

1 **a** $2 \times 3 \times 4$ **b** $3 \times 5 \times 2$ **c** $4 \times 6 \times 2$ **d** $8 \times 3 \times 2$
   e $5 \times 5 \times 2$ **f** $7 \times 3 \times 3$ **g** $6 \times 8 \times 2$

2 **a** $40 \times 30 = 4 \times 10 \times 3 \times 10$ **b** $50 \times 60 = 5 \times 10 \times 6 \times 10$
$= 4 \times 3 \times 100$ $= 5 \times 6 \times 100$
$= \underline{\hspace{2cm}}$ $= \underline{\hspace{2cm}}$
$= \underline{\hspace{2cm}}$ $= \underline{\hspace{2cm}}$

   c $200 \times 40 = 2 \times 100 \times 4 \times 10$
$= 2 \times 4 \times 1000$
$= \underline{\hspace{2cm}}$
$= \underline{\hspace{2cm}}$

   d $900 \div 3$ **e** $900 \div 30$ **f** $900 \div 300$
   g 300×60 **h** $600 \div 20$ **i** 800×70

3 Find these.
   The first few have been started.
   a $160 \times 4 = (100 \times 4) + (60 \times 4)$ **b** $260 \times 5 = 260 \times 10 \div 2$
$= \underline{\hspace{2cm}}$ $= \underline{\hspace{2cm}}$
   c $210 \times 6 = 210 \times 2 \times 3$ **d** 440×3 **e** 150×9
$= \underline{\hspace{2cm}}$
   f 230×8 **g** 124×4 **h** 282×5 **i** 514×6
   *j 16×25 *k 26×11 *l 25×14 *m 42×29

4 **a** $3{\cdot}8 \times 0{\cdot}1$ **b** $4 \times 0{\cdot}4$ **c** $5 \times 0{\cdot}6$
   d $8 \times 0{\cdot}3$ *e $4{\cdot}1 \times 0{\cdot}6$

Use partitioning or make the numbers into whole numbers or use patterns.

5 Calculate these.

   a $6{\cdot}2 \times 2 = 6 \times 2 + 0{\cdot}2 \times 2$ **b** $1{\cdot}4 \times 6 = 1 \times 6 + 0{\cdot}4 \times 6$

          = _____   = _____

   c $2{\cdot}4 \times 3$ **d** $4{\cdot}8 \times 5$ **e** $2{\cdot}9 \times 8$ **f** $3{\cdot}4 \times 7$ **g** $2{\cdot}5 \times 8$

   h $2{\cdot}5 \times 16$ **i** $6{\cdot}4 \times 4$ **j** $2{\cdot}6 \times 9$ **k** $3{\cdot}7 \times 8$ **l** $13 \times 1{\cdot}2$

6 a $3{\cdot}6 \times 7 = 25{\cdot}2$

     Write down another multiplication fact and two division facts that use the
numbers 3·6, 7 and 25·2.

   b Given that $42{\cdot}8 \times 1{\cdot}3 = 55{\cdot}64$, write down the answers to these.

     i $55{\cdot}64 \div 1{\cdot}3$ **ii** $55{\cdot}64 \div 42{\cdot}8$ **iii** $42{\cdot}8 \times 13$

7 a $550 \div 5 = (500 \div 5) + (50 \div 5)$ **b** $280 \div 4 = (200 \div 4) + (80 \div 4)$

            = _____   = _____

   c $270 \div 6 = (240 \div 6) + (30 \div 6)$

            = _____

   d $225 \div 5$ **e** $484 \div 4$ **f** $184 \div 8$

T

8

$\overline{\quad}$	$\overline{\quad}$		$\overline{\quad}$	$\overline{\quad}$	$\overline{\quad}$
2·8	45		2	1560	45

$\overline{90}$ $\overline{45}$ $\overline{15{\cdot}6}$ $\overline{2{\cdot}8}$ $\overline{35}$ $\overline{0{\cdot}02}$ $\overline{103}$ $\overline{20{\cdot}5}$ $\overline{400}$ $\overline{520}$

                                      H

$\overline{0{\cdot}02}$ $\overline{45}$ $\overline{45}$ $\overline{20{\cdot}5}$ $\overline{2}$ $\overline{\textbf{1600}}$ $\overline{400}$ $\overline{520}$ $\overline{103}$ $\overline{1200}$ $\overline{400}$

Use a copy of this box.

Write the letter beside each calculation above its answer in the box.

H $80 \times 20 = \textbf{1600}$ **M** 30×40 **T** $600 \div 300$ **S** 130×4 **W** 312×5

E 16×25 **R** $5{\cdot}2 \times 3$ **O** $2{\cdot}5 \times 18$ **N** $4 \times 0{\cdot}7$ **L** $0{\cdot}08 \div 4$

C $450 \div 5$ **A** $618 \div 6$ **F** $280 \div 8$ **K** $8{\cdot}2 \times 2{\cdot}5$

T

9 Use a copy of these number chains.
Fill in the missing numbers.

   a

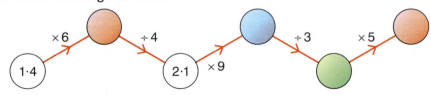

   b

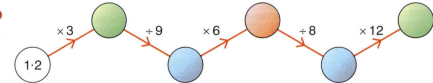

10 Do these by doubling and halving.
    a 6.4×5 **b** 12.6×5 **c** 32×20 **d** 15×16 **e** 246×50

11 **a** Find the **product** of 80 and 4.
    b $15 \times 49 = 735$. What is 15×4.9?
    ∗**c** What number divided by 8 gives the answer 1.2?
    ∗**d** Subtract 4.6 from the product of 2.4 and 6.

∗**12** Find ways of filling in the box and the circle.
    a $\square \times \bigcirc = 0.6$ **b** $\square \div \bigcirc = 0.4$

∗**13** Choose numbers from the box to make these true.
Use each number only once in each part.
    a $\square \times \square + \square = 6.4$ **b** $(\square + \square) \times \square = 20$
    c $(\square + \square) \div (\square - \square) = 3.1$

1·2	2	2·5
1·5	4	5

Solving problems mentally

**This exercise is to be done mentally.
You may use jottings.**

1 Write the answers to these as quickly as possible.
    a What is £27·95 to the nearest £10?
    b If $a + b = 20$, when a is 8 what is b?
    c Write a number that is bigger than zero point five but smaller than zero point six.
    d Add six to negative seven.
    e I am thinking of a number, n.
I subtract 3 from it.
Write an expression to show what I have now.
    f It takes 4 hours to travel from my home to the station.
I arrive at 2:30 p.m. What time did I leave home?

There are lots of links to other areas of maths in this exercise. Try to find them.

    g Shade one quarter of this diagram.
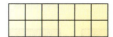
    h How much is half of £19·60?
    i A plan has a scale of 1 cm to 5 m.
On the plan a hall is 3 cm long.
How long is the hall in real life?
    j Write thirty-seven million in figures.
    k Copy this. Complete the diagram so the shaded part is the net of a cube.
    l A bag has 3 blue, 5 red and 12 purple counters in it.
If I take a counter at random from the bag, what is the probability it will be red?
    ∗**m** $34 \times 56 = 1904$
What is 35×56?

Number

Estimating answers to calculations

Jenna's school is putting more shelves in the library.
There are 580 books that need extra shelves.
Each shelf holds about 28 books.
Jenna wants to **estimate** how many extra shelves will be needed.

To **estimate the answer to a calculation** we round the numbers.
$\approx$ means 'is **approximately equal to**'.

Jenna worked out her estimate like this. $580 \div 28 \approx 600 \div 30$
$$= 20$$

When rounding numbers we try to round to 'nice numbers' which are easy to calculate mentally.

Example $\frac{58}{7} \approx \frac{56}{7} = 8$

> Approximate $\frac{58}{7}$ to $\frac{56}{7}$ rather than $\frac{60}{7}$.
> 56 is a multiple of 7.

Worked Example

Estimate the answers to these.   **a** 188×32 **b** $\frac{18 \cdot 6 - 5 \cdot 2}{3 \cdot 5}$

Answers

a $188 \times 32 \approx 200 \times 30$

$$= \mathbf{6000}$$

b $\frac{18 \cdot 6 - 5 \cdot 2}{3 \cdot 5} \approx \frac{20 - 5}{3}$

$$= \frac{15}{3}$$

$$= \mathbf{5}$$

> We choose 3 instead of 4 because $\frac{15}{3}$ is easier to do mentally than $\frac{15}{4}$.

Exercise 4

1 Choose the best estimate.

		A	**B**	**C**
a	33×57	30×60	40×50	40×60
b	82×78	80×70	100×80	80×80
c	$942 \div 32$	$1000 \div 30$	$900 \div 40$	$900 \div 30$
d	748×26	800×30	700×30	700×20
e	$324 \div 53$	$300 \div 60$	$300 \div 50$	$400 \div 50$
f	$20 \cdot 9 \times 39 \cdot 8$	20×39	20×30	20×40
g	$9 \cdot 18 \times 4 \cdot 81$	10×5	9×4	9×5

2 Estimate the answers to these.
Show how you found your estimate.

a 125×76 **b** 341×77 **c** $214 \div 19$ **d** $527 \div 48$
e $378 \div 37$ **f** 211×82

3 Estimate the answers to these.

a $81 \cdot 6 \times 5$ **b** $21 \cdot 6 \div 3$ **c** $48 \cdot 4 \div 5$ **d** $64 \cdot 2 \div 8$
e $5 \cdot 6 \times 3 \cdot 2$ **f** $8 \cdot 9 \times 7 \cdot 6$

4 For each of these write down a calculation you could do to estimate the answer.

a A club spent £248 on a new building and £684 on landscaping.
How much was spent altogether?

b Each class at Hendley School had 32 pupils.
How many pupils were in this school if there were 17 classes?

c 46 pupils each made 23 masks to be sold at the school fête.
How many masks did the pupils make altogether?

You could check your answer using a calculator.

d In a fire drill, the pupils at a school lined up in rows of 38.
How many rows of pupils would there be if there were 798 pupils at school on the day of a fire drill?

e Martha used 8 bottles of sauce.
Each bottle contained 16·75 mℓ.
How much sauce did she use altogether?

f 778 football fans travelled by coach from Derby to Liverpool.
Find the number of coaches needed if each could carry 43 passengers.

Written calculations – adding and subtracting

Remember
We **add** and **subtract** by lining up digits with the same place value.
For decimals, line up the decimal points.
Estimate the answer first.

Example   Dan wanted to know the total area of two of his posters.
They are 52·7 cm^2 and 138·4 cm^2.
52·7 + 138·4 is approximately equal to
50 + 140 = 190.
He added the areas.

$$\begin{array}{r} 52{\cdot}7 \\ 138{\cdot}4 \\ \hline 191{\cdot}1 \\ \hline \tiny{1\ 1} \end{array}$$

The total area is **191·1 cm^2**.

Worked Example

16 − 5·29

Answer

$$\begin{array}{r} {}^{5}\ {}^{1}9^{1} \\ 16{\cdot}00 \\ -\ \ 5{\cdot}29 \\ \hline 10{\cdot}71 \end{array}$$

The answer is **10·71**.

Number

Exercise 5

1 **a** 654 + 543 **b** 2356 − 1142 **c** 729 + 5496
   d 7092 − 354 **e** 5632 + 159 + 78 **f** 3145 + 68 − 154
   g 10 043 − 658 **h** 12 061 + 328 − 1516

T

2

| 9·55 | 105·38 | 62·74 | 244·34 | 82·87 | | 35·55 | 792·58 | 2·18 |

R
82·87   **67·1**   792·58   24·73   4·18   9·55

R
67·1   12·2   1·19   2·18   53·1   82·87   244·34   4·18

            R
4·18   792·58   **67·1**   82·87   244·34 $7\frac{1}{2}$

82·87   105·38   36·72   4·18   103·46   792·58

103·46   4·18   35·55   12·2   2·18   53·1

Use a copy of the box.
Find the answers to these.
Write the letter beside each question above its answer in the box.

R   62·4   **H**   61·7   **O**   33·4   **N**   4·9
     + 4·7   + 182·64   − 21·2   − 2·72
     67·1
      1

D 35·1 + 18   **V** 29·63 − 4·9   **I** 97 + 8·38   **G** 58·9 + 3·84
T 83·72 − 0·85   **U** 5·79 − 4·6   **E** 4·9 − 0·72   **A** 843 − 50·42
C 40·25 − 4·7   **L** 4·06 + 4·8 + 0·69   **M** 12·8 + 7·92 + 16   **S** 33·6 + 42·5 + 27·36

3 **a** Find the sum of 8·4 and 3·7 and subtract this from 20.
   b Find the difference between 536·04 and 7·8 then add 40 to the answer.
   c The sum of three numbers is 364·29.
      What might the numbers be?
   d The difference between two numbers is 12·73.
      What might the numbers be?

4 4·86 mm, 3·08 mm and 6·4 mm of rain fell in Manchester one Bank Holiday weekend.
  a How much rain fell altogether?
  b The same weekend 16 mm of rain fell in Minehead.
    How much more rain fell in Minehead?

5 Gareth sells fruit at a stall.
  One day he sold £268·40 worth of fruit.
  It had cost him £97·25.
  His rent for the day was £28·75.
  How much money did Gareth make that day?

6 An adult needs about 40 g of protein each day.
  Sasha works out she has eaten this much protein so far today.

  from vegetables 3·6 g   *from dairy products 6·8 g*   *from meat 14·4 g*

  a How much protein has she eaten so far today?
  b How many more grams does she need to eat to reach 40 g?

*7 Tim bought short ends of material to make a jacket for
  design and technology.
  These measured 1·55 m and 175 cm.
  Altogether Tim used 3 m to make his jacket.
  How many centimetres did Tim have left over?

  Link to design
  and technology.

*8 ☐☐·☐ and ☐·☐☐ and ☐☐·☐
  Choose a different digit from 1 to 9 for each box so that the three numbers
  a make the largest sum   **b** make the smallest sum.

*9 Put the digits 1 to 5 in the boxes to make each true.
  Use each digit only once in each part.
  a ☐·☐ + ☐·☐☐ = 4·74   **b** ☐·☐☐ − ☐·☐ = 3·33

Investigation

Ten pounds and eighty-nine pence

Follow these steps, writing **all** numbers as **decimals with two decimal places**.

Example

1 Choose an amount of money less than £10.   **£6·72**
2 Reverse the digits and write down the new amount.   **£2·76**
3 Find the difference.   **£6·72 − £2·76 = £3·96**
4 Reverse the digits in your answer to **3**.   **£6·93**
5 Add the answers to **3** and **4** together.   **£3·96 + £6·93 = £10·89**

Repeat steps **1** to **5**, starting with different amounts of money.
Do you always get £10·89? **Investigate.**

Multiplying

People need 0·6 g of protein a day for each kilogram they weigh.
Todd's father weighs 83 kg.
He needs $83 \times 0·6$ g of protein daily.

$83 \times 0·6$ is approximately $80 \times 0·6 = 48$ **because $80 \times 6 = 480$**

$83 \times 0·6$ is equivalent to $83 \times 0·6 \times 10 \div 10 = 83 \times 6 \div 10$

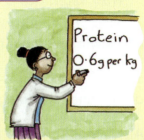

> Always estimate the answer first.

$$\begin{array}{r} 83 \\ \times \ \ 6 \\ \hline 498 \\ \scriptstyle 1 \end{array}$$ and $498 \div 10 = 49·8$

> We multiply 0·6 by 10 to get a whole number.
> ÷ 10 undoes × 10.
> We could also find $83 \times 0·6$ by
> $83 \times 6 \times 0·1 = 83 \times 6 \div 10$.

Todd's father needs 49·8 g of protein each day.

Worked Example
$16 \times 4·5$

Answer
$16 \times 4·5$ is approximately $20 \times 5 = 100$.
Two ways of finding the answer are shown below.

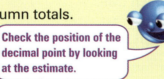

> Protein
> 0·6g per kg

Method 1

×	**10**	**6**	check
4	40	24	64
0·5	5	3	+ 8
	45	27	**72**

We add the column totals.

$45 + 27 = \mathbf{72}$

We check by adding the row totals.
$64 + 8 = 72$

> Check the position of the decimal point by looking at the estimate.

Method 2

$16 \times 4·5$ is equivalent to
$16 \times 4·5 \times 10 \div 10$ or $16 \times 45 \div 10$.

$$\begin{array}{r} 16 \\ \times \ \ 45 \\ \hline 640 \\ 80 \\ \hline 720 \\ \scriptstyle 1 \end{array}$$ 16×40
   16×5

Answer $720 \div 10 = \mathbf{72}$

> We multiply 4·5 by 10 to get a whole number. The ÷ 10 undoes this.

Exercise 6   **Always estimate the answer first.**

1 Use a copy of this cross number.
   Fill it in.

Across	Down
1. 460×29	**1.** 371×53
4. 852×16	**2.** 508×79
6. 673×45	**3.** 34×982
8. 343×25	**5.** 65×430
11. 423×18	**7.** 132×42
12. 84×782	**9.** 293×32
	10. 324×27

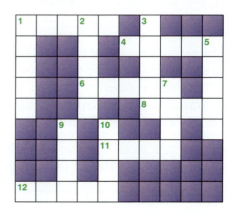

2 What goes in the boxes to make these true?
There is more than one answer for all of them.
How many different answers are there for each one?

a ☐☐
× ☐
‾‾‾‾‾‾‾
1 1 2

b ☐☐
× ☐
‾‾‾‾‾‾‾
1 4 0

c ☐☐
× ☐
‾‾‾‾‾‾‾
1 4 4

d ☐☐
× ☐
‾‾‾‾‾‾‾
1 3 2

3 A pencil can draw a line 56 kilometres long.
What length line could 128 of these pencils draw?

4 **a** $56 \times 0{\cdot}4$ **b** $42 \times 0{\cdot}6$ **c** $58 \times 0{\cdot}7$ **d** $42 \times 0{\cdot}9$ **e** $38 \times 0{\cdot}4$
    f $45 \times 0{\cdot}8$ **g** $324 \times 0{\cdot}3$ **h** $186 \times 0{\cdot}4$ **i** $143 \times 0{\cdot}7$ **j** $897 \times 0{\cdot}7$
   *__k__ $52{\cdot}6 \times 0{\cdot}9$ *__l__ $7{\cdot}94 \times 0{\cdot}8$ *__m__ $8{\cdot}63 \times 0{\cdot}5$ *__n__ $5{\cdot}72 \times 0{\cdot}6$

T

5

					A	
‾‾‾‾‾	‾‾‾‾‾	‾‾‾‾‾	‾‾‾‾‾	‾‾‾‾‾	‾‾‾‾‾	‾‾‾‾‾
307·38	188·33	188·33	128·8	41·6	**105·4**	4993·2

							A	
‾‾‾‾‾	‾‾‾‾‾	‾‾‾‾‾	‾‾‾‾‾	‾‾‾‾‾	‾‾‾‾‾	‾‾‾‾‾	‾‾‾‾‾
128·8	8384·8	188·33	832·3	832·3	285·48	188·33	**105·4**	685

Use a copy of the box.
Find the answers to these.
Write the letter beside each question above its answer in the box.

A $6{\cdot}2 \times 17 =$ **105·4**   **C** $3{\cdot}2 \times 13$ **S** $5{\cdot}6 \times 23$ **R** $25 \times 27{\cdot}4$
N $684 \times 7{\cdot}3$ **M** $892 \times 9{\cdot}4$ **F** $39 \times 7{\cdot}32$ **B** $94 \times 3{\cdot}27$
E $37 \times 5{\cdot}09$ **L** $406 \times 2{\cdot}05$

6 **a** A club is planning a trip.
The club hires 146 coaches.
Each coach holds 48 passengers.
How many passengers is that altogether?
Show your working.
b The club puts a first aid kit into each coach.
These are sold in boxes of 16.
How many boxes does the club need?

7 **a** Lee paid £15·50 each week to travel to school last year.
He went to school for 39 weeks.
How much did he pay to travel to school for the year?
Show your working.
b Lee could buy a special ticket which he could use for the 39 weeks.
It costs £585.
How much would that be each week?

Number

*8 Find the cost of these. You may have to round your answer to the nearest penny.

a Rose bought 15 m of rope.
b Peter bought 26 m of hose.
c Rhian bought 3·5 m of chain.
d Habib bought 17·2 m of plastic.
e Jake bought 3·2 m of rope and 15·6 m of plastic.

ROPE
£1·86/m

PLASTIC
£1·20/m

HOSE
£3·92/m

CHAIN
£8·20/m

*9 Find two numbers with a product of 875 and a difference of 10.

 Puzzle

In these sums, some of the digits got covered by ink blots. Work out what the missing digits are.

1

```
      35
    ×  ▨
    ─────
     770
```

*2

```
      ▨
    ×  3▨
    ─────
     ▨0
      1
    ─────
     52▨
```

Dividing

Worked Example

472 people are at a football dinner. They sit at tables which seat 22 people.

a How many tables are filled?
b How many people are at the table that is not filled?

Answer

We must divide 472 by 22 and find the remainder.

472 ÷ 22 is approximately 500 ÷ 20 = 25.

```
22 ) 472
    −440        22 × 20      because 22 × 2 = 44
    ─────
      32
    −  22        22 × 1
    ─────
      10
```

Answer 21 R 10
a **21** tables are filled.
b **10** people are at the table that is not full.

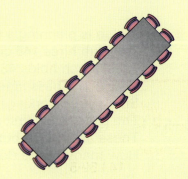

Exercise 7 **Always estimate the answer first.**

1 What is the remainder when
  a 747 is divided by 27
  b 864 is divided by 33
  c 573 is divided by 47?

2 680 test tubes are stored in frames which each hold 12 test tubes.
  a How many frames can be filled?
  b How many test tubes are left over?

3 784 boxes of baked beans were to be delivered to 15 shops in a supermarket chain. Each shop got the same number of boxes.
  a How many boxes did each shop get?
  b How many boxes were left over?

4 Voting papers are packed in bundles of 25.
  a How many bundles can be made from 873 voting papers?
  b How many are left over?

5 A box of chocolate bars, with 181 bars, was given to a children's holiday camp.
14 children each got the same number of bars.
  a How many bars did each child get?
  b How many were left over?

Worked Example
$59 \cdot 5 \div 14$

Answer

```
      14 ) 59·5
         − 56        14 × 4
           3·5
         −  2·8      14 × 0·2      because 14 × 2 = 28
            0·7
         −  0·7      14 × 0·05     because 14 × 5 = 70 and 14 × 0·5 = 7
            0·0
```

Answer **4·25**

Number

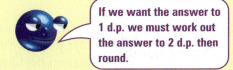

If we want the answer to 1 d.p. we must work out the answer to 2 d.p. then round.

Exercise 8   **Always estimate the answer first.**

1 Give the answers to these as decimals.
  a 364 ÷ 8   **b** 849 ÷ 5   **c** 192 ÷ 15   **d** 296 ÷ 16
  e 402 ÷ 15   **f** 570 ÷ 24

2 Hannah went on a cycling holiday.   **[SATs Paper 1 Level 5]**
  The table shows how far she cycled each day.

Monday	Tuesday	Wednesday	Thursday
32.3 km	38.7 km	43.5 km	45.1 km

  Hannah says: 'On average I cycled over 40 km a day'.
  Show that Hannah is wrong.

3 Give the answers to these to 1 d.p.
  a 487 ÷ 14   **b** 321 ÷ 17   **c** 843 ÷ 23   **d** 560 ÷ 19

4 296 people live in an area of 16 square miles.
  What is the mean number of people per square mile?

Link to geography.

5 Twelve friends shared these costs.
  How much did each pay for
  a the train   **b** the meals?

Train   **£75·60 total**
Meals   **£98·40 total**

6 Ben cut a 36·9 m rope into 8 lengths to make ropes for his tent.
  How long, to the nearest tenth of metre, is each piece?

80

*7   A supermarket sells soaps in boxes.

   12 for £4·32
   24 for £7·68
   36 for £12·60

Which is the best buy?
Show your working.

*8   ☐☐·☐ ÷ ☐☐ = 2·4
   Put the numbers 1, 2, 2, 8, 8 in the boxes to make this true.

Dividing by a decimal

Discussion

Jill   Madhu

Is Madhu right? **Discuss.**
How would you find the answers to these?

81 ÷ 0·4   35 ÷ 0·08   686 ÷ 4·9   616 ÷ 3·2   162 ÷ 3·6

To **divide by a decimal** we do an **equivalent calculation** that has a whole number as the **divisor**.

 The divisor is the number you are dividing by.

Worked Example
21 ÷ 0·4

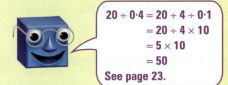
20 ÷ 0·4 = 20 ÷ 4 ÷ 0·1
        = 20 ÷ 4 × 10
        = 5 × 10
        = 50
See page 23.

Answer
21 ÷ 0·4 is approximately 20 ÷ 0·4 = 20 ÷ 4 ÷ 0·1
                                        = 50
21 ÷ 0·4 is equivalent to 210 ÷ 4.

```
    4 ) 210
      - 200        4 × 50
        10
       -  8        4 × 2
         2
       -  2        4 × 0·5
         0
```

Answer   **52·5**

Number

Exercise 9

1 Which of **A**, **B**, **C** or **D** is equivalent to the calculation given?

  a   42 ÷ 0·6   **A** 42 ÷ 0·06   **B** 4·2 ÷ 6   **C** 42 ÷ 6   **D** 420 ÷ 6

  b   58 ÷ 0·4   **A** 58 ÷ 4   **B** 580 ÷ 0·4   **C** 580 ÷ 40   **D** 580 ÷ 4

  c   300 ÷ 0·8   **A** 300 ÷ 8   **B** 3000 ÷ 8   **C** 30 ÷ 8   **D** 300 ÷ 80

  d   64 ÷ 0·03   **A** 64 ÷ 3   **B** 64·0 ÷ 3   **C** 64 000 ÷ 3   **D** 6400 ÷ 3

  ∗**e**   836 ÷ 0·05   **A** 836 ÷ 5   **B** 83 600 ÷ 5   **C** 83·6 ÷ 5   **D** 8360 ÷ 50

  ∗**f**   542 ÷ 0·08   **A** 542 ÷ 8   **B** 54 200 ÷ 8   **C** 542 ÷ 0·8   **D** 5420 ÷ 8

2 Write an equivalent division you could do to find the answer.
Do not do the calculation.

  a   82 ÷ 0·2   **b**   93 ÷ 0·3   **c**   72 ÷ 0·4   **d**   84 ÷ 0·09   ∗**e**   384 ÷ 0·04

3 Calculate these.

  a   18 ÷ 0·2   **b**   24 ÷ 0·8   **c**   32 ÷ 0·4   **d**   81 ÷ 0·3   **e**   42 ÷ 0·5

  f   22 ÷ 0·04   ∗**g**   273 ÷ 0·3   ∗**h**   482 ÷ 0·4   ∗**i**   405 ÷ 0·9   ∗**j**   301 ÷ 0·07

∗**4** Give the answers to these to 1 d.p.

  a   524 ÷ 0·6   **b**   388 ÷ 0·7   **c**   298 ÷ 0·3

  d   518 ÷ 0·09   **e**   342 ÷ 2·6

> Think about whether the answer should be bigger or smaller than the number you are dividing into.

Checking answers

We can **check an answer** using one of these methods.

1 Check that the answer is sensible

Example   142   168   151   133   147
Will found the mean of this data and got the answer 18·6.
This is not a sensible answer because the data values are all much higher than 18·6.

Example   Charlie worked out that the cost of the dinner he made was £216.
This is not sensible as it is far too expensive for a dinner.

2 Check the answer is the right order of magnitude

Example   The answer to 65 × 1·4 must be bigger than 65 because we are multiplying by a number bigger than 1.
The answer to 65 ÷ 1·4 must be smaller than 65 because we are dividing by a number bigger than 1.

3 Check using inverse operations

Example   Mark worked out 35 ÷ 25 as 1·4.
He could check this by working out 1·4 × 25.
The answer should be 35.

4 Check using an equivalent calculation

Example   12 × 14 = 168 can be checked by working out 24 × 7 or 6 × 28 or
12 × 2 × 7.

5 Check the last digits

Example   **18** × 2·**2** = 39·4 is wrong because 8 × 2 = 16.
The last digit should be 6.

Exercise 10   **except for questions 3, 6 and 9.**

1   Susie answered these questions for homework.
Her answers are shown.
Are her answers sensible? Explain.
  a   Toby spent £2·50 each week.
By March he had spent £35.
How many weeks had he taken to spend this?

  > 7

  b   Emily measured four equal pieces of ribbon.
Their total length was 105·8 cm.
How long was each piece?

  > 423·2 cm

  c   Nick had 7 m of climbing tape.
He cut off 2·34 m and 1·85 m to make two slings.
How much did he have left?

  > 2·81 m

  d   Neroli bought 25 chocolates for 65p each.
How much did these cost altogether?

  > £16·20

2   Paula did a science experiment.
She worked out the total mass of four pieces of calcium carbonate,
each weighing 21·7 g.
She got an answer of 18·8 g.
Is Paula's answer sensible? Explain.

3   Check the answers to these using inverse operations.
Which ones are wrong?
  a   75 × 1·2 = 90   **b**   294 ÷ 35 = 9·4   **c**   8·6 + 2·37 = 10·97
  d   4·2 × 9·6 = 40·36   **e** $\frac{2}{3}$ of 24·3 = 16·2   **f**   3·2² = 10·28
  g   √196 = 14

4   Janet entered the number 8653 into her calculator.
She multiplied by 7.
What should Janet then key to get 8653 displayed again?

5   Which of these is not equivalent to 24 × 6?
  A 12 × 2 × 3   **B** 48 × 3   **C** 12 × 2 × 6   **D** 6 × 4 × 6

6 Check these by doing an equivalent calculation.
Write down the calculation you did.
   a 128 ÷ 16 = 8   **b** 43 × 12 = 516   **c** 124 × 32 = 3968

7 Check these calculations by looking at the last digits.
Which ones are wrong?
   a 81 × 64 = 5148   **b** 72 × 38 = 2736   **c** 45 × 37 = 1656

8 Check the answers to these using a method of your choice.
Show how you did it.
   a 88 × 1·3 = 114·4   **b** 19·6 + 52·7 = 72·3   **c** 612 ÷ 12 = 51
   d 48 × 14 = 672   **e** √12·25 = 3·5

∗9 Without using a calculator, pick a possible answer to the calculation.
Explain your choice.
   a 48 × 89   **A** 4862   **B** 4272   **C** 4383
   b 53 × 32   **A** 1696   **B** 1486   **C** 1782
   c 79 × 1·5   **A** 118·5 **B** 142·5 **C** 120
   d 539 × 0·72 **A** 321·6 **B** 632   **C** 388·08

Using brackets on a calculator

When **brackets** are part of a calculation, we key them into the calculator as we come to them.

Always estimate the answer first.

Worked Example
a 6·4 − (3·7 + 4·8)   **b** $\frac{4·6}{8·4 - 5·7}$ **c** √11 + 6

Answer
a 6·4 − (3·7 + 4·8) is approximately 6 − (4 + 5) = 6 − 9 = ⁻3.

   Key 6·4 − ((3·7 + 4·8)) to get ⁻2·1.

b $\frac{4·6}{8·4 - 5·7}$ is approximately $\frac{5}{8-6} = \frac{5}{2} = 2·5$.

   Key 4·6 ÷ ((8·4 − 5·7)) = to get **1·7 (1 d.p.)**

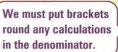

We must put brackets round any calculations in the denominator.

c √11 + 6 is approximately √16 = 4.

   Key √ ((11 + 6)) = to get **4·1 (1 d.p.)**

Exercise 11

1 **a** Jack keyed $\frac{13}{10-1}$ as

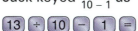

What mistake did Jack make?

b Jane keyed $\sqrt{5}+6$ as

√ 5 + 6 = .

What mistake did she make?

c Olivia keyed $4\cdot6(8\cdot3-2\cdot7)$ as

What mistake did Olivia make?

2 Round **f, g** and **h** to 1 d.p.
   a $8\cdot6-(4\cdot9+2\cdot3)$
   b $4\cdot7+(8\cdot6-3\cdot2)$
   c $3\cdot2\times(9\cdot7-2\cdot3)$
   d $39\cdot3-5\cdot7(6\cdot8-2\cdot3)$
   e $(16\cdot4-3\cdot2)\times(6\cdot8-5\cdot92)$
   f $\sqrt{36}-24$
   *g $\sqrt{11^2+5}$
   *h $\sqrt{5^2+16^2}$

3 **a** $\frac{96+30}{9}$ **b** $\frac{59+69}{12-4}$ **c** $\frac{132}{15-9}$ **d** $\frac{192}{6\times4}$ **e** $\frac{420}{5\times7}$

4 Give the answers to these to 1 d.p.
   a $\frac{4+6}{9}$ **b** $\frac{15}{12-5}$ **c** $\frac{25}{8\times3}$ **d** $\frac{4\cdot6+2\cdot3}{5\cdot8}$ **e** $\frac{8+6}{3\times2}$

5 $4\times6+3-3\times4$
   Put one set of brackets in this calculation to make
   a the largest possible answer   **b** the smallest possible answer.

It may take a long time to do these.

*6 Put decimal points, $+$, $-$, $\times$, $\div$ and brackets in these to make them true.
   a 19   36   2 = 2·62
   b 128   69   3 = 10·5
   c 124   4   18 = 1·3
   d 84   4   16   2 = 36·8
   e 4   116   5   22 = 14·4
   f 24   4   68   27 = 18·8
   g 69   24   3   18 = 9·5
   h 34   2   66   3 = 4·6
   i 27   11   5   7 = 0·7

Summary of key points

A   See page 62 for strategies that can be used to **add and subtract mentally**.

B   See page 66 for strategies that can be used to **multiply and divide mentally**.

C   We can use **mental strategies to solve problems**.

D To **estimate the answer to a calculation** we round the numbers.

Example $382 \times 24 \approx 400 \times 20 = 8000$

> ≈ means 'approximately equal to'.

Example $217 \div 25 \approx 200 \div 20 = 10$

> The best estimate is close to the actual answer but easy to do in your head.

Try to **round to 'nice numbers'** when estimating.

Examples   Approximate $\frac{61}{7}$ to $\frac{63}{7}$ rather than to $\frac{60}{7}$.

> 63 is a multiple of 7.

Approximate $\frac{397}{26}$ to $\frac{400}{25}$ rather than to $\frac{400}{30}$.

> 25 is a factor of 400.

 E When we **add and subtract** we line up digits with the same place value. For decimals, line up the decimal points.

Example $53 \cdot 8 - 6 \cdot 24$ is approximately equal to $54 - 6 = 48$.

$$\begin{array}{r} {}^{4\ 1\ 7\ 1}53.80 \\ -\ \ 6 \cdot 24 \\ \hline 47 \cdot 56 \end{array}$$

 F **Multiplication**

Example $28 \times 3 \cdot 4$

$28 \times 3 \cdot 4$ is approximately equal to $30 \times 3 = 90$.

Method 1	**Method 2**

Method 1

×	20	8	check
3	60	24	84
0·4	8	3·2	+ 11·2
	68	27·2	95·2

$68 + 27 \cdot 2 = \mathbf{95.2}$

Method 2

$28 \times 3 \cdot 4$ is equivalent to $28 \times 3 \cdot 4 \times 10 \div 10$
or $28 \times 34 \div 10$.

> Check the answer is about the right size by looking at the estimate.

$$\begin{array}{r} 28 \\ \times\ 34 \\ \hline 840 \quad 28 \times \mathbf{30} \\ 112 \quad 28 \times \mathbf{4} \\ \hline 952 \end{array}$$

Answer $952 \div 10 = \mathbf{95 \cdot 2}$

Division

Example   89·6 ÷ 23

89·6 ÷ 23 is approximately equal to 100 ÷ 25 = 4.

```
23 ) 89·6
      69·0        23 × 3
      20·6
      18·4        23 × 0·8
       2·20
       2·07       23 × 0·09
       0·13
```

Answer   3·89 R 0·13

3·9 (1 d.p.)

 When we **divide by a decimal** we do an equivalent division.

Examples   54 ÷ 0·9 is equivalent to 540 ÷ 9.

57·2 ÷ 3·4 is equivalent to 572 ÷ 34.

 We can **check the answer to a calculation** in one of these ways.

- Check the answer is sensible.
- Check the answer is about the right order of magnitude.
- Check using inverse operations, using an equivalent calculation or by checking the last digit.

For examples see page 82.

 When **brackets** are part of a calculation, we key them as we come to them.

Example   (5·6 + 2·3) × 8·4 is keyed as

 (5·6 + 2·3) × 8·4 = to get 66·36.

Sometimes we need to **add brackets to the calculation**.

The whole numerator must be divided by the whole denominator.

Example $\frac{5+11}{18-7}$ is keyed as

(5 + 11) ÷ (18 − 7) = to get 1·45 (2 d.p.)

Number

Find the answers to questions 1 to 7 mentally.

1 a $7 + 2 + 3$ **b** $14 - 3 - 8 + 4$ **c** $0.6 + 0.8 - 0.2 - 1.1$

T **2** In this pyramid, each number is the sum of the
two numbers below it.
Use a copy of this pyramid.
What number goes in the blue square?

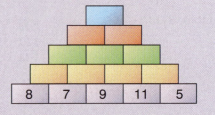

3 a $520 + 460$ **b** $630 - 380$ **c** $493 + 207$ **d** $836 - 429$

4 a $3 \times 5 \times 2$ **b** $6 \times 4 \times 2$ **c** $3 \times 7 \times 5$ **d** 50×70 **B**
   e $1800 \div 900$ **f** 510×8 **g** 23×9

T **5** Use a copy of this number chain.
Fill in the missing numbers.   **B**

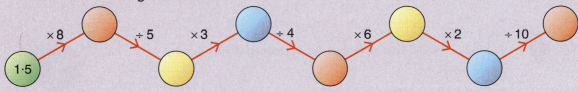

6 a 7×0.7 **b** 0.8×0.3 **c** $650 \div 5$ **d** $464 \div 4$ **e** $363 \div 3$ **f** $750 \div 15$

7 a I am counting back in zero point twos.   **C**
      Six point five, six point three, six point one, ...
      Write down the next two numbers.
   b Write a multiple of 3 that is bigger than 120.
   c Two angles join to make a straight line.
      One angle is 96°.
      What is the other?

8 Choose the best approximation for each calculation.   **D**
   a 343×22 **A** 400×20 **B** 300×20 **C** 300×30
   b $782 \div 38$ **A** $700 \div 40$ **B** $800 \div 30$ **C** $800 \div 40$
   c $3.86 \div 2.4$ **A** $4 \div 2$ **B** $3 \div 2$ **C** $4 \div 3$

9 Write down a calculation you could do to find an approximate answer.   **D**

   Renée bought 24 chocolate bears. Each one has a mass of 16·9 g.
   What is their total mass?

Check your answers to the calculations in questions 10 to 19 using one of the ways given in .

10 **a** $8{\cdot}36 + 52{\cdot}9$ **b** $16 - 5{\cdot}83$ **c** $186{\cdot}4 - 8{\cdot}03$

11 Amy bought books for £64·60 and pens for £5·65.
How much change did she get from £100?

12 **a** 65×306 **b** 42×528 **c** $52 \times 0{\cdot}8$ **d** $38 \times 4{\cdot}6$
   e $5{\cdot}7 \times 54$ **f** $3{\cdot}2 \times 11{\cdot}4$

13 Alex needed 5·4 m of braid.
   a How much did it cost, to the nearest penny?
   b How much change did she get from £30?

Braid
£3.80
per m

14 154 cakes of chocolate are packed in boxes of 15.
   a How many full boxes can be packed?
   b How many cakes of chocolate are left?

15 Give the answers to these as decimals.   **a** $207 \div 15$ **b** $342 \div 24$

16 Give the answers to these to 1 d.p.   **a** $587 \div 16$ **b** $429 \div 28$

17 Is it cheaper *per slice* to buy
36 pizza slices or 24 pizza slices?

Bag of
36 pizza slices
£39·60

Bag of
24 pizza slices
£27·20

18 **a** Which of these is equivalent to $84{\cdot}3 \div 0{\cdot}3$?
   A $843 \div 3$ **B** $8430 \div 0{\cdot}3$ **C** $8430 \div 3$ **D** $84{\cdot}3 \div 3$

   b Which two of these are equivalent to $605 \div 0{\cdot}07$?
   A $60500 \div 7$ **B** $6050 \div 7$ **C** $6050 \div 0{\cdot}7$ **D** $6{\cdot}05 \div 7$

19 **a** $75 \div 0{\cdot}3$ **b** $68 \div 0{\cdot}4$ **c** $426 \div 0{\cdot}06$

20 Use the brackets on your calculator to do these.
Give your answers to **c** and **d** to 1 d.p.
   a $(9{\cdot}4 - 4{\cdot}5) \times 2$ **b** $3{\cdot}6 \times (2{\cdot}1 - 0{\cdot}89)$ **c** $\frac{4{\cdot}3 + 6}{9}$ **d** $\frac{7}{15 - 6}$

You need to know

✓ fractions   page 7

✓ decimals   page 7

✓ percentages   page 7

Key vocabulary

cancel, convert, lowest terms, mixed number, recurring decimal, simplest form, tax

Line Up

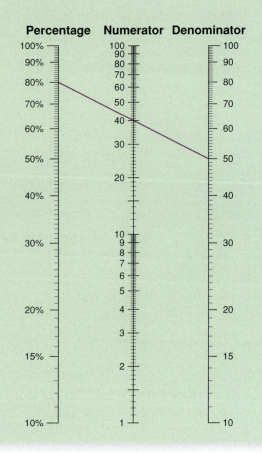

This is called a **nomogram**.

We can use it to change fractions to percentages.

To use it

1 join the numerator and denominator of the fraction with a line

2 extend the line to the percentage scale

3 read off the percentage scale to find out what the fraction is as a percentage.

Example $\frac{40}{50} = 80\%$ is shown with the purple line.

Write these as percentages.

1 a $\frac{5}{20}$ b $\frac{20}{50}$ c $\frac{10}{80}$

2 If the percentage is 75% and the denominator of the fraction is 60, what is the numerator?

Cancelling fractions

Remember

$\frac{4}{5}$ is a fraction in its simplest form.

4 and 5 have no common factors other than 1.

We write a fraction in its simplest form by **cancelling**.

$\frac{16}{20} = \frac{4}{5}$

We divide 16 and 20 by 4.
4 is the highest common factor of 16 and 20.

Example

$\frac{42}{49} = \frac{6}{7}$ **We divide 42 and 49 by 7.**

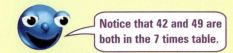

Notice that 42 and 49 are both in the 7 times table.

We can cancel in steps.

Examples

$\frac{36}{60} = \frac{18}{30} = \frac{9}{15} = \frac{3}{5}$ $\frac{24}{84} = \frac{6}{21} = \frac{2}{7}$

Exercise 1

1   What fraction of each shape is shaded?
Write your answer in its simplest form.

a   **b**   **c**

Each part must be the same size.

2   Look at the fractions in each of these diamonds.
Find the odd one out in each of them.

a   **b**   **c**   ***d**

3   Rose had these cards.
She made a fraction equivalent to $\frac{15}{25}$.
What might it have been?
There are two possible answers.

5   3   6   10

4 Heather cancelled each of these to its simplest form.
Did she do it correctly? Write yes or no for each.

a $\frac{^1\cancel{3}}{_4\cancel{12}} = \frac{1}{4}$ **b** $\frac{^2\cancel{8}}{_5\cancel{10}} = \frac{2}{5}$ **c** $\frac{^4\cancel{16}}{_{10}\cancel{40}} = \frac{4}{10}$ **d** $\frac{^5\cancel{35}}{_7\cancel{42}} = \frac{5}{7}$

5 Cancel these to their simplest form.

a $\frac{6}{8} = \frac{\square}{\square}$ (÷2 / ÷2)

b $\frac{4}{6} = \frac{\square}{\square}$ (÷2 / ÷2)

c $\frac{10}{15} = \frac{\square}{\square}$ (÷5 / ÷5)

d $\frac{18}{24} = \frac{\square}{\square}$ (÷6 / ÷6)

e $\frac{45}{60}$ **f** $\frac{32}{40}$ **g** $\frac{27}{63}$ **h** $\frac{36}{48}$ **i** $\frac{120}{180}$

6 Give the answer to each of these as a fraction in its lowest terms.
a 7 of the 28 pupils in Dick's class joined a rock climbing club.
What fraction is this?
b 10 of the 25 pupils in Jill's class joined an athletics club.
What fraction is this?

One number as a fraction of another

We can write **one number as a fraction of another**.

Example   To write 36 as a fraction of 60, we write

we are finding the fraction of this number → $\frac{36}{60} = \frac{3}{5}$

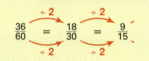
$\frac{36}{60} = \frac{18}{30} = \frac{9}{15}$ (÷2 / ÷2)

36 is $\frac{3}{5}$ of 60.

Worked Example
This bar chart shows the amount 5 pupils gave to charity on charity day.
What fraction of the total amount did Ben give?

Answer
Total amount given = £4 + £8 + £2 + £12 + £6
= £32

Fraction given by Ben = $\frac{£8}{£32}$
= $\frac{1}{4}$

We have to find the total first.

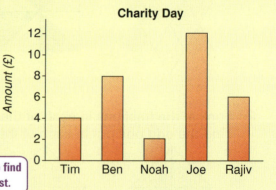

Charity Day

Amount (£): values shown 0, 2, 4, 6, 8, 10, 12
Tim = 4, Ben = 8, Noah = 2, Joe = 12, Rajiv = 6

Exercise 2

1 What fraction of an hour are these?
There are 60 minutes in an hour so for **a**, **30 minutes** $= \frac{30}{60} = \frac{1}{2}$ **of an hour**.

a 30 minutes   **b** 20 minutes
c 45 minutes   **d** 27 minutes

Write your answer in its simplest form.

2 a What fraction of 1 metre is 42 centimetres?
b What fraction of 1 kg is 250 g?
c What fraction of £2 is 75p?
d What fraction of 2 cm is 10 mm?
∗e What fraction of one metre is 350 mm?

> **Remember:**
> 1 kg = 1000 g
> 1 m = 100 cm
> 1 cm = 10 mm

3 Work these out.
The first two have been started.
a What fraction of 12 is 8?
$\frac{8}{12} =$ ___
b What fraction of 24 is 16?
$\frac{16}{24} =$ ___
c What fraction of 12 is 9?
d What fraction of 80 is 20?
e What fraction of 45 is 15?
f What fraction of 36 is 30?
∗g What fraction of 42 is 36?

4 What fraction of the adult price do students pay for sailing lessons?

SAILING LESSONS
Adults £25
Students £20

5 Sixteen out of the thirty pupils in Ella's class have ridden horses before.
What fraction is this?

∗6 Sue was doing a survey to find what proportion of cars didn't stop at a 'stop' sign.
2 didn't stop and 8 did stop.
a What fraction didn't stop?
b What fraction did stop?

> For 6, 7, 8 and 9 you need to find the total first.

∗7 In Nazir's class there are 14 pupils who were born in England and 18 who were born in other places.
What fraction was born in England?

∗8 In one litre of air there is about
210 mℓ of oxygen
780 mℓ of nitrogen
10 mℓ of argon
About what fraction of the air is oxygen?

> Link to science.

∗9 Rowena asked her friends which colour they would like the new school tracksuit to be.
This bar chart shows her results.
What fraction wanted the tracksuit to be
a red   **b** blue   **c** black
d green or black?

School Tracksuit

Number (y-axis: 1, 2, 3, 4)
Red = 3, Blue = 4, Green = 4, Black = 1
Colour (x-axis: Red, Blue, Green, Black)

Number

Practical

1 Write some fractions for your class.
   For example, the fraction that are left-handed, like pizza,
   have blue eyes, have curly hair, wear glasses,
   walk to school, ...

2 Write some fractions for your friends, family or school.

3 Choose a topic you are interested in.
   Write some fractions about it.
   For example, you might choose cars and write
   '$\frac{1}{3}$ of all cars are made in Japan.'

Fractions and decimals

Remember

$\frac{15}{4}$ is an **improper fraction**.

It is the same as $15 \div 4$.

$15 \div 4 = 3$ with 3 left over.

$\frac{15}{4} = 3\frac{3}{4}$

$2\frac{2}{5}$ is a **mixed number**.

It is the same as $2 + \frac{2}{5}$.

$2 + \frac{2}{5} = \frac{10}{5} + \frac{2}{5}$

$= \frac{12}{5}$

To **convert a decimal to a fraction** write it with a denominator of 10 or 100 or 1000.

Examples

$0.8 = \frac{8}{10}$

$= \frac{4}{5}$

$0.54 = \frac{54}{100}$

$= \frac{27}{50}$

$0.839 = \frac{839}{1000}$

Cancel the fraction to its lowest terms.

Exercise 3

1 Write these as fractions in their lowest terms.

a $0.7 = \frac{7}{\square}$

b $0.2 = \frac{\square}{10} = \frac{\square}{5}$ (÷ 2)

c $0.60 = \frac{60}{\square} = \frac{\square}{10} = \frac{\square}{5}$

d $0.4 = \frac{\square}{10} = \frac{\square}{5}$

e 0.49

f 0.73

g 0.25

h 0.85

i 0.03

j 0.157

k 0.087

l 0.007

*m 1.7

T 2

$2\frac{27}{1000}$	$\frac{49}{1000}$	$\frac{9}{10}$	$\frac{9}{1000}$	$\frac{103}{1000}$	$\frac{37}{100}$	$2\frac{3}{10}$	$\frac{13}{20}$		$\frac{9}{20}$	$\frac{37}{100}$	$2\frac{323}{1000}$	$\frac{2}{25}$
		S										
$\frac{421}{1000}$	$\frac{1}{4}$	$\frac{1}{2}$	$2\frac{3}{100}$	$2\frac{3}{10}$	$\frac{37}{100}$		$\frac{9}{10}$	$\frac{13}{20}$		$2\frac{3}{100}$	$\frac{49}{1000}$	$2\frac{3}{10}$
S		**2**										
$\frac{1}{2}$	$\frac{83}{1000}$	$\frac{37}{100}$	$\frac{9}{10}$	$\frac{13}{20}$	$\frac{9}{20}$							

Use a copy of this box.
Write the letter beside each question above its answer in the box.
Convert these to fractions in their simplest form.

S	$0.5 = \frac{1}{2}$	**I**	0.9	**W**	0.08	**R**	0.37
A	0.25	**G**	0.45	**F**	0.421	**D**	0.103
L	0.009	**H**	0.049	**P**	0.083	**N**	0.65
E	2.3	**T**	2.03	**C**	2.027	**O**	2.323

'Simplest form' and 'lowest terms' mean the same.

Remember

$3 \div 8 = \frac{3}{8}$ and $\frac{4}{9} = 4 \div 9$

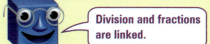

Division and fractions are linked.

To write a **fraction as a decimal** we can:

1 use known facts

Example   We know that $\frac{1}{8} = 0.125$
$$\frac{3}{8} = 0.125 \times 3$$
$$= 0.375$$

$\frac{1}{2} = 0.5$		$\frac{1}{4} = 0.25$	
$\frac{3}{4} = 0.75$		$\frac{1}{8} = 0.125$	
$\frac{1}{5} = 0.2$		$\frac{1}{10} = 0.1$	
$\frac{1}{3} = 0.\dot{3}$			

2 make the denominator 10 or 100

Examples

$$\overset{\times 4}{\frac{9}{25}} = \underset{\times 4}{\frac{36}{100}}$$
$$= 0.36$$

$$\overset{\div 3}{\frac{12}{30}} = \underset{\div 3}{\frac{4}{10}}$$
$$= 0.4$$

$$\frac{17}{20} = \frac{85}{100}$$
$$= 0.85$$

$$6\frac{4}{5} = 6\frac{8}{10}$$
$$= 6.8$$

Use this method when it is easy to change the denominator to 10 or 100 using equivalent fractions.

Exercise 4

1   Write these fractions as decimals.

a $\frac{3}{10}$ **b** $\frac{17}{100}$ **c** $\frac{1}{4}$ **d** $\frac{3}{5} = \frac{\square}{10} = $ ___

e $\frac{1}{20} = \frac{\square}{100} = $ ___   **f** $\frac{3}{20} = \frac{\square}{100} = $ ___   **g** $\frac{4}{25}$ ***h** $\frac{5}{8} = 5 \times \frac{1}{8} = $ ___

***i** $1\frac{3}{5}$ ***j** $\frac{36}{40}$

Number

2 How much of these shapes is shaded?
Match your answer with a decimal in the box.

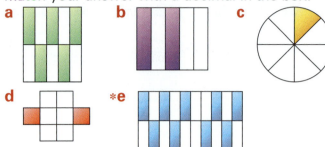

a   b   c

d   *e

0·375	
0·5	0·25
0·025	
0·05	0·04
0·125	0·4	0·45

3 a Molly visited her sister who lived 8 km away.
She walked 6 km and ran the rest.
What fraction did she walk?
Write this as a decimal.
  b A bag of mixed lollies weighed 250 g.
There was 150 g of mints in the bag.
What fraction was mints?
Write this as a decimal.

We can **divide the numerator by the denominator** to write a fraction as a decimal.

Examples

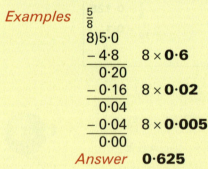

$\frac{5}{8}$

```
  8)5·0
  - 4·8      8 × 0·6
  ──────
    0·20
  - 0·16     8 × 0·02
  ──────
    0·04
  - 0·04     8 × 0·005
  ──────
    0·00
```

Answer **0·625**

$\frac{3}{7}$

```
  7)3·0
  - 2·8      7 × 0·4
  ──────
    0·20
  - 0·14     7 × 0·02
  ──────
    0·060
  - 0·056    7 × 0·008
  ──────
    0·004
```

Answer 0·428 R 0·004
0·43 (2 d.p.)

> We use this way when it is difficult to use the first two ways.

Sometimes we **use a calculator** to divide the numerator by the denominator.

Example $\frac{27}{39}$ key 27 ÷ 39 = to get 0·69 (2 d.p.)

or key 27 a^b/c 39 = a^b/c to get 0·69 (2 d.p.)

> **Remember:**
> $\frac{27}{39} = 27 ÷ 39$

> Keying this again gives the answer as a decimal.

In some decimals, one or more of the digits repeat.
These are called **recurring decimals**.

Examples $\frac{1}{3} = 0·3333...$ $\frac{2}{3} = 0·6666...$ $\frac{3}{11} = 0·272727$

          $= 0·\dot{3}$ $= 0·\dot{6}$ $= 0·\dot{2}\dot{7}$

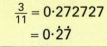

> The dot above the 3 shows that 3 repeats.

> The dots above the 2 and 7 show that these both repeat.

1 Use a copy of this.
Finish the division and write the answer as a decimal.
Round to 2 d.p.
Check your answers are correct using the $a^{b/c}$ key on your calculator.

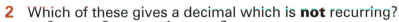

a $\frac{1}{7}$ 7)1·0
$-0·7$

b $\frac{5}{9}$ 9)5·0
$-4·5$

c $\frac{2}{7}$ 7)2·0
$-1·4$

d $\frac{26}{3}$ 3)26
-24

e $\frac{25}{6}$ 6)25
-24

f $\frac{13}{7}$ 7)13
-7

g $\frac{7}{6}$ 6)7
-6

2 Which of these gives a decimal which is **not** recurring?

A $\frac{3}{7}$ **B** $\frac{5}{9}$ **C** $1\frac{1}{3}$ **D** $\frac{7}{8}$

3 Use your calculator to write these as decimals.
Round **d**, **e**, **f** and **g** to 1 d.p.

a $\frac{13}{25}$ **b** $\frac{5}{16}$ **c** $\frac{21}{32}$ **d** $\frac{25}{41}$ **e** $\frac{104}{163}$ **f** $\frac{37}{29}$ **g** $\frac{147}{109}$

4 Write $\frac{1}{3}$ as a recurring decimal.

*5 What fraction of these shapes is shaded?
Give the answers as recurring decimals.

a   **b**   **c**    **d**

Investigation

Recurring decimals

1 What is the recurring decimal for $\frac{1}{3}$?
What about $\frac{2}{3}$? $\frac{4}{3}$? $\frac{5}{3}$?
Use these answers to predict what decimal you would get for $\frac{7}{3}$, $\frac{8}{3}$.

2 What is the recurring decimal for $\frac{1}{11}$?
What about $\frac{2}{11}$? $\frac{3}{11}$? $\frac{4}{11}$?

Describe the pattern.
Use these answers to predict what decimal you would get for
$\frac{5}{11}$, $\frac{6}{11}$, $\frac{7}{11}$ and $\frac{8}{11}$.

Converting fractions, decimals and percentages

Remember

We convert a **percentage to a fraction or decimal** by writing it as the number of parts per hundred.

Examples $19\% = \frac{19}{100}$

$= \mathbf{0 \cdot 19}$

$137\% = \frac{137}{100}$

$= \mathbf{1 \cdot 37}$

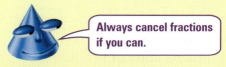

We can write this as $1\frac{37}{100}$.

Examples

$35\% = \frac{35}{100}$ $= \frac{7}{20}$

$130\% = \frac{130}{100}$

$= \frac{13}{10}$

$= \mathbf{1\frac{3}{10}}$

Always cancel fractions if you can.

Fractions and decimals can be converted to percentages by writing them with denominators of 100.

Examples

$\frac{8}{25} = \frac{32}{100}$

$= \mathbf{32\%}$

$0 \cdot 63 = \frac{63}{100}$

$= \mathbf{63\%}$

You should know these.

$\frac{1}{4} = 0 \cdot 25 = 25\%$ $\frac{3}{4} = 0 \cdot 75 = 75\%$ $\frac{1}{8} = 0 \cdot 125 = 12\frac{1}{2}\%$

Worked Example

Al got these marks in two tests.

Science	Maths
Al	Al
67%	18 out of 25

In which test did Al do better?

Answer

Change the fraction to a percentage.

18 out of 25 $= \frac{18}{25} = 72\%$ in Maths

$\frac{18}{25} = \frac{72}{100}$

Al did better in Maths.

Exercise 6   Except for question 11.

1 Write these fractions as percentages.

a $\frac{54}{100}$ **b** $\frac{7}{100}$ **c** $\frac{45}{100}$ **d** $\frac{70}{100}$

e $\frac{8}{10} = \frac{\square}{100} = \underline{\quad}\%$ **f** $\frac{7}{10}$ **g** $\frac{1}{10}$ **h** $\frac{3}{10}$

2 Convert these to percentages.
The first two are started.

a $\dfrac{4}{50}$ (×2) $= \dfrac{\square}{100} = \underline{\quad}\%$

b $\dfrac{7}{20} = \dfrac{\square}{100} = \underline{\quad}\%$

c $\dfrac{17}{50} = \dfrac{\square}{\square} = \underline{\quad}\%$

d $\dfrac{17}{20}$ (×5) $= \dfrac{\square}{100} = \underline{\quad}\%$

e $\dfrac{3}{25}$ (×4) $= \dfrac{\square}{100} = \underline{\quad}\%$

f $\dfrac{13}{20}$

g $\dfrac{33}{50}$

h $\dfrac{3}{5}$

i $\dfrac{16}{25}$

*j $\dfrac{42}{60}$

*k $\dfrac{12}{30}$

T

3 Do this question mentally.
Use a copy of this table.
Fill in the gaps.

Fraction	Decimal	Percentage
$\dfrac{3}{10}$		
	0·7	
		60%
	0·45	
		35%
	0·06	
	2·5	250%
		275%

T

4

$\overline{\quad}$ 8% $\quad$ $\overline{\quad}$ 30% $\overline{\quad}$ 65% $\overline{\quad}$ 8% $\overline{\quad}$ 265% $\overline{\quad}$ 140% $\overline{\quad}$ 90·5% $\overline{\quad}$ 30% $\overline{\quad}$ 7%

$\overline{\quad}$ 24% $\overline{\quad}$ 8% $\overline{\quad}$ 23% $\quad$ $\overline{\quad}$ 65% $\overline{\quad}$ 75% $\overline{\quad}$ 265% $\overline{\quad}$ 23%

$\overline{\quad}$ 90·5% $\overline{\quad}$ 65% $\overline{\quad}$ 30% $\quad$ $\overline{\quad}$ 30% $\overline{\quad}$ 65% $\overline{\quad}$ 46% $\overline{\quad}$ 402% $\overline{\quad}$ 8% $\overline{\quad}$ 24% $\overline{\quad}$ 7%

$\overset{\textbf{E}}{\overline{\quad}}$

$\overline{\quad}$ 90·5% $\overline{\quad}$ 23% $\overline{\quad}$ 30% $\overline{\quad}$ 90·5% $\overline{\quad}$ 7·5% $\overset{\textbf{E}}{\overline{\textbf{56\%}}}$ $\quad$ $\overline{\quad}$ 46% $\overline{\quad}$ 75% $\overline{\quad}$ 65%

Use a copy of this box.
Write the letter beside each decimal or fraction above its answer in the box.
Convert these to percentages.

E $0.56 = \dfrac{56}{100} = \mathbf{56\%}$

U 0.75

O 0.46

A 0.08

N 0.23

R 2.65

D 0.075

I 0.905

M 4.02

H $\dfrac{7}{100}$

S $\dfrac{3}{10} = \dfrac{30}{100} = \underline{\quad}\%$

F $1\dfrac{2}{5}$

C $\dfrac{6}{25}$ (×4) $= \dfrac{\square}{100} = \underline{\quad}\%$

T $\dfrac{13}{20}$ (×5) $= \dfrac{\square}{100}$

5

| 40% | 50% | 30% | 25% | 12½% | 60% | 12% |

Find a percentage from the box for each sentence.

a One out of four horses in a race lost its rider.

b Three out of five people in a town are over 20.

c Twelve out of forty people who applied for a job were interviewed.

d Out of every two hundred and fifty calories eaten, at least one hundred is usually carbohydrate.

e One out of every eight people at a concert left early.

6 An 80 g bar of chocolate contains 20 g of fat.
What percentage of fat is this?

7 Each diagram below was drawn on a square grid. **[SATs Paper 2 Level 5]**

a Write what **percentage** of each diagram is shaded.
The first one is done for you.

75%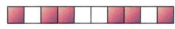

b Explain how you know that 12½% of the diagram is shaded.

c Shade 37½% of the diagram.

8 What percentage of 1 hour is 12 minutes?

Write these as fractions first then convert.

9 A 500 g rock contains 25 g of gold.
What percentage of the rock is gold?

10 Last month, Alex had tests in three of her subjects.
She got   74% for geography
          16 out of 20 for science
          0·72 of the total possible marks in English.
Which subject did Alex do best in?
What percentage did she get in this subject?

∗11 Charlie changed $\frac{4}{7}$ to a percentage using his calculator.

$\frac{4}{7} = 4 \div 7$ key 4 ÷ 7 = to get 0·57142857i

$= 0·57$ He rounded the answer to 2 d.p.

$= \frac{57}{100}$

$= 57\%$

Use Charlie's way to change these to percentages.

a $\frac{3}{7}$ **b** $\frac{4}{9}$ **c** $\frac{7}{17}$ **d** $\frac{13}{19}$ **e** $\frac{19}{17}$

*12 The human body has 206 bones.
This table shows how many are in each body part.

Head	Back	Chest	Both arms	Both legs
29	26	25	64	62

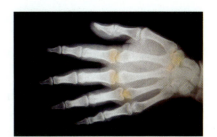

Give the percentage answers to 1 d.p.
a What fraction is in the head?
b What percentage is in the back?
c What fraction is in the chest?
d Write the answer to **c** as a decimal to 2 d.p.
e What percentage is in the arms?
f What percentage is in each leg?
g What percentage is in the legs and arms together?

*Discussion

Megan did a survey on summer and winter. She gave her results as percentages rounded to the nearest per cent.

    Like summer best   68%
    Like winter best   20%
    Like both the same   13%

These add to 101%.
Does this mean Megan's results are wrong? **Discuss**.

* Practical

1 Carry out a survey to find one of these percentages.
   ● percentage of cars with personalised number plates in school car park
   ● percentage of fat, protein, carbohydrate, etc. in one type of tinned or packaged food
   ● percentage of students in your class who don't have any brothers

Adding and subtracting fractions

Fractions can be **added or subtracted** easily if they have the same denominator.

Examples
$$\frac{5}{16} + \frac{7}{16} = \frac{5+7}{16}$$
$$= \frac{12}{16}$$
$$= \frac{3}{4}$$

$$\frac{7}{12} + \frac{11}{12} = \frac{7+11}{12}$$
$$= \frac{18}{12}$$
$$= 1\frac{6}{12}$$
$$= 1\frac{1}{2}$$

Always write the answers in their simplest form.

Number

To **add or subtract fractions with different denominators**

1 find the LCM (lowest common multiple) of the denominators
2 write equivalent fractions with this LCM as the common denominator
3 add or subtract the equivalent fractions.
  Give the answer in its simplest form.

Example $\frac{4}{5}$ and $\frac{3}{4}$ have different denominators.

1 Find the lowest common multiple of 4 and 5.

  The multiples of 5 are 5, 10, 15, 20, 25, ...

  The multiples of 4 are 4, 8, 12, 16, 20, ...

The lowest common multiple of 4 and 5 is 20.
So the lowest common denominator is 20.

2   $\frac{4}{5} = \frac{16}{20}$ $\frac{3}{4} = \frac{15}{20}$

3 $\frac{4}{5} + \frac{3}{4} = \frac{16}{20} + \frac{15}{20}$

$= \frac{16+15}{20}$

$= \frac{31}{20}$

$= 1\frac{11}{20}.$

We can show this using a diagram.

 + 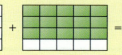 =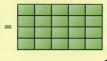

$\frac{4}{5}$ + $\frac{3}{4}$ = $1\frac{11}{20}$

> Find the LCM to find the number of parts to have in each diagram.

Worked Example
In Baysdown, $\frac{3}{8}$ of the population is under 5 and $\frac{1}{3}$ is aged from 5 to 16.
What fraction of the population is aged 16 or under?

Answer
Fraction aged 16 or under $= \frac{3}{8} + \frac{1}{3}$ LCM of 8 and 3 is 24.

$= \frac{9}{24} + \frac{8}{24}$ **multiples of 8 are 8, 16, 24, ...**

$= \frac{9+8}{24}$ **multiples of 3 are 3, 6, 9, 12, 15, 18, 21, 24, ...**

$= \frac{17}{24}$

Exercise 7

1 What goes in the boxes?

a $\frac{3}{12} + \frac{5}{12} = \frac{\square + \square}{12}$ b $\frac{17}{20} - \frac{9}{20} = \frac{\square - \square}{20}$

$= \frac{\square}{12}$ $= \frac{\square}{20}$

$= \frac{\square}{\square}$ $= \frac{\square}{\square}$

2 Calculate these.

   a $\frac{1}{5} + \frac{2}{5}$ **b** $\frac{3}{14} + \frac{5}{14}$ **c** $\frac{7}{8} - \frac{5}{8}$ **d** $\frac{11}{15} + \frac{14}{15}$

3 What goes in the boxes?

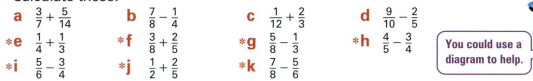

   a $\frac{3}{4} - \frac{5}{12} = \frac{\square}{12} - \frac{\square}{12}$

           $= \frac{\square - \square}{12}$

           $= \frac{\square}{12}$

           $= \frac{\square}{\square}$

   b $\frac{5}{6} + \frac{1}{3} = \frac{\square}{6} + \frac{\square}{6}$

           $= \frac{\square + \square}{6}$

           $= 1\frac{\square}{6}$

4 Calculate these.

   a $\frac{3}{7} + \frac{5}{14}$ **b** $\frac{7}{8} - \frac{1}{4}$ **c** $\frac{1}{12} + \frac{2}{3}$ **d** $\frac{9}{10} - \frac{2}{5}$

   ***e** $\frac{1}{4} + \frac{1}{3}$ ***f** $\frac{3}{8} + \frac{2}{5}$ ***g** $\frac{5}{8} - \frac{1}{3}$ ***h** $\frac{4}{5} - \frac{3}{4}$

   ***i** $\frac{5}{6} - \frac{3}{4}$ ***j** $\frac{1}{2} + \frac{2}{5}$ ***k** $\frac{7}{8} - \frac{5}{6}$

> You could use a diagram to help.

5 **a** $\frac{1}{4} + \frac{3}{4} + \frac{2}{4}$ **b** $\frac{7}{10} + \frac{5}{10} - \frac{3}{10}$ **c** $\frac{14}{16} - \frac{3}{16} - \frac{5}{16}$ ***d** $\frac{1}{3} + \frac{1}{4} + \frac{5}{12}$

   ***e** $\frac{9}{10} - \frac{2}{5} - \frac{1}{2}$ ***f** $\frac{3}{5} + \frac{7}{10} - \frac{3}{20}$ ***g** $\frac{5}{8} + \frac{3}{4} - \frac{13}{24}$ ***h** $\frac{2}{3} + \frac{1}{8} + \frac{3}{4}$

6 Mr Brown left all his money to his three children.
The eldest got $\frac{1}{2}$, the youngest got $\frac{1}{5}$.
What fraction did these two get altogether?

***7** A netball team won $\frac{2}{5}$ of their games and lost $\frac{1}{8}$ of them.
What fraction of their games did they draw?

***8** Two fractions with different denominators add to 1.
What might they be?

? **Puzzle**

What digits could ▲ be to make these true?

   a $\frac{▲}{4} + \frac{▲}{▲} = \frac{3}{▲}$

   ***b** $\frac{▲}{5} + \frac{▲}{▲▲} = \frac{7}{▲▲}$

Investigation

Egyptian fractions

The old Egyptian number system used these for numbers.

I	II	III	IIII	II	III	III	IIII	III	○	ꝰ
one stick	two sticks								heel bone	coiled rope
1	2	3	4	5	6	7	8	9	10	100

They used ⬭, called 'ro', to show fractions.

$\frac{⬭}{||} = \frac{1}{2}$ $\frac{⬭}{|||} = \frac{1}{5}$ $\frac{⬭}{∩} = \frac{1}{10}$ $\frac{⬭}{ꝰ|||} = \frac{1}{103}$

All Egyptian fractions have a numerator of 1.

Other fractions, such as $\frac{3}{4}$, were written as the sum of two **different** fractions with a numerator of 1.

$\frac{3}{4} = \frac{1}{2} + \frac{1}{4}$.   It was written as ⬭ ⬭ .
                                          || ||||

$\frac{2}{3} = \frac{1}{2} + \frac{1}{6}$.   It was written as ⬭ ⬭ .
                                          || |||
                                             |||

$\frac{2}{5}$ could not be written as ⬭ ⬭ because the fractions had to be different.

$\frac{2}{5}$ was written as ⬭ ⬭ ($\frac{1}{3} + \frac{1}{15}$).

1 What fraction do each of these show?

a ⬭ ⬭   **b** ⬭ ⬭   **c** ⬭ ⬭   **d** ⬭ ⬭

2 How would the Egyptians have written these?

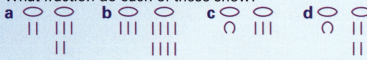

a $\frac{5}{6}$ **b** $\frac{3}{10}$ ***c** $\frac{5}{8}$ ***d** $\frac{2}{9}$ ***e** $\frac{7}{12}$ ***f** $\frac{8}{15}$

Percentage of – mentally

Discussion

Sam worked out 35% of 60 like this.

$$100\% \text{ of } 60 = 60$$
$$10\% \text{ of } 60 = 6$$
$$5\% \text{ of } 60 = 3$$
$$30\% \text{ of } 60 = 18$$
$$35\% \text{ of } 60 = 18 + 3$$
$$= 21$$

Discuss Sam's way.
Can you think of another way to work this out?

Remember

$10\% = \frac{1}{10}$ $5\% = $ half of 10% $1\% = \frac{1}{100}$ $25\% = \frac{1}{4}$

Worked Examples

a A 40 g top is 35% polyester.
How many grams are polyester?

b Find 125% of £160.

Answers

a 10% of $40 = \frac{1}{10}$ of 40
           $= 4$
     5% of $40 = 2$
    30% of $40 = 3 \times 4$
           $= 12$
    35% of $40 = 12 + 2$
           $= 14$

14 g are polyester.

b 25% of £160 $= \frac{1}{4}$ of £160
               $= £40$
    125% of £160 $= £160 + £40$
               $= \textbf{£200}$

> Remember: 100% is one whole.

Exercise 8

**This exercise is to be done mentally.
You may use jottings.**

1 a 20% of 50   **b** 40% of 20   **c** 60% of 40
   d 30% of 80   **e** 5% of 150   **f** 15% of 120
   g 45% of 150 km   **h** 65% of 180 g   **i** $33\frac{1}{3}$% of 96 kg
   *j 125% of £80   ***k** 120% of 220 cm

> 15% = 10% + 5%

2 How much would Perry save on a shirt in the sale?

Shirts **£60**   SALE 25% off this price

3 How much would be saved on a watch today that usually costs
    a £50   **b** £60   **c** £80?

4 Jane said that 30% of £45 is the same as 45% of £30.
    Is she correct?

*5 A file was 120 kbytes.
    How much of it had been downloaded in each of these?

a

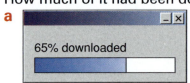

65% downloaded

b

85% downloaded

*6 Carla worked out $17\frac{1}{2}$% of 120 mentally like this.

        10% of 120 = 12
        5% of 120 = 6
        $2\frac{1}{2}$% of 120 = 3
    So $17\frac{1}{2}$% of 120 = 12 + 6 + 3 = 21

    $17\frac{1}{2} = 10 + 5 + 2\frac{1}{2}$

Use her method to work these out.
    a $17\frac{1}{2}$% of 360   **b** $12\frac{1}{2}$% of 240

*7 Find four different ways to fill in these boxes. $\square$% of $\square$ = 30

Percentage of – written and calculator methods

To find 27% of £652 **using a calculator**, we can use one of these methods.

27% of 652 = $\frac{27}{100} \times 652$ **Key** 27 ÷ 100 × 652 = to get **£176·04**.	27% of 652 = 0·27 × 652 **Key** 0·27 × 652 = to get **£176·04**.	1% of 652 = 6·52 27% of 652 = 27 × 6·52 **Key** 27 × 6·52 = to get **£176·04**.
Using fractions	**Using decimals**	**Finding 1% first**

To find 16% of 65 we can use one of these **written methods**.

Using fractions	Using decimals	Finding 1% first
16% of 65 = $\frac{16}{100} \times 65$ $\begin{array}{r} 65 \\ \times\ 16 \\ \hline 650 \\ 390 \\ \hline 1040 \end{array}$ $65 \times 16 = 1040$ $= \frac{16 \times 65}{100}$ $= \frac{1040}{100}$ $= \mathbf{10\cdot4}$	16% of 65 = 0·16 × 65 $\begin{array}{r} 65 \\ \times\ 16 \\ \hline 650 \\ 390 \\ \hline 1040 \end{array}$ $= 16 \times 65 \div 100$ $= 1040 \div 100$ $= \mathbf{10\cdot4}$	1% of 65 = 0·65 16% of 65 = 16 × 0·65 $\begin{array}{r} 65 \\ \times\ 16 \\ \hline 650 \\ 390 \\ \hline 1040 \end{array}$ $= 16 \times 65 \div 100$ $= 1040 \div 100$ $= \mathbf{10\cdot4}$

Note To key $12\frac{1}{2}$% key either ⌈12·5⌉ ⌈÷⌉ ⌈100⌉ **or** key ⌈0·125⌉

Exercise 9

1 Find the answer to these using a calculator.
  a 17% of £265 using fractions.
  b 38% of £136 using decimals.
  c 55% of £72·50 by finding 1% first. Round to the nearest penny.
  d $12\frac{1}{2}$% of £56.

2 Last year Bill earned £26 500.
   He paid 24% of this in tax.
   How much did Bill pay in tax?

3 Samantha borrowed £800 from her sister.
   Her sister charged her 4·5% interest per year.
   How much interest did she owe after 1 year?

4 Round the answers to these sensibly.
  a A 2p coin is 95% copper, 3·5% tin and 1·5% zinc.
     Five hundred 2p coins have a mass of 3·56 kg.
     What mass of copper, tin and zinc are needed to make five hundred 2p coins?
  b It is predicted that by 2021 the population of the United Kingdom will be about 63 640 000.
     How many people will live in each of England, Wales, Scotland and Northern Ireland in 2021?

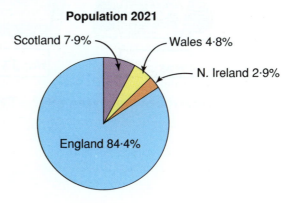

Population 2021

Scotland 7·9% Wales 4·8% N. Ireland 2·9% England 84·4%

5 Work out the new price of these.

a   b   c

6 Calculate   [SATs Paper 2 Level 5]
  a 8% of £26·50
  b $12\frac{1}{2}$% of £98.

7 Find the answer to these using a written method.
  a 16% of 200 m using fractions. $\frac{16}{100} \times 200 =$
  b 26% of 800 ℓ using decimals.   **0·26 × 800**
  c 37% of 200 cm by finding 1% first.
 *d 13% of £68 using fractions.
 *e 24% of 84 km using decimals.
 *f 19% of 120 g by finding 1% first.

8 A dress weighs 320 g.
  How many grams are lycra?

*9 a William bought a stereo in the sale.
    How much did he save?

  b How much was taken off the TV in the sale?

Fraction of

Exercise 10

1

$\overline{}$ 24		$\overline{}$ 100	$\overline{}$ 24	$\overline{}$ 8	**H** $\overline{}$ **5**	$\overline{}$ 24	$\overline{}$ 45	**32**

| $\overline{}$ 124 | $\overline{}$ 16 | $\overline{}$ 45 | $\overline{}$ 100 | $\overline{}$ 12 | $\overline{}$ 32 | $\overline{}$ 45 | | $\overline{}$ 9 | $\overline{}$ 21 | | $\overline{}$ 32 | $\overline{}$ 24 | $\overline{}$ 100 | **H** $\overline{}$ **5** |

| $\overline{}$ 32 | $\overline{}$ 24 | $\overline{}$ 4 |

Use a copy of the box. Find the answer to these.
Write the letter beside the question above its answer in the box.

H $\frac{1}{8}$ of 40 = **5** **R** $\frac{1}{6}$ of 24 **I** $\frac{1}{5}$ of 45 **T** $\frac{1}{7}$ of 56

U $\frac{2}{5}$ of 40 **A** $\frac{3}{5}$ of 40 **N** $\frac{7}{10}$ of 30 **L** $\frac{3}{8}$ of 32

E $\frac{2}{3}$ of 48 **S** $\frac{3}{4}$ of 60 **C** $\frac{4}{5}$ of 125 **M** $\frac{4}{7}$ of 217

2 Use a copy of these.
Find the fraction of the number in the green part.
Write the answer in the space at the top.

a   **b**   **c**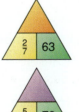

d   **e**   **f**

3 Find the missing numbers.   **[SATs Paper 1 Level 5]**

 a $\frac{1}{2} \times 20 = \frac{1}{4}$ of ___ **b** $\frac{3}{4}$ of 100 = $\frac{1}{2}$ of ___ **c** $\frac{1}{3}$ of 60 = $\frac{2}{3}$ of ___

Fraction of – multiplying an integer by a fraction

$\frac{1}{4}$ of 16, $\frac{1}{4} \times 16$, $16 \times \frac{1}{4}$, $16 \div 4$ are all equivalent.

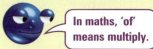

In maths, 'of' means multiply.

Worked Examples

a $5 \times \frac{3}{8}$

b Steff used $\frac{4}{5}$ of a 12 m length of rope to tie up his boat. How much rope was this?

Answers

a $5 \times \frac{3}{8} = 5 \times 3 \times \frac{1}{8}$

$= \frac{15}{8}$

$= 1\frac{7}{8}$

It is not easy to find $\frac{1}{8}$ of 5 or 3 so we multiply 5 and 3 to get $\frac{15}{8}$.

b $\frac{4}{5}$ of $12 = \frac{4}{5} \times 12$

$= 4 \times \frac{1}{5} \times 12$

$= 4 \times 12 \times \frac{1}{5}$

$= \frac{48}{5}$

$= 9\frac{3}{5}$ **m**

Whenever possible find the answers mentally.

Worked Example

About $\frac{2}{5}$ of Toby's burger patties are protein. If a burger pattie has a mass of 200 g, how much of this is protein?

Answer

$\frac{2}{5}$ of 200 g $= \frac{2}{5} \times 200$

$= 2 \times \frac{1}{5} \times 200$

$= \frac{400}{5}$

$= 80$

80 g is protein.

Exercise 11

1 Find the answer to these mentally.
a $\frac{2}{5}$ of 40   **b** $\frac{3}{8}$ of 64   **c** $\frac{7}{25}$ of 125   **d** $\frac{5}{9}$ of 81   **e** $1\frac{1}{2}$ of 18

2 Write true or false for these.
a $\frac{5}{9} \times 30 = 5 \times \frac{1}{9} \times 30$ **b** $\frac{3}{8} \times 13 = 13 \times \frac{3}{8}$ **c** $24 \times \frac{7}{16} = \frac{7}{16} \times \frac{24}{1}$ **d** $\frac{5}{6} \times 9 = \frac{54}{5}$

3 Give the answers as mixed numbers.
a $\frac{3}{4}$ of 9   **b** $\frac{5}{8}$ of 7   **c** $\frac{2}{3}$ of 16   **d** $\frac{5}{6}$ of 21 $*$**e** $\frac{2}{9}$ of 80

4 **a** Askra had a 32 cm length of ribbon.
          She was asked to cut $\frac{3}{8}$ off it.
          How much should she cut off?

   b A flea can long jump 35 cm.
          It can high jump $\frac{3}{5}$ of this distance.
          How high can a flea jump?

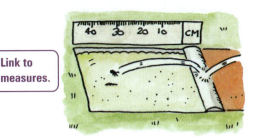

Link to measures.

*__*__5__ Write the next two lines for these.

   a $\frac{1}{7} \times 1 = \frac{1}{7}$ **b** $\frac{2}{7} \times 1 = \frac{2}{7}$ **c** $\frac{4}{5} \times 1 = \frac{4}{5}$
          $\frac{1}{7} \times 2 = \frac{2}{7}$ $\frac{2}{7} \times 2 = \frac{4}{7}$ $\frac{4}{5} \times 2 = \frac{8}{5}$
          $\frac{1}{7} \times 3 = \frac{3}{7}$ $\frac{2}{7} \times 3 = \frac{6}{7}$ $\frac{4}{5} \times 3 = \frac{12}{5}$
          $\frac{1}{7} \times 4 = \frac{4}{7}$ $\frac{2}{7} \times 4 = \frac{8}{7}$ $\frac{4}{5} \times 4 = \frac{16}{5}$

*__*__6__ Bea said 'multiplying **always** makes a number larger'.
       Give an example to show that Bea is wrong.

* Dividing an integer by a fraction

$4 \div \frac{1}{3}$ means 'How many $\frac{1}{3}$s are there in **4**?'

There are 3 lots of $\frac{1}{3}$ in 1. So there are $3 \times$ **4** lots of $\frac{1}{3}$ in 4. The answer is 12.

We could write this question as $4 = \square \times \frac{1}{3}$.

Discussion

8P were asked to find the answer to $4 \div \frac{2}{5}$.
Jolene drew these diagrams.

Each circle has been divided into fifths.
How could she use this to find the answer to $4 \div \frac{2}{5}$? **Discuss.**

Sirah drew this number line.

How could she use this to find the answer to $4 \div \frac{2}{5}$? **Discuss.**

Tim knew that $10 \times \frac{2}{5} = 4$.

How could he use this to find the answer to $4 \div \frac{2}{5}$? **Discuss.**

Use each of Jolene's, Sirah's and Tim's methods to find the answer to these.

   $3 \div \frac{1}{7}$ $4 \div \frac{1}{5}$ $5 \div \frac{1}{6}$ $6 \div \frac{3}{5}$ $9 \div \frac{3}{4}$ $8 \div \frac{2}{3}$

Number

Worked Example

a $3 \div \frac{1}{6}$ **b** $4 \div \frac{2}{3}$

Answers

a We read this as 'How many $\frac{1}{6}$s in 3?'
From the diagram we can see that there are 18 sixths in 3.
$3 \div \frac{1}{6} =$ **18**

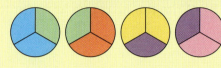

3 circles divided into sixths

b We read this as 'How many $\frac{2}{3}$s in 4?'
From the diagrams we can see that there are
6 two-thirds in 4.
$4 \div \frac{2}{3} =$ **6**

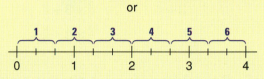

or

Exercise 12

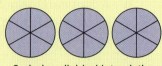

1 $60 \times \frac{1}{4} = 15$

     $30 \times \frac{2}{4} = 15$

     $20 \times \frac{3}{4} = 15$

     Use this to find the answers to these.

     a $15 \div \frac{1}{4}$ **b** $15 \div \frac{2}{4}$ **c** $15 \div \frac{3}{4}$

2 $120 \times \frac{1}{6} = 20$

     $60 \times \frac{2}{6} = 20$

     $40 \times \frac{3}{6} = 20$

     $30 \times \frac{4}{6} = 20$

     Find these.   **a** $20 \div \frac{1}{6}$ **b** $20 \div \frac{1}{3}$ **c** $20 \div \frac{1}{2}$ **d** $20 \div \frac{2}{3}$

3   Write a division for each of these.
     The answer to **a** is $3 \div \frac{1}{4}$.

     a   How many quarters are there in 3 oranges?
     b   How many thirds are there in 7 pizzas?
     c   How many fifths are there in 4 pies?
     d   How many eighths are there in 6 apples?

4 **a** $4 \div \frac{1}{4}$ **b** $6 \div \frac{1}{5}$ **c** $8 \div \frac{1}{3}$ **d** $5 \div \frac{1}{8}$

5 What goes in the box?

a $32 \times \frac{3}{4} = 24$ so $24 \div \frac{3}{4} = \square$

b $55 \times \frac{3}{5} = 33$ so $33 \div \frac{3}{5} = \square$

6 Use the diagrams to help find the answers to these.

a $6 \div \frac{2}{3}$

b $6 \div \frac{3}{4}$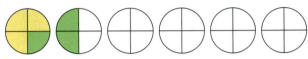

*c $4 \div \frac{4}{5}$

*d $10 \div \frac{5}{6}$

*7 What goes in the gap, increases or decreases?
Dividing a number by a fraction smaller than 1 _____ the number.

When we divide by a whole number we get an answer smaller than the number we divided into.

Summary of key points

 A We **cancel fractions** to write them in their simplest form.

Examples

 B We can write **one number as a fraction of another**.

Example 8 as a fraction of 24 is $\frac{8}{24} = \frac{1}{3}$.

Always cancel fractions to their lowest terms.

C To write a **decimal as a fraction** write it with a denominator of 10, 100 or 1000.
Then cancel to the simplest form.

Examples $0 \cdot 7 = \frac{7}{10}$ $0 \cdot 82 = \frac{82}{100}$ $0 \cdot 317 = \frac{317}{1000}$
$= \frac{41}{50}$

113

To write a **fraction as a decimal** we can

1 use known facts

Example $\frac{3}{5} = 3 \times \frac{1}{5} = 3 \times 0 \cdot 2 = 0 \cdot 6$

2 make the denominator 10 or 100.

Example $\frac{7}{25} = \frac{28}{100} = 0 \cdot 28$

3 divide the numerator by the denominator ⟶

Example $\frac{47}{5}$

$$
\begin{array}{r}
5\overline{)47} \\
-45 \qquad 5 \times \mathbf{9}\\
\hline
2 \cdot 0 \\
-2 \cdot 0 \qquad 5 \times \mathbf{0 \cdot 4}\\
\hline
0 \cdot 0
\end{array}
$$

4 use a calculator

Example $\frac{36}{42}$

Answer **9·4**

key [36] [÷] [42] [=] to get 0·86 (2 d.p.)

or key [36] [aᵇ/c] [42] [=] [aᵇ/c] to get 0·86 (2 d.p.)

In some decimals, one or more digits repeat.

These are **recurring decimals**.

Examples $\frac{1}{3} = 0 \cdot 3333\ldots \qquad \frac{5}{11} = 0 \cdot 454545\ldots$

$\qquad\qquad = 0 \cdot \dot{3} \qquad\qquad\quad = 0 \cdot \dot{4}\dot{5}$

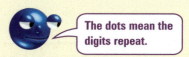

The dots mean the digits repeat.

 We can write **fractions and decimals as percentages** by writing with denominator 100

Examples

$$\frac{3}{25} \xrightarrow{\times 4} \frac{12}{100}$$
$$= 12\%$$

$$\frac{24}{40} \xrightarrow{\div 4} \frac{6}{10} \xrightarrow{\times 10} \frac{60}{100}$$
$$= 60\%$$

$$0 \cdot 72 = \frac{72}{100}$$
$$= 72\%$$

 To **add and subtract fractions** with the same denominator, add the numerators.

Write the answers in their simplest form.

Examples

$$\frac{4}{15} + \frac{7}{15} = \frac{4 + 7}{15}$$
$$= \frac{11}{15}$$

$$\frac{17}{20} + \frac{11}{20} = \frac{17 + 11}{20}$$
$$= \frac{28}{20}$$
$$= 1\frac{8}{20}$$
$$= 1\frac{2}{5}$$

To **add and subtract fractions** which do not have the same denominator we find equivalent fractions.

Examples

$$\frac{5}{8} + \frac{2}{3} = \frac{15}{24} + \frac{16}{24}$$
$$= \frac{15 + 16}{24}$$
$$= \frac{31}{24}$$
$$= \mathbf{1\frac{7}{24}}$$

$$\frac{5}{8} - \frac{3}{5} = \frac{25}{40} - \frac{24}{40}$$
$$= \frac{25 - 24}{40}$$
$$= \frac{1}{40}$$

LCM of 8 and 5 is 40.

multiples of 8 are 8, 16, 24, 32, **40**, ...
multiples of 5 are 5, 10, 15, 20, 25, 30, 35, **40**, ...

We can find **'percentage of'** mentally.

Always try to use a mental method first.

Example   Find 65% of £480.

50% of £480 = £240

10% of £480 = £48

5% of £480 = £24

65% of £480 = £240 + £48 + £24   **65% = 50% + 10% + 5%**

= **£312**

We can find **'percentage of'** using a calculator.

Example   Find 17% of 84.

Using fractions	Using decimals	Finding 1% first
17% of 84 = $\frac{17}{100} \times 84$	17% of 84 = 0·17 × 84	1% of 84 = 0·84
Key (17)(÷)(100)(×)(84)(=)	Key (0·17)(×)(84)(=)	17% of 84 = 17 × 0·84
to get **14·28**.	to get **14·28**.	Key (17)(×)(0·84) to get **14·28**.

We can find **'percentage of'** using a written method.

Example   Find 16% of 750.

Using fractions	Using decimals	Finding 1% first
16% of 750 = $\frac{16}{100} \times 750$	16% of 750 = 0·16 × 750	1% of 750 = 7·5
75 $\quad = \frac{16 \times 750}{100}$	= 16 × 750 ÷ 100	16% of 750 = 16 × 7·5
×16 $\quad = \frac{12\,000}{100}$	= 12 000 ÷ 100	= 16 × 75 ÷ 10
750	= **120**	= 1200 ÷ 10
450 $\quad$ = **120**		= **120**
1200		

 To find a **fraction of**, we divide.

To find $\frac{1}{9}$ we divide by 9. $\frac{1}{9}$ of 54 = 54 ÷ 9 = 6

To find $\frac{2}{3}$ of 36, we find one third then multiply by 2.

$\frac{1}{3}$ of 36 = 36 ÷ 3

= 12

$\frac{2}{3}$ of 36 = 2 × 12

= 24

 $\frac{1}{8}$ of 24, $\frac{1}{8} \times 24$, $24 \times \frac{1}{8}$,   24 ÷ 8 are all equivalent.

We can **multiply integers and fractions**.

Example $4 \times \frac{5}{8} = 4 \times 5 \times \frac{1}{8}$

$= \frac{20}{8}$

$= 2\frac{4}{8}$

$= \mathbf{2\frac{1}{2}}$

115

Number

Test yourself **Except for questions 7 and 14.**

1 Cancel these fractions to their simplest form.   **A**
   a $\frac{2}{12}$ **b** $\frac{12}{24}$ **c** $\frac{6}{24}$ **d** $\frac{15}{25}$ **e** $\frac{27}{36}$

2 **a** What fraction of 30 is 5?   **B**
   b What fraction of 60 is 20?

3 At a dance class there were 20 girls and 5 boys.   **B**
   a How many were in the class altogether?
   b What fraction of the class was girls?

4 Write these as fractions.   **C**
   a 0·4   **b** 0·25   **c** 0·43   **d** 0·039   *e 1·27

> Simplify the fractions.

5 Write these as decimals.   **D**
   a $\frac{9}{10}$ **b** $\frac{19}{100}$ **c** $\frac{3}{100}$ **d** $\frac{7}{20}$ **e** $\frac{3}{25}$

6 Write the fractions in question **5** as percentages.   **E**

 7 Write these as recurring decimals.   **a** $\frac{1}{3}$ *b $\frac{4}{11}$ **D**

8 Use a copy of the table.   **C D E**
Fill it in.

Decimal	Percentage	Fraction
0·5		
	30%	
		$\frac{7}{20}$
0·03		

9 **a** $\frac{2}{5} + \frac{1}{5}$ **b** $\frac{9}{12} - \frac{4}{12}$ **c** $\frac{2}{9} + \frac{3}{9} + \frac{2}{9}$ **d** $\frac{5}{6} + \frac{2}{6} - \frac{3}{6}$ **F**

10 **a** What goes in the gap? $\frac{2}{3} \overset{\times 4}{\underset{\times 4}{=}} \frac{\square}{12}$ **b** Find $\frac{7}{12} + \frac{2}{3}$.   **A F**

11 **a** What goes in the gap? $\frac{1}{4} = \frac{\square}{12}$ **b** Find $\frac{7}{12} - \frac{1}{4}$.

12 **a** What goes in the gaps? 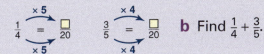 $\frac{1}{4} = \frac{\square}{20}$ $\frac{3}{5} = \frac{\square}{20}$ **b** Find $\frac{1}{4} + \frac{3}{5}$.

13 Work these out mentally.
   a 40% of £60   **b** 35% of 80 m   *c 21% of 70 m

14 About 70% of our body mass is water.
   About how much water is in a person who weighs 55 kg?

15 Use a written method to find these. Each is started for you.
   a 17% of 400 m $\frac{17}{100} \times 400 =$ **b** 36% of 650 mℓ
   $0.36 \times 650 =$

16 Hakan asked 30 pupils which subject they liked best.   [SATs Paper 1 Level 5]

Subject	Number of boys	Number of girls
Maths	4	7
English	2	4
Science	3	3
History	0	1
French	1	5
	total 10	total 20

a Which subject did **20%** of **boys** choose?
b Which subject did **35%** of **girls** choose?
c Hakan said:
    'In my survey, Science was equally popular with boys and girls'.
    Explain why Hakan was wrong.
d Which subject was equally popular with boys and girls?

17 Find these mentally.
   a $\frac{1}{6}$ of 48 m   **b** $\frac{1}{5}$ of 55 kg   **c** $\frac{3}{4}$ of 16 m   **d** $\frac{2}{3}$ of 21 km

18 Jill's frog can jump 125 cm.
   Troy's frog can jump $\frac{4}{5}$ of this distance.
   How far can Troy's frog jump?

19 Give the answers to these as fractions.
   a $2 \times \frac{3}{5}$ **b** $5 \times \frac{3}{8}$ **c** $7 \times \frac{5}{6}$

117

Number

20 Write a division for these.
Find the answers.
   a How many eighths in 6?   **b** How many thirds in 5?

21 $84 \times \frac{3}{4} = 63$ so $63 \div \frac{3}{4} = \square$.
What goes in the box?

22 Use this diagram to find the answer to $5 \div \frac{5}{6}$.

5 Ratio and Proportion

You need to know

✓ ratio and proportion   page 9

Key vocabulary

direct proportion, proportion, ratio

Golden Rectangles

Measure the length and width of this rectangle with your ruler.

Use your calculator to find $\frac{\text{length}}{\text{width}}$.

Your answer should be very close to 1·6.

This is called the **golden ratio**.

Rectangles with this ratio are said to be 'nice to look at'.

Buildings such as the one in the picture (in Greece) are built to this ratio.

Construct a golden rectangle as follows.

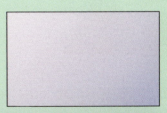

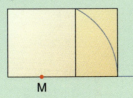

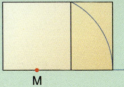

Draw a square (any size). Mark the mid-point, M, as shown.

Make the base longer. Put your compass point on M and draw an arc as shown.

Draw the rectangle so that it just encloses the arc.

119

Writing ratios in their simplest form

Ian made a drink using 2 parts juice and 5 parts lemonade.
The ratio of juice to lemonade is 2 : 5.

Jackie made a drink using 4 parts juice and 10 parts lemonade.
The ratio of juice to lemonade is 4 : 10.
We can simplify this ratio by cancelling.

Ian's and Jackie's drinks were mixed in the **same ratio**.

Examples

Exercise 1

1 For each of these diagrams, write the ratio of orange parts to purple parts.

a   **b**   **c**

d   **e**   **f**

2 Write these ratios in their simplest form.

a 4 : 8   **b** 3 : 12   **c** 6 : 18   **d** 20 : 5

= __ : __   = __ : __   = __ : __   = __ : __

e 21 : 7   **f** 6 : 9   **g** 20 : 25   **h** 16 : 24

i 30 : 24   ***j** 6 : 4 : 10   ***k** 8 : 16 : 4   ***l** 12 : 9 : 15

3 A fuel mix is 10 parts petrol to 4 parts oil.
What is the ratio, in its simplest form, of petrol to oil?

4 There are 27 pupils and 6 teachers on a school camp.
Write the ratio of
  a pupils to teachers
  b teachers to pupils.

*5 Robbie painted his room blue and purple in the ratio 5 : 2.
Was there more blue or more purple in his room?

*6 Write as a ratio in its simplest form
  a carbohydrate to fat to sugar
  b protein to carbohydrate to sugar.

	No. of grams
Carbohydrate	24
Fat	8
Protein	16
Sugar	12

*7 Belinda had £15. She spent £10.
  a How much did she have left?
  b Write what Belinda spent to what she had left as a ratio in its lowest terms.

Investigation

Squares and triangles

You will need 1 cm squared paper,
            1 cm triangle dotty paper.

a On the 1 cm squared paper, draw six
squares with sides of 1 cm, 2 cm,
3 cm, 4 cm, 5 cm and 6 cm.
Use a copy of this table.
Fill it in.

Length of side (cm)	Perimeter (cm)	Ratio of side : perimeter
1	4	1 : 4
2	8	2 : 8 = __ : __

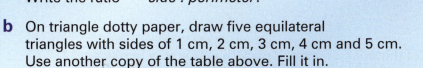

Simplify the ratios.

A square has sides of x cm.
Write an expression for its perimeter.
Write the ratio   *side : perimeter*.

b On triangle dotty paper, draw five equilateral
triangles with sides of 1 cm, 2 cm, 3 cm, 4 cm and 5 cm.
Use another copy of the table above. Fill it in.

An equilateral triangle has sides of y cm.
Write an expression for its perimeter.
Write the ratio   *side : perimeter*.

Number

All parts of a ratio must have the same units.
Once the ratio is simplified we do not include the units.

Worked Example
Write the ratio *adult cost to child cost* in its simplest form.

Rides

Adult £1·60
Child   70p

Answer
The ratio is £1·60 : 70p.
We must write both parts either in pounds or in pence.

Using pence, the ratio is
160 : 70 = 16 : 7

Using pounds the ratio is
1·60 : 0·70
× 10 ( ) × 10
= 16 : 7

Exercise 2

1 Write these ratios in their simplest form.

   a   50p : £1
      = 50p : ___ p
      = ___ : ___

   b   5 mm : 1 cm
      = 5 mm : ___ mm
      = ___ : ___

   c   20 min : 1 hour
      = 20 min : ___ min
      = ___ : ___

   d   750 mℓ : 1 litre
      = 750 mℓ : ___ mℓ
      = ___ : ___

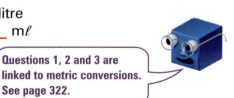

Questions 1, 2 and 3 are linked to metric conversions. See page 322.

2 Write these ratios in their simplest form.

   a 2 mm : 5 mm   **b** 3 mm : 1 cm   **c** 53 cm : 2 m   **d** 13 min : 1 hour
   e 60c : $3   ***f** $4·50 : $3   ***g** 450 g : 2 kg   ***h** 1 litre : 340 mℓ

3 A recipe uses 1·5 kg of flour and 500 g of sugar.
   Write, in its simplest form, the ratio of flour to sugar.

*4

CRISPS
Small   35p
Family £1·05

   What is the ratio of the price of a small pack of crisps to a family pack?

*5 Sarah and Kate bought a raffle ticket which cost £2.
   Sarah paid 89p and Kate paid the rest.
   What was the ratio of the amount Kate paid to the amount Sarah paid?

A ratio in its simplest form does not have fractions.

Worked Example

Hamish put $1\frac{1}{2}$ litres of juice and 1 litre of lemonade in a jug.
Write the amount of juice to the amount of lemonade as a ratio in its simplest form.

Answer

The ratio is $1\frac{1}{2}$: 1.

$$1\frac{1}{2} : 1$$
$$\times 2 \left(\right) \times 2 \qquad \text{\textbf{Multiply both parts of the ratio by 2 to get whole numbers.}}$$
$$= 3 : 2$$

The ratio in its simplest form is **3 : 2**.

Exercise 3

1 a $\frac{1}{2} : 1$
$\times 2 \left(\right) \times 2$
$= \_ : \_$

b $\frac{1}{3} : 2$
$\times 3 \left(\right) \times 3$
$= \_ : \_$

c $\frac{1}{5} : 3$
$\times 5 \left(\right) \times 5$
$= \_ : \_$

d $1\frac{1}{3} : 3$
$\times 3 \left(\right) \times 3$
$= \_ : \_$

2 Write these ratios in their simplest form.
   a $2\frac{1}{2} : 1$ **b** $1\frac{1}{2} : 2$ **c** $\frac{1}{2} : 1\frac{1}{2}$ **d* $2\frac{1}{5} : 3$ **e* $3\cdot5 : 2$

3 Samantha used $\frac{1}{4}$ kg of red peppers and 2 kg of green peppers to make a pickle.
   Write, in its simplest form, the ratio of red to green peppers.

Ratio and proportion

Proportion compares part to whole.
Ratio compares part to part.

Example   There are 3 pink squares and 7 purple squares.
The ratio of pink to purple is 3 : 7.

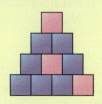

There are 10 squares in total.

The proportion of pink squares is 3 out of 10 or $\frac{3}{10}$.

The proportion of purple squares is 7 out of 10 or $\frac{7}{10}$ or 70% or $0\cdot7$.

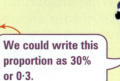

We could write this proportion as 30% or $0\cdot3$.

Number

Exercise 4

1 Bernie planted these 36 plants.

    16 tomato

    20 broccoli

  a Write the ratio of tomato to broccoli plants.

  b What proportion of the plants were tomato?

  c What proportion were broccoli?

Always give the answers in the simplest form.

2 This table shows the points scored in a game.

Name	Nicola	Robyn
Points	15	20

  a Write these as ratios in their simplest form.

    i Nicola's to Robyn's points

    ii Robyn's to Nicola's points

  b How many points were scored in total?

  c What proportion of points did these score?

    i Nicola   **ii** Robyn

Give the answers to c as fractions in their simplest form.

3 **a** What is the ratio of rolls to crisps sold?

  b What is the ratio of pies to crisps sold?

  c What proportion of items sold were crisps? Give your answer as a percentage.

  d What proportion of items sold were ice creams? Give your answer as a fraction.

  e What proportion of items sold were pies? Give your answer as a decimal.

Item	Number sold
Pies	25
Rolls	20
Crisps	35
Ice creams	20
Total	**100**

*4 a   What is the ratio of cotton to polyester?
    b   What is the ratio of lycra to polyester to cotton?
    c   What proportion of the trousers were lycra?
    d   Give the proportion of cotton in the trousers as a
        percentage.

Trousers	
Material	**Grams**
Cotton	20
Polyester	45
Lycra	15

Dividing in a given ratio

When things are shared, they are not always shared equally.
Instead they are sometimes **shared in a given ratio**.

Example   Some money was shared between two
            sisters in the ratio 2 : 1.
            This means that one sister got twice as
            much as the other.
            We could say that out of every £3, one
            sister got £2 and the other got £1.

Worked Example
Two friends, Rachel and Paul, paid for a house in the ratio **2 : 3**.
They sold the house for £80 000. How much should each get?

Answer
2 + **3** = 5
There are 5 shares altogether.
1 share is $\frac{£80\,000}{5}$ = £16 000.
2 shares are 2 × £16 000 = £32 000.
3 shares are 3 × £16 000 = £48 000.

Rachel should get £32 000 and Paul £48 000.

Notice that you are
given the total.

Check that these
add up to £80 000.

Exercise 5

1 Brian's mother won £160.
   She gave it to Brian and his sister Daphne in the ratio 5 : 3.
   Brian got the larger amount.
   a   How many shares are there altogether?
   b   How much is one share?
   c   How much did Brian get?
   d   How much did Daphne get?

2 Faye and her sister Margaret bought a house.
It cost £125 000.
They shared the cost in the ratio 2 : 3.
Margaret paid the larger amount.
  a How many shares are there altogether?
  b How much is one share?
  c How much did Faye pay?
  d How much did Margaret pay?

3 **a** Share £50 in the ratio 3 : 2.
  b Share £2400 in the ratio 7 : 5.
  c Share £128 in the ratio 3 : 5.
  d Share £72 000 in the ratio 5 : 4.

4 The ratio of sunny days to rainy days last June was 2 : 3.
  a How many days were sunny?
  b How many days were rainy?

You will need to know how many days are in June.

5

Moira and Anna bought a boat for £5400.
They paid for it in the ratio 4 : 5.
Moira paid the smaller amount.
How much did each pay?

*6 Share
  a £128 in the ratio 1 : 3 : 4.
  b £72 000 in the ratio 2 : 3 : 4.

*7 A fertiliser mixture contains 2 parts nitrogen,
2 parts potash and 3 parts lime.
How much of each is in a 70 kg bag?

Fertiliser
2 parts nitrogen
2 parts potash
3 parts lime
70 kg

*8 The three angles of a triangle are in the ratio 2 : 3 : 4.
What is the size of each angle of this triangle?

Proportion

Daphne wants to make some purple paint to paint her room.
She mixes 5 spoonfuls of blue paint with 2 spoonfuls of red paint.
She then finds out this isn't enough paint!
She needs to mix up more.

How will she get the same colour paint?

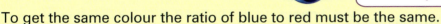

To get the same colour the ratio of blue to red must be the same.
The amounts must be in **direct proportion**.
If we double one we double the other.
If we have three times as much of one we must multiply the other by three.

For 10 spoonfuls of blue paint she needs 4 of red.   10 : 4 = 5 : 2
For 15 spoonfuls of blue paint she needs 6 of red.   15 : 6 = 5 : 2
For 20 spoonfuls of blue paint she needs 8 of red.   20 : 8 = 5 : 2

Worked Example
2 pies cost £2·40
How much will these cost?
a 4 pies   b 8 pies

Answers
a 4 is twice as many as 2.
   4 pies will cost twice as much
   as 2 pies.
   4 pies will cost 2 × £2·40 = £4·80

b 8 is four times 2.
   8 pies will cost four times
   as much as 2 pies.
   8 pies will cost 4 × £2·40 = £9·60

Exercise 6

1 To make a drink you mix 1 ℓ of juice concentrate with 3 ℓ of water.
   a How much water would you add to 2 ℓ of juice?
   b How much water would you add to 5 ℓ of juice?
   ∗c How much water would you add to 500 mℓ of juice?

2 One kilogram of apples cost £2·50.
   What would these cost?
   a 3 kg   **b** 10 kg   **c** $\frac{1}{2}$ kg

3 To make orange paint you mix 5 ℓ of yellow paint with 3 ℓ of red paint.
   a How much yellow paint would you mix with
      i 6 ℓ of red paint
      ii 9 ℓ of red paint?
   ∗b How much red paint would you mix with $2\frac{1}{2}$ ℓ of yellow paint?

6 Steph and Becky share £800 in the ratio 5 : 3.
How much does each of them get?

C

7 a Divide 450 g in the ratio 4 : 5.
   b Divide 175 g in the ratio 3 : 2.

C

8 A recipe for 5 people needs 250 g of mushrooms.
   a What mass of mushrooms would be needed for 1 person?
   b What mass of mushrooms would be needed for 3 people?
   c What mass of mushrooms would be needed for 11 people?

D

9 This recipe shows the ingredients needed to
make 24 bonbons.
   a How many ounces of chocolate hail are
needed for 48 bonbons?
   b How many ounces of butter are needed
for 12 bonbons?
   c How many ounces of mixed fruit are
needed for 12 bonbons?
   d How many ounces of butter are needed for 9 bonbons?
  *e How many ounces of mixed fruit are needed for 15 bonbons?

E

> **BONBONS**
>
> 4 oz butter
> 8 oz mixed fruit
> 1 tbsp cocoa
> 7 oz icing sugar
> 3 oz chocolate hail

10 At an art show there were water colours and oil paintings in the ratio 2 : 3.
There were 12 oil paintings.
How many water colours were there?

E

11 The ratio of the number of cars to the number of motorbikes in a rally is 7 : 2.
There are 63 cars. How many motorbikes are there?

E

12 Screenwash is used to clean car windows.   **[SATs Paper 2 Level 5]**
To use screenwash you mix it with water.

B **E**

> **Winter mixture**
> Mix 1 part screenwash
> with 4 parts water

> **Summer mixture**
> Mix 1 part screenwash
> with 9 parts water

   a In winter, how much water should I mix with 150 mℓ of screenwash?
   b In summer, how much screenwash should I mix with 450 mℓ of water?
   c Is this statement correct?

> 25% of winter mixture is screenwash

Explain your answer.

Algebra Support

Expressions

Writing expressions

$n + 6$, $3n$ and $2n + 3$ are all **expressions**.

Example   Paul has n coins in his hand.
Lala has 3 times as many.
Lala has $3 \times n$ or $3n$ coins.

We usually write expressions without $\times$ and $\div$.

Letters **follow the same rules as numbers**.

Examples $4 + 5 = 5 + 4$ $a + b = b + a$

$7 \times 8 = 8 \times 7$ $xy = yx$

$5 \times (8 \times 3) = (5 \times 8) \times 3$ $m(pq) = (mp)q$

Simplifying expressions

We can **collect like** terms.

Examples $x + x = 2x$ $3m + 2m = 5m$ $6y - 2y = 4y$

We can **simplify expressions by cancelling**.

Example $\frac{^1x}{_1x} = 1$ $\frac{^13m}{_13} = m$

Divide the numerator and denominator by the same thing.

Substituting into expressions

We can find the value of an expression by **substituting** values for the unknown.
We follow the **order of operations** rules.

Remember BIDMAS.

Example   If $n = \mathbf{4}$ then
$$3n = 3 \times \mathbf{4} \qquad\qquad 3n + 6 = 3 \times \mathbf{4} + 6 \qquad\qquad \frac{12}{n} = \frac{12}{\mathbf{4}}$$
$$= 12 \qquad\qquad\qquad\qquad\quad = 12 + 6 \qquad\qquad\qquad\quad = 3$$
$$\qquad\qquad\qquad\qquad\qquad\qquad\qquad = 18$$

Practice Questions 1, 2, 5, 6, 7, 9, 13, 18

Formulae

A **formula** is a rule for working out one unknown when you know others.
In a formula each unknown stands for something.

Example $P = \mathbf{8}T$ is a formula for the pressure, P, in a container at temperature, T.
If $T = \mathbf{20}$ then $P = 8 \times \mathbf{20}$
$= 160$.

Practice Questions 3, 10

Algebra

Equations

$n + 7 = 12$ is an **equation**. n has a particular value.

$x + y = 15$ is an equation. x and y can take many different values but they must always add to 15.

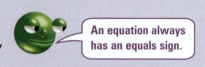

An equation always has an equals sign.

We can **solve equations** that have just one unknown **using inverse operations**.

Examples $y + 12 = 19$
$y = 19 - 12$ **The inverse of adding 12 is subtracting 12.**
$y = \mathbf{7}$

$y \rightarrow \boxed{+12} \rightarrow 19$

$7 \leftarrow \boxed{-12} \leftarrow 19$

$\frac{x}{5} = 4$
$x = 4 \times 5$ **The inverse of dividing by 5 is multiplying by 5.**
$x = 20$

Practice Questions 20, 27, 28, 29

Sequences

A **sequence** is a set of numbers in a given order.
Each number is called a **term**.

Examples   3, 8, 13, 18, 23, ...   **These dots mean the sequence continues forever. Its is infinite.**

1st term   5th term

8, 16, 24, 32, 40, ... , 80   **This sequence if finite. It starts at 8 and ends at 80.**

We can **write a sequence** by **counting on** or **counting back**.

Examples   Starting at 5 and counting on in steps of 4 we get 5, 9, 13, 17, 21, ...

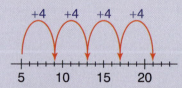

Starting at 6 and counting back in steps of 2 we get 6, 4, 2, 0, ⁻2, ⁻4, ...

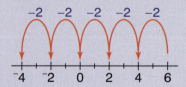

We can **write a sequence** if we know the first term and the **rule for finding the next term**.

Example   **1st term** 3   **rule** add 10   gives   3, 13, 23, 33, 43, ...

136

We can write a sequence if we know the rule for the *n*th term.

Example   If the rule for the *n*th term is $3n + 1$ the sequence is

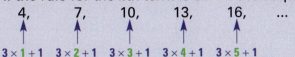

4,   7,   10,   13,   16,   ...

$3 \times 1 + 1$ $3 \times 2 + 1$ $3 \times 3 + 1$ $3 \times 4 + 1$ $3 \times 5 + 1$

> We find the terms of the sequence by substituting $n = 1$, $n = 2$, $n = 3$ and so on into $3n + 1$.

We can write **sequence for practical situations**.

Example   Les makes bags.
He puts patterns on the bags using strips of tape.

1 bag	2 bags	3 bags
4 strips	8 strips	12 strips

The sequence of the number of strips is 4, 8, 12, ...
There are 4 strips on each bag.
The number of strips on *n* bags will be $4 \times n$.

Practice Questions 4, 8, 14, 21, 22, 24

Functions

$x \rightarrow$ [multiply by 2] $\xrightarrow{2x}$ [add 4] $\xrightarrow{2x+4} y$ This is a **function machine**.

input   output

The rule for the function machine is $y = 2x + 4$ or $x \longrightarrow 2x + 4$.

We can find the **output** if we are given the input.

Example   2 → [multiply by 2] → 4
5 → → 10
3 → → 6

We can show the function on a **mapping diagram**.

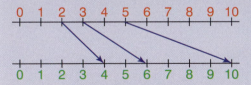

0 1 2 3 4 5 6 7 8 9 10

0 1 2 3 4 5 6 7 8 9 10

We can **find the rule** if we are given the input and output.

Example   5 → [?] → 20 $5 \times 4 = 20$
0 → → 0 $0 \times 4 = 0$
2 → → 8 $2 \times 4 = 8$

The rule for this function machine is 'multiply by 4'.

Practice Questions 15, 16, 17

Graphs

(5, 6) is a **coordinate pair**.

Each of these coordinate pairs satisfy the rule $y = x - 1$.

(0, $^-$1), (1, 0), (2, 1), (3, 2)

x-coordinate y-coordinate

Each y-coordinate is found by subtracting 1 from the x-coordinate.

To draw a **straight-line graph** of $y = x + 2$ find three or more coordinate pairs.

x	working	y
3	$3 + 2 = 5$	5
2	$2 + 2 = 4$	4
0	$0 + 2 = 2$	2
$^-$1	$^-1 + 2 = 1$	1

Then plot the coordinate pairs (3, 5), (2, 4), (0, 2) and ($^-$1, 1) on a grid.
Label your line.

$y = mx$ is the equation of a straight line through (0, 0).
m can have any value.
The greater the value m, the steeper the slope.

Example $y = 2x$ is steeper than $y = x$.

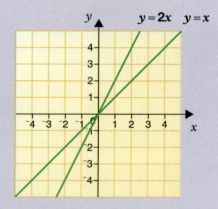

If m is positive the slope is positive, e.g. $y = 3x + 2$.

If m is negative the slope is negative, e.g. $y = {}^-2x + 2$.

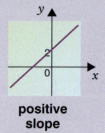

positive slope

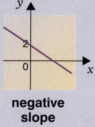

negative slope

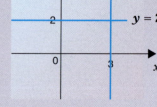

Lines **parallel to the** x-axis have equations $y = a$ where a is any number.
Lines **parallel to the** y-axis have equations $x = b$ where b is any number.

Practice Questions 11, 12, 19

Graphs of real-life situations

We often draw straight-line graphs to show real-life situations.
We can read information from these graphs.

Example This graph shows the charges for plumbing.
The points (0, £40), (10, £70) and (20, £100) have been plotted and joined with a straight line.

We can estimate other values from the graph.

Example If the plumber spend 5 hours on a job the charge is about £55.

Practice Questions 23, 25, 26

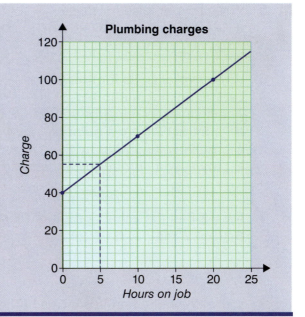

Plumbing charges

Practice Questions

T

1 Use a copy of this.
Write the expression on the left without a multiplication sign.
Join it to the answer on the right with a line.

$3 \times n$ $5n$

$4 \times n$ $3n + 2$

$n \times 5$ $4n$

$3 \times n + 2$ $4n + 3$

$4 \times n + 3$ $3n$

2 Write each of these without a multiplication or division sign.
a $3 \times n$ **b** $y \times 4$ **c** $a \div 5$ **d** $3 \times y + 2$ **e** $4 \times (x + 3)$

3 A formula for finding the cost of sweets is:
    cost = cost of one bag × number of bags
Find the cost if
a there are 4 bags and each bag costs £2
b there are 5 bags and each bag costs £3.

4 Write down the first six terms of these sequences.
You may use the number lines to help.

a Start at 3 and count on in steps of 4.

b Start at 4 and count back in steps of 2.

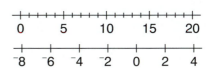

139

Algebra

5 Write true or false for these.

   a $a + b = b + a$ **b** $ab = ba$ **c** $pqr = p + q + r$ **d** $a - b = b - a$

6 Write expressions for these.
Let the unknown number be n.
The first one is done for you.

   a subtract 2 from a number $n - 2$ **b** add 3 to a number
   c multiply a number by 5   **d** divide a number by 3
   e multiply a number by 3 then add 1

7 If $x = 3$ find the value of each of these expressions.

   a $2x = 2 \times 3$ **b** $x + 4 = 3 + 4$ **c** $x - 2$ **d** $2x + 1$ **e** $20 - x$
       $=$ $=$

8 Write down the first five terms of these.
The first term and the rule for finding the next term is given.

   a **1st term** 2,   **rule** add 3
   b **1st term** 1,   **rule** add 5
   c **1st term** 20,   **rule** subtract 5
   d **1st term** 1·5,   **rule** add 0·5

9 Write an expression for the number of marbles Brad has.
The answer to **a** is $r - 4$.

   a Brad had r marbles. He gave 4 away.
   b Brad had s marbles. He was given 6 more.
   c Brad had t marbles. He lost 3.
   d Brad had p marbles. He gave 7 to a friend.

10 $C = 4n$ is a formula for the cost, in pence, of n jelly jubes.
Find the cost of

   a 5 jelly jubes   **b** 20 jelly jubes   **c** 35 jelly jubes.

Remember the units.

T **11** Use a copy of these tables.
Fill them in.

   a $y = 2x$ **b** $y = x - 3$ **c** $y = 10 - x$

x	Working	y
1	2×1	2
2		
3		

x	Working	y
6		
3		
0		

x	Working	y
10		
8		
5		

T **12** Use a copy of this.

x	Working	y
1		
0		
⁻2		

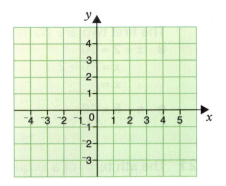

a Complete the table for $y = x + 2$.
b Write down the coordinate pairs for the points.
c On the grid, plot these three points.
    Draw and label the line with equation $y = x + 2$.

13 Simplify these by collecting like terms.
   a $y + y$ **b** $x + x + x$ **c** $3a + 4a$ **d** $7x + 3x$
   e $8y - y$ **f** $12p - 8p$ **g** $3b + b + 2b$ **h** $4m - 2m + 7m$
   i $7a + 2a + 3b + 4b$ **j** $4x - x + 5b - 2b$ ✳**k** $7y + 2y + 2z - 3z$

14 Match the descriptions with the sequences given in the box.
   a the even numbers from 4 to 14
   b the sequence begins at 3 and increases in steps of 4
   c each term is one more than a multiple of 5
   d the sequence begins at 7 and decreases in steps of 3

> A. 4, 6, 8, 10, 12, 14
> B. 6, 11, 16, 21, ...
> C. 7, 4, 1, ⁻2, ⁻5, ...
> D. 7, 10, 13, 16, ...
> E. 3, 7, 11, 15, 19, ...

15 Find the outputs for these.
   a 3 → [add 5] →
       6 → →
       2 → →
   b 1 → [multiply by 2] → [add 1] →
       0 → → →
       5 → → →

16 Finish these rules for the function machines in question **15**.
   a $y = $ _____ or $x \rightarrow$ _____
   b $y = $ _____ or $x \rightarrow$ _____

T **17** Use a copy of this.
    Complete the mapping diagram for the function machine in question **15a**.

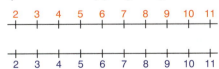

18 Simplify these expressions by cancelling.
   a $\frac{m}{m}$ **b** $\frac{y}{y}$ **c** $\frac{2n}{2}$ **d** $\frac{5x}{5}$ **e** $\frac{6p}{p}$ **f** $\frac{25p}{5}$

19 $y = $ ___ x
    What number could go in the gap to give the equation of a line which is steeper
    than $y = x$?

Algebra

*28 Solve each of the equations in question **27**.

*29 **a** A yard is 15 m long.
A fence 50 m long goes right around the
perimeter of the yard.
If w is the width of the yard, which of
these is correct?

A $30 + 2w = 50$ **B** $15 + w = 50$
C $50 + 2w = 30$ **D** $w = 50 - 30$

b Solve the correct equation to find the
width of the yard.

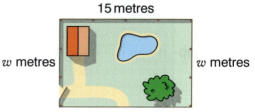

15 metres

w metres w metres

15 metres

6 Algebra and Equations

You need to know
✓ equations   page 136

Key vocabulary

brackets,   collect like terms,   equals,   equation,   formula,
function,   inverse operations,   substitute

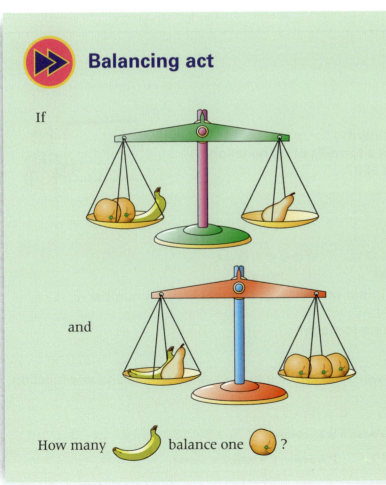

Balancing act

If

and

How many 🍌 balance one 🍊 ?

Understanding algebra

In **algebra** we use letters to stand for numbers.

$n + 8 = 4$ is an **equation**.

> *n* can have just one value.

$V = lbh$ is a **formula** for the volume of a cuboid.

V = volume, l = length, b = breadth, h = height

$y = 3x + 2$ is a **function**.
It tells us the rule for this function machine.

$$x \rightarrow \boxed{\begin{array}{c}\text{multiply}\\\text{by 3}\end{array}} \xrightarrow{x \times 3} \boxed{\begin{array}{c}\text{add}\\2\end{array}} \rightarrow x \times 3 + 2 \ (\text{or } 3x + 2)$$

Discussion

Which of these are true? **Discuss.**

a An equation *always* has an equals sign.
b In a formula, the letters *always* stand for something, such as length, volts, etc.
c A function tells us how to find the output, y, of a function machine if we know the input, x.

Exercise 1

1 Mr Jones put this on the board.
Bobbie was asked if it was a formula or an equation.
Can he tell just by looking at it?
Explain.

2

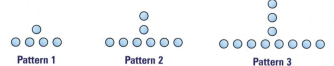

Pattern 1   Pattern 2   Pattern 3

The rule for finding the number of circles, C, from the pattern number, n, is
   $C = 3n + 2$.
Is $C = 3n + 2$ an equation or a formula?
Explain.

3 In each of the following rows, there is *one* equation, *one* formula and *one* function.
Which is which?

a $y = 2x - 4$, $a + b = 7$, $C = 4b - T$

b $F = ma$,   *F* is force, *m* is mass, *a* is acceleration $5x + 3 = 18$, $y = \frac{1}{2}x - 3$

c $\frac{5a + 21}{9} = 17$, $P = VI$,   *P* is power, *V* is volts, *I* is current $y = \frac{3x - 2}{4}$

4 The input and output of a function machine were plotted on a graph.
$y = 2x - 7$ is the equation of the straight line.
Is $y = 2x - 7$ also a function?

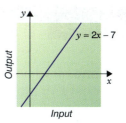

Discussion

Sanjay wrote this.

$$28 + 46 = 28 + 40$$
$$= 68 + 6$$
$$= 74$$

What is wrong with what Sanjay wrote? **Discuss.**

Does 'equals' mean the same as 'makes'? **Discuss.**

Writing equations

Remember
An **equation** always has an equals sign.

Discussion

Janet took x pounds to the fair.
Rashida took £15 to the fair.

Discuss how to write an equation for each of the following.

- Together Janet and Rashida have £40.
- Twice Janet's amount plus Rashida's amount add to £65.
- Janet's amount less Rashida's amount equals £10.

Worked Example
Louise bought 2 frisbees for £y.
She also bought a sunhat for £3.
In total she spent £7.
Write an equation for what Louise spent.

Answer
Louise spent $2 \times £y$ and another £3.
The equation for what Louise spent is
$2y + 3 = 7$

Algebra

1   Fiona hired a video camera.
    It cost £15 plus £5 a day for d days.
    She paid £30 in total.
    Finish the equation for this. $5 \times d$_____ = _____

2   Ben is x years old. His sister, Amy, is 5 years older. —— Amy's age
    Together their ages add to 21 years.
    Finish the equation for this. $x +$ _____ $= 21$

3   f people went to the school play on the first night.
    700 went on the second night.
    1600 went on the third night.
    Write an equation for each of these.
    a   Altogether 1500 went on the first two nights.
    *b   Twice as many went on the third night as on the
        first night.

4   Write an equation for these.
    a   I think of a number.   b   I think of a number.
        I add 7.   I multiply it by 2 then add 5.
        The answer is 16.   The answer is 17.
    c   I think of a number.   d   I think of a number.
        I subtract 3 then multiply this by 4.   I divide by 4 then add 11.
        I get the answer 8.   The answer is 15.

> Use n for the number.

Writing and solving equations

Remember
Inverse operations can be used to **solve equations**.
We work backwards and do the inverse operations.

Worked Example
Solve these.
a $\frac{y}{2} = 7$ b $4p + 3 = 11$

Answer

a $y \rightarrow$ [divide by 2] $\rightarrow 7$

    Work backwards
    14 $\leftarrow$ [multiply by 2] $\leftarrow 7$ doing inverse
                                        operations

    $y = 14$

b $4p + 3 = 11$
    $4p = 11 - 3$ Subtracting 3 is the inverse of adding 3.
    $4p = 8$
    $p = \frac{8}{4}$ Dividing by 4 is the inverse of multiplying by 4.
    $p = 2$

Always check your answer.
Substitute your answer into the equation to see if it makes it true.
a $\frac{y}{2} = \frac{14}{2}$ b $4p + 3 = 4 \times 2 + 3$
    $= 7$ ✓ $= 8 + 3$
                                               $= 11$ ✓

Exercise 3

T

1 Use a copy of these.
Solve the equations by filling in the gaps.

a $4x = 20$

b $x + 3 = 10$

c $m - 5 = 7$

d $\frac{p}{3} = 5$

T

2 Use a copy of these.
Solve the equations by filling in the gaps.

a $2x + 3 = 11$

b $5y - 2 = 13$

c $4(y - 5) = 20$

d $\frac{m}{3} + 4 = 8$

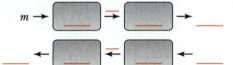

3 There are two small tins and one big tin on these scales.   **[SATs Paper 1 Level 4]**

The two small tins each have the same mass.
The mass of the big tin is 2·6 kg.
What is the mass of one small tin?
Show your working.

4 Solve these equations.

a $3n = 15$

b $\frac{p}{5} = 4$

c $x + 6 = 11$

d $y - 3 = 7$

e $2p - 1 = 10$

f $5x + 6 = 21$

g $3(a - 2) = 18$

h $5(p + 7) = 50$

i $5(x + 1) = 22$

j $\frac{g}{5} - 3 = 7$

Algebra

5 Solve these.
Write an equation first.

a I think of a number.
I multiply it by 5 then add 3.
The answer is 43.
What was the number I thought of?

b I think of a number.
I subtract 6 then multiply by 5.
The answer is 30.
What was the number I thought of?

*6 Write and solve equations for these.

a Eva bought 3 bags of Easter eggs.
Each cost £n.
She also bought a magazine for £2.
Altogether she spent £8.
How much is a bag of Easter eggs?

b A coffee shop has 8 tables altogether.
5 of the tables seat 4 people.
The other 3 tables seat n people each.
The coffee shop seats 38 people altogether.
How many people do each of the 3 other
tables seat?

c Grace's grandmother gave her £x.
She spent £6 on gifts.
She banked half of what was left.
Grace banked £7.
How much did Grace's grandmother give her?

Discussion

● 9P was asked to solve this equation. $3(x + 4) = 27$

Dan solved it like this.

$3(x + 4) = 27$
$3x + 12 = 27$ **multiply out the brackets first**
$3x = 27 - 12$ **inverse of +12 is −12**
$3x = 15$
$x = \frac{15}{3}$ **inverse of × 3 is ÷ 3**
$x = \mathbf{5}$

Rhian solved it like this

$3(x + 4) = 27$
$(x + 4) = \frac{27}{3}$ **inverse of × 3 is ÷ 3**
$x + 4 = 9$
$x = 9 - 4$ **inverse of + 4 is − 4**
$x = \mathbf{5}$

Discuss Dan's and Rhian's methods.

● **Discuss** how to write and solve an equation for
this.

Supreme pizzas cost £3 more than regular pizzas.
Regular pizzas cost £x each.
Peter bought 5 supreme pizzas.
They cost £40 altogether.
How much does a regular pizza cost?

Exercise 4

T

1 Use a copy of these.
Fill in the gaps to help you solve the equations.

a $3(x + 1) = 12$
$3 \times x + 3 \times 1 = 12$
$3x + 3 = 12$
$3x = \underline{} - 3$
$3x = \underline{}$
$x = \underline{}$

b $6(y - 3) = 18$
$6 \times y - 6 \times 3 = 18$
$6y - \underline{} = 18$
$6y = \underline{} + \underline{}$
$= \underline{}$
$y = \underline{}$

2 Solve these equations.
a $4(p - 2) = 24$ **b** $8(w - 6) = 32$ **c** $10(d + 4) = 40$ **d** $5(x + 2) = 20$
e $2(a - 7) = 18$ **f** $7(y + 3) = 42$ **g** $4(f + 3) = 28$ **h** $9(y - 4) = 36$

Collecting like terms first

Before solving an equation we must multiply out any brackets, then **collect any like terms**.

Worked Example

Paula: I have £n

Rosa: I have £(n + 2)

Ralph: I have £2n

Paula   Rosa   Ralph

Altogether Paula, Rosa and Ralph have £34.
How much does each have?

Answer

$$\overset{\text{Paula}}{n} + \overset{\text{Rosa}}{(n + 2)} + \overset{\text{Ralph}}{2n} = 34$$
$$n + n + 2 + 2n = 34$$
$$n + n + 2n + 2 = 34$$
$$4n + 2 = 34 \qquad \text{collect like terms}$$
$$4n = 34 - 2 \qquad \text{The inverse of +2 is −2}$$
$$4n = 32$$
$$n = \frac{32}{4} \qquad \text{The inverse of ×4 is ÷4}$$
$$n = £8$$

Paula has £8.
Rosa has £(8 + 2) = £10.
Ralph has £(2 × 8) = £16.

Algebra

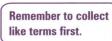

1 Solve these equations.
  a $3x + 6 + 2x = 26$ **b** $4n - 3 + 5n = 15$
  c $6p + 8 - 4p + 3p = 28$ **d** $3x + 3 + 4x + 1 = 39$

> Remember to collect like terms first.

2 a The perimeter of this window is $w + 4 + w + $ _____ $+$ _____.
  b Collecting the like terms, the perimeter is _____.
  c The perimeter is 44 m so _____ $= 44$.
  d Solve the equation to find w.

T

3 Use a copy of this.
  a In this pyramid, each number is the sum of the two numbers above it.
    Amanda wrote this.

  > $6 + n$ goes in the empty box
  > $6 + n + n + 8 = 34$ adding the 2 boxes

  Finish Amanda's working to find n.

  b Fill in the empty boxes.
    Then write an equation and solve it to find n.

  i 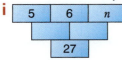   **ii** 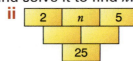   **iii**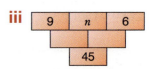

4 a Write an expression for the perimeter of this shape.
  b The perimeter of the shape is 33 cm.
    Write an equation for the perimeter.
    Solve the equation to find x.

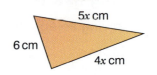

5 a Write an expression for the perimeter of this shape.
  b The perimeter of the shape is 27 mm.
    Write an equation for the perimeter.
    Solve the equation to find y.

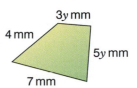

*6 Write and solve an equation to find x.
  a is started for you.

  a 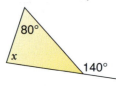   **b** 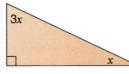   **c**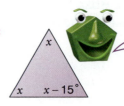

> Remember the angles of a triangle add to 180°.

  > $x + 80° = 140°$
  > $x = 140° - 80°$

*7 a Cakes cost n pence. Muffins cost $(n + 1)$ pence.
    Rachel buys 6 cakes and 6 muffins.
    Write an expression for the total cost of these.
    Simplify your expression.
  b Rachel pays 966 pence.
    Write an equation for this.
  c Solve your equation to find the cost of a cake.

Balancing equations

Discussion

- 4 oranges and 3 bananas balance
2 oranges and 4 bananas.
Mel takes a banana from the right-hand pan.
What will happen?
What would she need to do to the left-hand
pan to make the scales balance? **Discuss.**

- This set of scales is balanced.

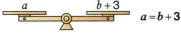

$$a \quad\quad b+3$$
$$a = b+3$$

If we subtract **3** from both sides, will the
new equation be true?

$$a-3 \quad\quad b+3-3$$
$$a-3 = b+3-3$$

$$a-3 \quad\quad b$$
$$a-3 = b$$

What if we now add **7** to both sides?
Discuss.

$$a-3+7 \quad\quad b+7$$

$$a+4 \quad\quad b+7$$
$$a+4 = b+7$$

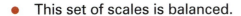

$$2x-3 \quad\quad 7 \quad\quad\quad 2x \quad\quad\quad\quad x$$

add 3 to both sides   **divide both sides by 2**

Discuss what goes on the empty pans.

Draw balance diagrams to solve these equations.

$$3x = 12 \qquad \frac{x}{2} = 5 \qquad 4x - 1 = 7$$

How did you work out in what order to do things? **Discuss.**

We can solve equations by **doing the same to both sides** of the equation.

Worked Example
Solve $4x - 7 = 15$.

Answer

We could say that we *transform* both sides of the equation.

$$4x - 7 = 15$$ We add 7 first, before dividing by 4. Why?

$$4x-7 \quad\quad 15$$

$$4x - 7 + \mathbf{7} = 15 + \mathbf{7} \qquad \text{add 7 to both sides}$$

$$4x-7+7 \quad\quad 15+7$$

$$4x = 22$$

$$4x \quad\quad 22$$

$$\frac{4x}{4} = \frac{22}{4} \qquad \text{divide both sides by 4}$$

$$\frac{4x}{4} \quad\quad \frac{22}{4}$$

$$x = \mathbf{5\cdot5}$$

$$x \quad\quad 5\cdot5$$

Algebra

T

1 Use a copy of this.

$\overline{17}$	$\overline{6}$	$\overline{3}$	$\overline{22}$	$\overline{6}$	$\overline{3}$		$\overline{16}$	$\overline{33}$	$\overline{11}$

R

$\overline{30}$ $\overline{14}$ $\overline{100}$ $\overline{0}$ $\overline{4}$ $\overline{\textbf{12}}$ $\overline{11}$ $\overline{30}$

$\overline{24}$ $\overline{4}$ $\overline{30}$ $\overline{2}$ $\overline{4}$ $\overline{3}$ $\overline{30}$ $\overline{14}$ $\overline{100}$

R

$\overline{16}$ $\overline{6}$ $\overline{\textbf{12}}$ $\overline{17}$ $\overline{22}$ $\overline{30}$ $\overline{6}$ $\overline{14}$ $\overline{33}$ $\overline{9}$ $\overline{100}$

R

$\overline{6}$ $\overline{9}$ $\overline{100}$ $\overline{\textbf{12}}$ $\overline{6}$ $\overline{3}$ $\overline{100}$

$\overline{\frac{1}{2}}$ $\overline{4}$ $\overline{17}$ $\overline{17}$ $\overline{4}$ $\overline{6}$ $\overline{3}$

$\overline{48}$ $\overline{100}$ $\overline{6}$ $\overline{48}$ $\overline{17}$ $\overline{100}$

Solve these equations by doing the same to both sides.

R $r + 8 = 20$ $\quad r = \textbf{12}$ **S** $s + 23 = 34$ **T** $t - 14 = 16$ **A** $a - 18 = 15$

L $l + 9 = 26$ **D** $d - 5 = 17$ **H** $26 = h + 12$ **E** $111 = e + 11$

W $w + 16 = 32$ **F** $f + 17 = 17$ **Y** $3y = 6$ **I** $8i = 32$

V $7v = 63$ **M** $4m = 2$ **C** $\frac{c}{3} = 8$ **P** $\frac{p}{4} = 12$

O $\frac{3x}{2} = 9$ **N** $\frac{4n}{3} = 4$

2 Solve these equations using the balance method.

a $3a + 2 = 20$ **b** $4b - 7 = 25$ **c** $3c + 5 = 32$

d $5d + 1 = 26$ **e** $4e - 3 = 15$ **f** $2f + 3 = 24$

g $7g + 4 = 32$ **h** $9h - 6 = 66$ **i** $6i - 20 = 19$

T

3 Use a copy of the cross number.
Fill it in by solving the equations.

Across

1. $n - 7 = 4$

2. $\frac{n}{7} = 3$

5. $n - 20 = 12$

9. $n - 10 = 3$

10. $\frac{n}{3} = 14$

13. $2n - 1 = 41$

15. $\frac{n}{3} + 4 = 5$

Down

1. $\frac{n}{4} + 1 = 4$

3. $n + 3 = 16$

4. $3n = 12$

6. $\frac{2n}{3} = 16$

7. $2 + n = 15$

8. $n - 1 = 13$

11. $2n = 44$

12. $1 + 2n = 21$

14. $2n - 3 = 19$

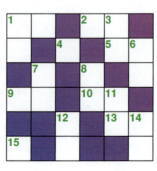

Solving equations using a graph

Discussion

T

One bag of peanuts costs £3.
How much do 2, 3, 4, 5, ... bags cost? **Discuss.**

We could write a table for this.
Is the cost directly proportional
to the number of packets bought?
Discuss.

Number of bags	1	2	3	4	5	...
Cost (£)	£3					...

+1 +1 +1 +1

Remember if it is directly
proportional then $\frac{cost}{number\ of\ bags}$
will be the same for every set
of values in the table.

We could draw a graph of Cost versus
Number of bags.
Will the graph be a straight line?
Discuss.
Use a copy of this grid and plot the
points to check.

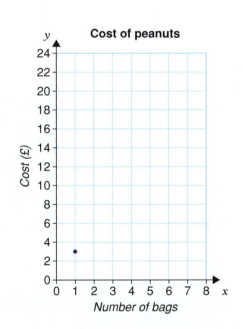

Cost of peanuts

How could you use the graph to find the
cost of 8 bags? **Discuss.**

In the above discussion, the ratio

   Cost : Number of bags

is the same for each set of values.

   3 : **1** = 6 : 2 = 9 : 3 = 12 : 4 = 15 : 5
         3 × 2 = 6   3 × 3 = 9   3 × 4 = 12   3 × 5 = 15
         1 × 2 = 2   1 × 3 = 3   1 × 4 = 4   1 × 5 = 5

So cost is **directly proportional** to the number of bags.

Algebra

T 1 Use a copy of this.
  a One pizza costs £4.
    Fill in your table.
  b The ratio *Cost : Number of pizzas*
    for the first pair of values in the
    table is £4 : 1.
    Fill these in for the other pairs of
    values.

    £ 8 : 2   = ___ : 1
    £__ : 3   = ___ : 1
    £__ : 4   = ___ : 1

    What do you notice?
  c Is the cost directly proportional to
    the number of pizzas bought?
    Explain.
  d Use a copy of the grid.
    Plot the points in the table.
    Do the points lie in a straight line?
  e Find the cost of 8 pizzas.

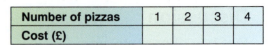

Number of pizzas	1	2	3	4
Cost (£)				

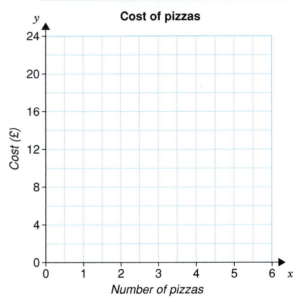

Cost of pizzas

T 2 Use a copy of this.
  Mobile phone covers were reduced
  from £4 each to £3 each in a sale.
  a Fill in this table

Number	1	2	3	4	5	...
Original price	£4	£8				...
Sale price	£3					...

  b Work out the ratio
    Sale price : Original price
    for each pair of values on the table.
    What do you notice?
  c Is the sale price directly
    proportional to the original price?
  d Plot the points in the table on the
    grid.
    Do the points lie in a straight line?
    Is this what you would expect?
    Explain.
  e Find the sale price of a number
    of phone covers that originally
    cost £48.

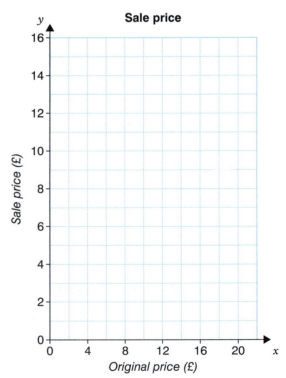

Sale price

T

*3 Use a copy of this grid.
Rachel runs at a speed of 250 metres every 2 minutes.
She runs at a constant speed.

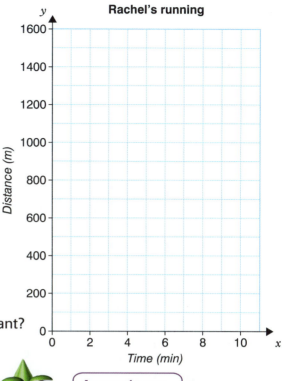

Rachel's running

a Write down five pairs of numbers for distance and time.
For example (2, 250).

b Is the ratio
Distance travelled : Time taken constant?
What is it?

c Plot the points in **a**.

d Use the graph to find the distance she would run in
i 18 minutes **ii** 9 minutes.

Assume she runs at the same speed.

Summary of key points

 A $x + 3 = 5$ is an **equation**. x has a specific value.
An equation always has an equals sign.

$A = lb$ is a **formula**. Each letter stands for something specific. A is area, l is length, b is breadth.

In algebra letters stand for numbers.

$y = 2x - 5$ is a **function**. If we know x we can find y.

B We can write **equations**.

Example Toby has £n. Angus has 3 times as much as Toby.
Together they have £20.
This can be written as the equation.

$$3 \times n + n = 20$$

Angus's money   Toby's money

Algebra

 We can **solve equations using inverse operations**.

Example $3x - 2 = 13$

$3x = 13 + 2$

$3x = 15$

$x = \frac{15}{3}$

$x = \mathbf{5}$

Adding 2 is the inverse of subtracting 2.

Dividing by 3 is the inverse of multiplying by 3.

$x \rightarrow$ | multiply by 3 | $\xrightarrow{3x}$ | subtract 2 | $\rightarrow 3x - 2$

$5 \leftarrow$ | divide by 3 | $\xleftarrow{15}$ | add 2 | $\leftarrow 13$

Start with the answer.

Note Always check your answer by substituting it into the equation.

$3\,x - 2 = 3 \times \mathbf{5} - 2$

$= 15 - 2$

$= 13 \quad \checkmark$

If an equation has **brackets**, we usually multiply them out first.

Example $2(x - 3) = 14$

$2x - 6 = 14$ **Multiply brackets first.**

$2x = 14 + 6$ **Inverse of subtracting 6 is adding 6.**

$2x = 20$

$x = \frac{20}{2}$ **Inverse of multiplying by 2 is dividing by 6.**

$x = \mathbf{10}$

 We need to **collect like terms** before solving an equation.

Example $2(n + 1) + 3n = 17$

$2n + 2 + 3n = 17$

$5n + 2 = 17$ **Collecting like terms.**

$5n = 17 - 2$ **The inverse of adding 2 is subtracting 2.**

$5n = 15$

$n = \frac{15}{5}$ **The inverse of multiplying by 5 is dividing by 5.**

$n = \mathbf{3}$

 We can **solve equations by doing the same to both sides** of the equation.

Example $\frac{p}{2} - 3 = 5$

$\frac{p}{2} - 3 + \mathbf{3} = 5 + \mathbf{3}$

$\frac{p}{2} = 8$

$\frac{p}{2} \times \mathbf{2} = 8 \times \mathbf{2}$

$p = \mathbf{16}$

F   We can **solve equations using a graph**.

Example   Lee biked at a steady speed.

Time (min)	1	2	3	4	...
Total distance cycled (m)	400	800	1200	1600	...

Distance is directly proportional to time because

$$400 : 1 = 800 : 2 = 1200 : 3 = 1600 : 4$$

When we draw a graph of distance against time we get a straight line.

We can read values off the graph.

After 6 minutes Lee will have cycled 2400 m.

Distance cycled

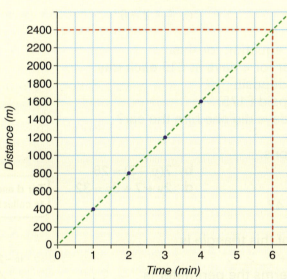

Test yourself

1   In each of the following, there is *one* equation, *one* formula and *one* function.   **A**
     Which is which?

   a $y = 3x + 7$, $2x + 4 = 7$, $v = u + at$ (v = final velocity, u = initial velocity, a = acceleration, t = time)

   b $P = nRT$,   (P = pressure, T = temperature, n = number of moles, R = a constant) $2p + 7 = {}^-3$, $y = \frac{x + 3}{7}$

2   Juliet earns £p each week and Nina earns £400 each week.   **B**
     Write an equation for each of these. Choose from the box.
     a   Together Juliet and Nina earn £900 each week.
     ∗b   If Juliet earned £100 more each week, she would
            earn the same as Nina.

   $p + 400 = 900$
   $p - 400 = 900$
   $p + 100 = 400$

7 Expressions and Formulae

You need to know

Key vocabulary

collect like terms, expression, formula, formulae, index, multiply out a bracket, simplest form, substituting

There's a skeleton in the cupboard

When a skeleton is found, forensic scientists can estimate the height, in cm, this person was from the length of various bones.

For **males**, the formulae are:

> Height = 3·08 × length of humerus bone + 70·45
> Height = 3·7 × length of ulna bone in cm + 70·45
> Height = 2·52 × length of tibia bone in cm + 75·79

For **females**, the formulae are:

> Height = 3·36 × length of humerus bone + 57·97
> Height = 4·27 × length of ulna bone in cm + 57·76
> Height = 2·90 × length of tibia bone in cm + 59·24

1 A scientist found a female tibia bone of 39·9 cm. What was her estimated height?

*2 Work out the length your humerus, ulna and tibia should be for your height.

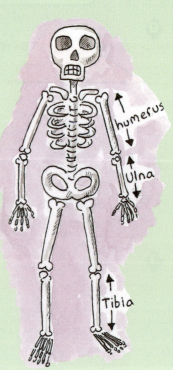

Writing expressions

We **write expressions** without multiplication or division signs.

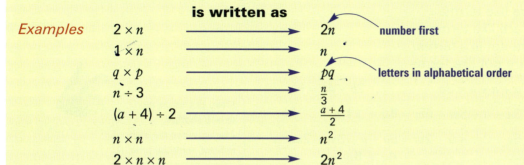

Examples

is written as

$2 \times n$ ⟶ $2n$ number first

$1 \times n$ ⟶ n

$q \times p$ ⟶ pq letters in alphabetical order

$n \div 3$ ⟶ $\frac{n}{3}$

$(a + 4) \div 2$ ⟶ $\frac{a+4}{2}$

$n \times n$ ⟶ n^2

$2 \times n \times n$ ⟶ $2n^2$

Exercise 1

1 Write these without multiplication or division signs.

a $4 \times n$	**b** $3 \times y$	**c** $2 \times m$	**d** $5 \times q$	**e** $p \times 7$
f $5 \times (x + 4)$	**g** $7 \times (y - 3)$	**h** $a \times b$	**i** $s \times t$	**j** $7 \times (p + q)$
k $a \times b \times c$	**l** $3 \times n \times m$	**m** $(n + m) \div p$	**n** $(p + 4) \div q$	**o** $5y \div 7$
p $a \times a$	**q** $p \times p$	**r** $2 \times x \times x$	**s** $4 \times y \times y$	

2 Write these **with** multiplication signs.

a $4p$ **b** $4(x + 3)$ **c** $8(n - 2)$ **d** x^2 **e** $5b^2$ **f** $4ab$

3 Match each of these with an expression in the box.

a 5 more than x

b 7 less than x

c 8 times x

d one third of x

e x multiplied by y

f twice x plus y

g half the sum of x and y

h three times the sum of x and y

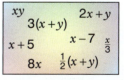

xy $2x + y$ $3(x + y)$ $x + 5$ $x - 7$ $\frac{x}{3}$ $8x$ $\frac{1}{2}(x + y)$

*4   **area of a rectangle = length × width**.

Sam found the area of this rectangle as:

length × width $= 6 \times x$

$= 6x$ cm^2

6 cm

x cm

Use Sam's way to write an expression for the area of these.

a
4 metres / x metres

b 3n centimetres / 4 centimetres

c
4p kilometres / 3p kilometres

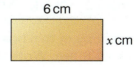

163

Algebra

Collecting like terms

We can simplify expressions by **collecting like terms**.

$3x$ and $4x$ are like terms.
$7x^2$ and $4x^2$ are like terms.
$5a$ and $7b$ are not like terms.
$5x$ and $2x^2$ are not like terms.

Examples $\quad a + a + a = 3a \qquad\qquad 2b + 4b = 6b$

Examples $\quad 3a + a + 5m + 6m = \mathbf{4a + 11m}$

$\qquad\qquad 5a + 3m - a + 2m = 5a - a + 3m + 2m$
$\qquad\qquad\qquad\qquad\qquad = \mathbf{4a + 5m}$

Move like terms next to one another. Always move the sign in front, with the term.

Examples $\quad 3x^2 + 2x^2 = 5x^2$
$\qquad\qquad 4x^2 + 3x^2 + x = 7x^2 + x$

Sometimes we are asked to write and simplify an expression.

Worked Example
Write an expression for the perimeter of this park.
Simplify your expression.

$2x + 4$

$3x$

Answer

$\qquad$ Perimeter $= 2x + 4 + 2x + 4 + 3x + 3x$
$\qquad\qquad\qquad\quad = 2x + 2x + 3x + 3x + 4 + 4$
$\qquad\qquad\qquad\quad = \mathbf{10x + 8}$

Exercise 2

T | **1**

| $\overline{7a + 1}$ | $\overline{5a + 5b}$ | $\overline{10a + b}$ | $\overline{9b + 8}$ | $\overline{10a + b}$ | | $\overline{15b + 1}$ | $\overline{9b + 8}$ | $\overline{10a + b}$ |

N

| $\overline{\mathbf{6a + 9b}}$ | $\overline{6a + 4b}$ | | $\overline{9b + 8}$ | $\overline{5a + 3b}$ | $\overline{3a + b}$ | $\overline{10a + b}$ | $\overline{9b + 8}$ | $\overline{{}^-2b + 2}$ |

N

| $\overline{5a + 3b}$ | $\overline{\mathbf{6a + 9b}}$ | | $\overline{{}^-2b + 2}$ | $\overline{15b + 1}$ | $\overline{6a + 2}$ | $\overline{a + 11}$ | $\overline{5a + 3b}$ |

| $\overline{15b + 1}$ | $\overline{9b + 8}$ | $\overline{15b + 1}$ | $\overline{11a - b}$ | $\overline{5a + 3b}$ | $\overline{15b + 1}$ |

Use a copy of this box.
Simplify these. Write the letter beside each, above its answer in the box.

N $\quad 4a + 3b + 2a + 6b = \mathbf{6a + 9b}$
E $\quad 8a + 2a + 4b - 3b$
T $\quad 5a + 3 + 2a - 2$
A $\quad 12b - 7 + 3b + 8$
*S $\quad 6b + 7 - 8b - 5$

H $\quad 9a - 4a + 3b + 2b$
O $\quad 5a + 6b + a - 2b$
R $\quad 3b + 11 + 5b - 3 + b$
U $\quad 4a + 6 + 3a - 4 - a$

I $\quad 3a + 2a + 4b - b$
V $\quad 7a + 2b - 4a - b$
D $\quad 9a + 7 - 8a + 4$
B $\quad 5a - 3b + 6a + 2b$

164

2 Simplify these.
  a $4x^2 + 2x^2$ **b** $7x^2 - 5x^2$ **c** $3x^2 + 9x^2$
  d $10x^2 - 4x^2$ **e** $8x^2 - x^2$ **f** $5x^2 + 2x^2 + 3x^2$
  g $3x^2 + 4x^2 - 2x^2$ **h** $7x^2 + 3x^2 - 4x^2$ **i** $6x^2 - x^2 - 2x^2$

3 Write expressions for the perimeters of these gardens.
Simplify your expressions.

  a   **b**   **c**   **d**   ***e**

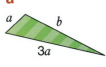 

4 This diagram shows a park.
Write an expression for the perimeter of the park.
Simplify the expression.

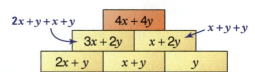

*5 The expression in each box is found by adding
the expressions in the two boxes beneath it.
Find the missing expressions in these.
Simplify them.

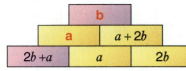 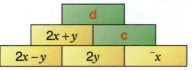

? **Puzzle**

1 Use a copy of these.
Write possible expressions for the empty boxes.

  a   **b**

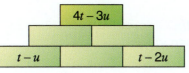 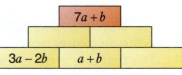

*2 Show this is a magic square by adding the
expressions in each row, column and diagonal.

Remember: in a magic
square each row, column
and diagonal add to the
same number.

$a - b$	$a + b - c$	$a + c$
$a + b + c$	a	$a - b - c$
$a - c$	$a - b + c$	$a + b$

Algebra

Simplifying expressions – multiplying and dividing

Multiplying

n stands for a number of sandwiches.

Yesterday Ben bought $3 \times n$ or $3n$
sandwiches for his family.
Today friends are coming and he needs
4 times as many.
He buys $4 \times 3n$ sandwiches.
This is 4 lots of $3n$ which is **12n**.

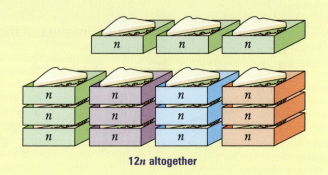

12n altogether

Example

$$n \times n^2 = n \times n \times n \qquad n^2 = n \times n$$
$$= n^3$$

Exercise 3

1 Simplify these expressions.

a $6 \times 4x$	**b** $3 \times 2n$	**c** $5 \times 6p$	**d** $3 \times 2y$	**e** $8 \times 4w$
f $5 \times 2t$	**g** $8y \times 3$	**h** $7w \times 4$	**i** $7x \times 3$	**j** $5 \times 3p$
k $6 \times 8z$	**l** $9p \times 9$	**m** $7 \times 8m$	**n** $9x \times 8$	**o** $4p \times 10$
p $12w \times 2$	**q** $3 \times 9n$	**r** $15t \times 2$	**s** $20 \times 4z$	

2 Write an expression which is the same as each of these.

a $a \times a \times a$	**b** $c \times c$	**c** $d \times d \times d \times d$	**d** $b \times b$
e $y \times y \times y \times y$	**f** $x \times x \times x \times x$	**g** $e \times e \times e$	**h** $f \times f \times f \times f$
i $g \times g$	**j** $p \times p \times p$	**k** $a \times a \times a \times a$	

Remember
$m \times m \times m = m^3$

3 Explain the difference between
a n^2 and $2n$ **b** m^3 and $3m$.

4   *area of a rectangle = length × width*.
Write an expression for the area of these yards.
The first one is done for you.

a   3t   **b**   3   **c**   7y   **d**   5b   **e**   5

Area $= 4 \times 3t$
       $= 12t$

*5 Simplify these.

a $a \times a^2$	**b** $p^2 \times p$	**c** $d^2 \times d^2$	**d** $n \times n^3$	**e** $m^2 \times m$
f $x^3 \times x^3$	**g** $p^2 \times 2$	**h** $y^3 \times 4$	**i** $2y \times 3y$	**j** $3x \times 2x$
k $4a \times a$	**l** $3b \times 5b$	**m** $6m \times 2m$	**n** $8b^2 \times 2b$	

166

Dividing

Discussion

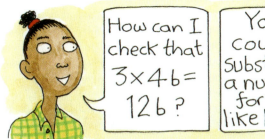

How can I check that $3 \times 4b = 12b$?

You could substitute a number for b, like b = 2.

What might Menna do to check this? **Discuss.**

We can often **simplify an expression by cancelling**.

Example

$$\frac{m^3}{m^2} = \frac{{}^1\!\not{m} \times {}^1\!\not{m} \times m}{{}_1\!\not{m} \times \not{m}_1}$$
$$= m$$

To cancel, we have divided both the numerator and the denominator by m.
$m \div m = 1$

Worked Example

Simplify these. **a** $\frac{12x}{8}$ **b** $\frac{5p^2}{5}$

Answer

a $\frac{{}^3\!\not{12}x}{{}_2\!\not{8}} = \frac{3x}{2}$

Divide both the numerator and denominator by 4.

b $\frac{5p^2}{5} = \frac{{}^1\!\not{5}p^2}{\not{5}_1}$
$= p^2$

Divide both the numerator and denominator by 5.

Worked Example

Simplify these. **a** $\frac{m^6}{m^3}$ **b** $t^4 \div t^2$ **c** $\frac{18p^2}{9}$

Answer

a $\frac{m^6}{m^3} = \frac{{}^1\!\not{m} \times {}^1\!\not{m} \times {}^1\!\not{m} \times m \times m \times m}{{}_1\!\not{m} \times {}_1\!\not{m} \times {}_1\!\not{m}}$
$= m \times m \times m$
$= m^3$

b $t^4 \div t^2 = \frac{{}^1\!\not{t} \times {}^1\!\not{t} \times t \times t}{{}_1\!\not{t} \times {}_1\!\not{t}}$
$= t^2$

c $\frac{18p^2}{9} = \frac{{}^2\!\not{18}p^2}{\not{9}_1}$
$= 2p^2$

Exercise 4

1 Simplify these by cancelling. The first two are started.

a $\frac{4x}{4} = \frac{{}^1\!\not{4}x}{{}_1\!\not{4}}$
$=$ ___

b $\frac{7y}{7} = \frac{{}^1\!\not{7}y}{{}_1\!\not{7}}$
$=$ ___

c $\frac{5p}{5}$

d $\frac{11m}{11}$

e $\frac{3b}{3}$

f $\frac{8n}{4}$

g $\frac{15x}{3}$

h $\frac{18q}{6}$

i $\frac{12y}{4}$

j $\frac{5x}{x}$

k $\frac{3a}{a}$

l $\frac{7n}{n}$

m $\frac{10p}{p}$

*n $\frac{12x}{8}$

*o $\frac{25n}{15}$

Algebra

2

$\overset{\text{I}}{\overline{a^3}}$ $\overline{a^4}$ $\overset{\text{I}}{\overline{a^3}}$ $\overline{5a}$ $\overline{3b}$ $\overline{2a}$ $\overset{\text{I}}{\overline{a^3}}$ $\overline{4a}$ $\overline{3b}$ $\overline{5a}$

$\overline{30}$ $\overline{5b^2}$ $\overline{12n^2}$ $\overline{4a}$ $\overline{1}$ $\overline{a^4}$ $\overline{b^2}$ $\overline{30}$

$\overline{5a}$ $\overline{3b^2}$ $\overline{5b^2}$ $\overline{a^4}$ $\overline{5a}$ $\overline{30}$ $\overline{2a}$ $\overset{\text{I}}{\overline{a^3}}$ $\overline{a}$ $\overline{5b^2}$

$\overline{a^2}$ $\overline{1}$ $\overline{5a}$ $\overline{5a}$ $\overline{b^2}$ $\overline{5b^2}$ $\overline{3b}$ $\overline{1}$ $\overline{2a}$

$\overline{\frac{3b^2}{2}}$ $\overline{1}$ $\overline{\frac{3b^2}{2}}$ $\overline{12n^2}$ $\overline{\frac{3b^2}{2}}$ $\overline{1}$ $\overline{b^2}$ $\overline{12n^2}$ $\overline{3b}$ $\overline{1}$ $\overline{b^2}$ $\overline{8a}$

Use a copy of this box. Simplify these expressions.
Write the letter beside each expression above its answer in the box.

I $a \times a^2 = a^3$ **N** $a^2 \times a^2$ **F** $\frac{8a}{4}$ **R** $\frac{20a}{5}$ **Y** $\frac{30b}{b}$ **D** $\frac{64a}{8}$

B $\frac{a^5}{a^3}$ **V** $a^6 \div a^5$ **E** $\frac{10b^2}{2}$ **W** $\frac{12b^2}{4}$ **T** $\frac{45a}{9}$ **S** $\frac{48b}{16}$

L $b^4 \div b^2$ **O** $a^3 \div a^3$ **C** $\frac{21b^2}{14}$ **A** $3n \times 4n$

Brackets

Remember
$8 \times 36 = 8(30 + 6) = 8 \times 30 + 8 \times 6$
$5 \times 39 = 5(40 - 1) = 5 \times 40 - 5 \times 1$

When we multiply a bracket by a number like this it is called 'multiplying out' the brackets.

When we **multiply out** a bracket we use the **distributive law**.

Examples $3(x + 2) = 3x + 6$ $4(y - 3) = 4y - 12$

Exercise 5

1 Write without brackets.
Use a copy of the grids to help you.

a $4(a + 2)$

	a	2
4		

b $3(x + 5)$

	x	5
3		

c $5(y + 3)$

	y	3
5		

d $7(b - 2)$

	b	$^-2$
7		

e $9(a + 4)$

f $8(m - 6)$

g $7(n - 1)$

h $4(p - 2)$

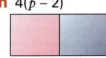

i $6(b - 5)$ **j** $17(c + 1)$

T

2 Use a copy of this.
Draw lines from column 1 to column 2 to column 3 to show what terms you get when you multiply out the brackets.
The first two are done.

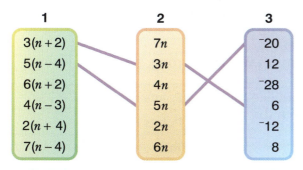

1	2	3
$3(n+2)$	$7n$	$^-20$
$5(n-4)$	$3n$	12
$6(n+2)$	$4n$	$^-28$
$4(n-3)$	$5n$	6
$2(n+4)$	$2n$	$^-12$
$7(n-4)$	$6n$	8

3 Paula multiplied out these brackets.
She made some mistakes.
Give her a ✓ if she is correct or a ✗ if she is wrong.

a $8(b+2) = 8b + 16$ **b** $3(y+4) = 3y + 4$ **c** $7(x-2) = 7x + 14$
d $5(y-4) = 5y - 20$ **e** $9(c+3) = 9c + 27$ **f** $10(d-4) = 4d - 40$

4 Write these expressions without brackets.

a $4(x+2)$ **b** $5(y+3)$ **c** $3(b-2)$ **d** $4(x-5)$
e $4(y-2)$ **f** $6(a+4)$ **g** $4(3y+2)$ **h** $5(3m+4)$
i $3(2p-5)$ **∗j** $3(2a+3b)$ **∗k** $2(3a-4b)$ **∗l** $6(3m-2n)$
∗m $4(4y-3z)$ **∗n** $5(4a-7b)$ **∗o** $3(2a-3b)$ **∗p** $9(9p-4q)$

Brackets – a game for a group

You will need three sets of these cards.

 $3n$ $2n$ $3x$ $2x$ $4x$ $4n$ $^-5$ $^-4$ $^-8$

$^-6$ 6 $5n$ $5x$ 5 $4(n-1)$ $2(x+3)$ $4(x-2)$ $3(n+2)$

$2(n-3)$ $2(n+3)$ $3(n-2)$ $3(x+2)$ $3(x-2)$ $5(n+1)$ $5(x-1)$

To play
- Choose a dealer.
- Give three cards face down to each player.
- Put the next card face up on the table.
- The rest of the pack is put face down in a pile.
- Take turns.
- The aim of the game is to get a set of three cards that make a correct statement.
  For example, this is a winning set of three cards since $3(n-2) = 3n - 6$.

$^-6$ $3(n-2)$ $3n$

continued

- At your turn you can do *one* of these:
  1 use the three cards in your hand to make a set
  2 pick up the card on the table and throw out one card from your hand
  3 pick up a card from the pile and throw out one card from your hand.
  Thrown out cards are put on the bottom of the pile.
- The first person to get a winning set of three cards gets 1 point and is the dealer for the next round.
- If a player puts out a winning set which is wrong, the player loses 1 point.
  You could keep a score sheet like this.
- The winner is the person with the most points after a set time.

Points scored	✓✓✓✓✓
Points lost	✓✓
Total	3

Multiplying out brackets then simplifying

Sometimes we have to **multiply out a bracket** before we **simplify**.

Worked Example
Simplify this $8(x + 2y) + 4(x + y)$

$$8(x + 2y) + 4(x + y) = 8x + 16y + 4x + 4y$$ **Multiply out the brackets.**
$$= 8x + 4x + 16y + 4y$$ **Write like terms together.**
$$= \mathbf{12x + 20y}$$ **Add or subtract like terms.**

Exercise 6

1 Multiply out the brackets and then simplify.
   a $3(x + 4) + 2(x + 1)$ **b** $4(a + 3) + 3(a + 2)$
   c $2(x + 3) + 3(x + 4)$ **d** $3(a + 5) + 2(a + 1)$
   e $3(a - 4) + 2(a - 1)$ **f** $4(b + 5) + 3(b + 1)$
   g $2(x - 1) + 3(x + 4)$

Practical

You will need a spreadsheet package.

Ask your teacher for the **Brackets** ICT worksheet.

Substituting into expressions

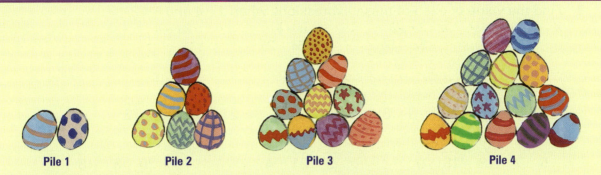

Pile 1   Pile 2   Pile 3   Pile 4

An expression for the number of Easter eggs in pile n is $4n - 2$.
To find the number of Easter eggs in pile **8** we **substitute 8** for n.

$$4n - 2 = 4 \times \mathbf{8} - 2$$
$$= 32 - 2$$
$$= \mathbf{30}$$

There are 30 eggs in pile 8.

We find the value of an expression by **substituting** values for the unknown into the expression.

Algebraic operations follow the same rules as **arithmetic operations**.

Order of operations
We carry out algebraic operations in the same **order** as arithmetic operations.
In arithmetic we do operations in this order.

Brackets
Indices
Division and **M**ultiplication
Addition and **S**ubtraction.

There is more about BIDMAS on page 5.

In algebra we do the same.

Example   In $5 - 3n$ $3 \times n$ is worked out first.
            In $7(y - 3)$ the value of the expression in the brackets is worked out first.
            In $8 - b^2$,   the square (index) is worked out first.

Worked Example
Find the value of   **a** $2a + 6$ when $a = \mathbf{3}$ **b** $4x - 2y$ when $x = \mathbf{3}$ and $y = \mathbf{6}$
                c $3n^2 - 1$ when $n = \mathbf{2}$.

Answer

a $2a + 6 = 2 \times a + 6$ **b** $4x - 2y = 4 \times \mathbf{3} - 2 \times \mathbf{6}$ **c** $3n^2 - 1 = 3 \times \mathbf{2}^2 - 1$
$\quad\quad = 2 \times \mathbf{3} + \mathbf{6}$ $\quad\quad\quad = 12 - 12$ $\quad\quad\quad = 3 \times 4 - 1$
$\quad\quad = 6 + 6$ $\quad\quad\quad = \mathbf{0}$ $\quad\quad\quad = 12 - 1$
$\quad\quad = \mathbf{12}$ $\quad\quad\quad = \mathbf{11}$

Algebra

Exercise 7 **except for questions 4, 5 and 6.**

1 Find the value of each of these expressions when $a = 4$ and $b = 2$.
  a $a + b$ **b** $3a + b$ **c** $a - b$ **d** $2a - b$
  e $3a - 2b$ **f** $5a - 5b$ **g** $5a - 2b$

T

2

| $\overline{22}$ | $\overline{20}$ | $\overline{4}$ | $\overline{16}$ | $\overline{3}$ | $\overline{3}$ | $\overline{18}$ | $\overline{5}$ | | $\overline{21}$ | $\overline{14}$ | $\overline{7}$ | $\overline{12}$ |

| $\overline{5}$ | $\overline{7}$ | $\overline{18}$ | $\overline{18}$ | $\overline{9}$ | | $\overline{3}$ | $\overline{21}$ | $\overline{4}$ | | $\overline{3}$ | $\overline{20}$ | $\overline{8}$ | $\overline{18}$ |

U

| $\overline{6}$ | $\overline{20}$ | $\overline{14}$ | $\overline{\mathbf{32}}$ | $\overline{1}$ | $\overline{18}$ | $\overline{5}$ | | $\overline{16}$ | | $\overline{10}$ | $\overline{16}$ | $\overline{12}$ |

Use a copy of this box.
If $n = 4$, $m = 2$ and $p = 3$, evaluate these. Write the letter that is beside each above the answer in the box.
U $8n = \mathbf{8} \times n = \mathbf{8} \times \mathbf{4} = \mathbf{32}$ **M** $n + m$ **T** $p - m$ **D** $2n + m$
L $3p - m$ **V** $20 - 3n$ **P** $15 - 2p$ **E** $6n - 2p$ **Y** $2(n + m)$
N $7(n - m)$ **O** $3(9 - m)$ **S** $\frac{n}{2} + p$ **F** $\frac{n + m}{2}$ ***A** n^2
*R m^2 ***I** $5m^2$ ***G** $n^2 + 2p$

3 If $a = 2$ and $b = 4$ find the value of these.
Shade the answer in the diagram.
You should get a path from the start to the finish.
  a $a + b$ **b** $b - a$ **c** $2a$
  d $2a + b$ **e** b^2 **f** b^3
  g $2b^2$ **h** $3a^2 + 2$ **i** $2b^2 + 1$

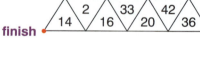

4 Use your calculator if you need to.
Find the value of each of these when $n = 1.5$.
  a $2n + 3$ **b** $4n - 1$ **c** $3(n + 1.5)$ **d** $5(8 - 2n)$ **e** $10 - 4n$

5 Use your calculator if you need to.
Find the value of each of these when $x = {}^-4$.
  a $2x + 12$ **b** $3x + 2$ **c** $20 - x$ **d** $3(x - 2)$

6

      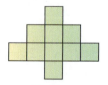   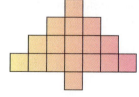

Shape 1   Shape 2   Shape 3   Shape 4

The number of squares in shape n is given by $n^2 + 1$.
Use your calculator if you need to.
Find the number of squares in
  a shape 10   **b** shape 9   **c** shape 13   **d** shape 24.

Investigation

Substitution

Remember A magic square adds to the same total no matter which way you add.

Find the value of each expression in the boxes if $a = 1$, $b = 3$ and $c = 5$.
Have you made a magic square?

Do you make a magic square if $a = 2$, $b = 4$, $c = 8$?

Choose some other numbers for a, b and c. Make sure c is bigger than a and b added together.
Do you always get a magic square?

$c + b$	$c - a - b$	$c + a$
$c + a - b$	c	$c - a + b$
$c - a$	$c + a + b$	$c - b$

Substituting into formulae

The area of this sail is

$A = \frac{b \times h}{2}$.

To find the area we need to know the values of b and h.
If $b = 20$ mm and $h = 12$ mm, we **substitute** these into the formula to find A.

$A = \frac{b \times h}{2}$ Write the formula down first.

$= \frac{20 \times 12}{2}$ Substitute the values.

$= \frac{240}{2}$

$= \mathbf{120}$ **mm**^{2} Remember the units.

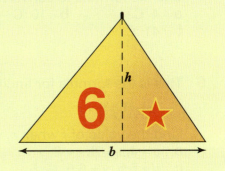

Exercise 8   **Only use a calculator if you need to.**

1 The cost, B pounds, of a bookcase with s shelves is calculated using this formula.

$B = \mathbf{25} + \mathbf{15}s$

Calculate the cost of a bookcase with these numbers of shelves.
a 3 shelves   **b** 5 shelves   **c** 4 shelves

Algebra

2 The cost, C pence, of a fish bowl and fish is worked out using the formula

$$C = 30f + 650$$

where f is the number of fish.
a Amy bought a fish bowl and 6 fish.
How much did this cost in pounds?
b Frank bought a fish bowl and 11 fish.
How much did this cost in pounds?

3 The perimeter of this sandpit is given by

$$P = 2l + 2w.$$

Find the perimeter if
a $l = 3$ m, $w = 2$ m   **b** $l = 3 \cdot 4$ m, $w = 3$ m
c $l = 3 \cdot 5$ m, $w = 2 \cdot 5$ m.

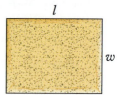

4 $V = IR$ gives the voltage, V, in an electrical circuit with current I and resistance R.
Find V if
a $I = 2 \cdot 5$, $R = 12$ **b** $I = 0 \cdot 6$, $R = 3$ **c** $I = 1 \cdot 2$, $R = 8$.

5 The formula for converting temperature in °C to K (Kelvin) is $K = C + 273$.
Use the formula to convert these to degrees Kelvin.
a 4 °C   **b** 10 °C   **c** 50 °C   **d** 29 °C   **e** 19 °C
f 41 °C

6 This is the formula for finding the distance, s, an object has travelled.

$$s = \left(\frac{v + u}{2}\right) \times t$$

Find the value of s when v, u and t have these values.
a $v = 4$, $u = 6$, $t = 2$ **b** $v = 8$, $u = 10$, $t = 3$
c $v = 5 \cdot 5$, $u = 6 \cdot 5$, $t = 5$ 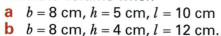**d** $v = 5$, $u = 8$, $t = 2 \cdot 5$

7 The volume, in cm^3, of this prism is given by the
formula $V = \frac{b + h}{2} \times l$.
Find the volume when
a $b = 8$ cm, $h = 5$ cm, $l = 10$ cm
b $b = 8$ cm, $h = 4$ cm, $l = 12$ cm.

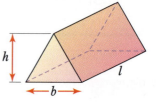

*8 The distance travelled, S metres, by a car is given by the formula $S = 20 + 0 \cdot 5t^2$.
t is the time in seconds.
How far has the car travelled after
a 1 sec   **b** 2 sec   **c** 4 sec   **d** $\frac{1}{2}$ sec?

Investigation

Perimeters and areas

You will need some square dotty paper.

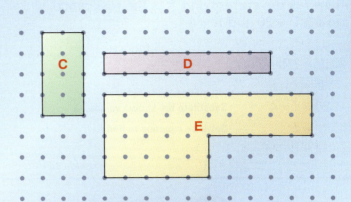

Leah drew this table for shapes **C**, **D** and **E**.

Shape	Number of dots on the perimeter, P	Number of dots inside the shape, I	Area (in squares), A
C	12	3	8
D	18	0	8
E	28	17	30

Leah worked out that the formula for finding the area, A, of a shape drawn on square dotty paper was

$$A = \frac{P}{2} + I - 1$$

where P is the number of dots on the perimeter and I is the number of dots inside the shape.

Draw lots more shapes on square dotty paper.
Draw a table like Leah's for your shapes.
Count the dots on the perimeter, and inside each of your shapes.
Find the areas by counting squares.

Does Leah's formula work for all the shapes you drew on square dotty paper?

 ## Puzzle

Replace each of the letters by one of the digits, 0, 1, 2, 3, 4, 5, 6, 7, 8, 9 so that the subtractions are correct.

a
```
  F I V E
– F O U R
─────────
    O N E
```

E = 2
O = 3

b
```
  S E V E N
–   F O U R
─────────
  T H R E E
```

N = 8
H = 9

Is there more than one possible answer?

Algebra

Sometimes we must **solve an equation** to find the value of the unknown.

Worked Example

The voltage, V, in a circuit is given by $V = IR$
where I is the current and R is the resistance.
Find I if $V = 20$ and $R = 4$.

Answer

$V = IR$

$V = I \times R$

$20 = I \times 4$ **Substitute the known values.**

$\frac{20}{4} = I$ **Solve an equation to find I.**

$I = \textbf{5}$

Exercise 9

1 The formula for finding force, F, from mass, m, and acceleration, a, is $\textbf{F = ma}$.
   Find
      a a if $F = 60$ and $m = 20$ **b** m if $F = 30$ and $a = 5$ **c** a if $F = 60$ and $m = 100$.

2 The formula for finding speed, S, from distance, D, and time, T, is $\textbf{S} = \frac{\textbf{D}}{\textbf{T}}$.
   Find
      a D if $S = 30$ and $T = 2$ **b** D if $S = 6$ and $T = 7$.

3 The voltage, V, in an electrical circuit, with current I and resistance R, is given by
   the formula $\textbf{V = IR}$.
   Find
      a R when $V = 6$ and $I = 2$ **b** I when $V = 3{\cdot}6$ and $R = 4$
      c R when $V = 13{\cdot}8$ and $I = 4{\cdot}6$ **d** I when $V = 8{\cdot}1$ and $R = 3{\cdot}6$.

*4 The formula for the change £C from £40 for n pieces of pie is $\textbf{C = 40 - 5n}$.
   Find n if C is
      a   10   **b**   15   **c**   0   **d**   5.

Writing expressions

    $n + 4$, $x - 7$, $\frac{p}{5}$ and $3m$ are all expressions.
The letters n, x, p, and m stand for numbers.

Example   George won x matches in a tennis tournament.
                 Tim won 3 more matches
                 Tim won $x + 3$.

Example   Pamela has *n* points in a game.
She picks up the card shown.
She now has $3 \times n$ points.
Note We write $3 \times n$ as $3n$.

Exercise 10

1   Bonnie planted some apple trees.
a stands for the number of apple trees she planted.
   a   Bonnie planted three times as many plum trees as apple trees.
   Write an expression for the number of plum trees she planted.
   b   Bonnie planted twice as many cherry trees as apple trees.
   Write an expression for the number of cherry trees she planted.
   c   Bonnie planted four times as many pear trees as apple trees.
   Write an expression for the number of pear trees she planted.

2   Slide 1 is *t* metres high.
   Slide 2 is $3t$ metres high.
   Which of these are true?
   a   Slide 2 is $3 \times t$ metres high.
   b   Slide 2 is 3 metres higher than slide 1.
   c   Slide 1 is 3 times as high as slide 2.
   d   Slide 2 is 3 times as high as slide 1.

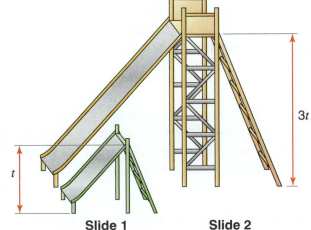

Slide 1   Slide 2

3   Lisa, Alice and Pru each have a bag of sweets.
   Call the number of sweets in Lisa's bag *m*.
   a   Lisa put the sweets in her bag into 3 equal piles.
   Write an expression for the number of sweets in each pile.
   b   Alice has half as many sweets as Lisa in her bag.
   Write an expression for the number of sweets in Alice's bag.
   c   Pru has 4 more sweets than Alice.
   Write an expression for the number of sweets in Pru's bag.

Algebra

4 Rex has 5 boxes of chocolates and 4 extra chocolates.
c is the number of chocolates in each box.
Which of these gives the total number of chocolates shown?
There is more than one answer.

A $5c$ **B** $5c + 4$ **C** $5 \times c + 4$ **D** $4 + 5c$

5 Pam, Clive, Maya and Jill play a game.
They each start with some bags of pegs.
Each bag has n pegs in it.
Write an expression for the number of pegs each of these has at the end of the game.
  a Pam had 3 bags at the start and she lost 6 pegs.
  b Clive had 4 bags at the start and he won 8 pegs.
  c Maya had 5 bags at the start and she won 20 pegs.
  d Jill had 1 bag at the start and she won 10 pegs.

6 Catherine takes n minutes to eat her breakfast each day.
She takes twice as long to eat her lunch.
  a Write an expression for the time it takes Catherine to eat her lunch.
  b Write an expression for the time it takes Catherine to eat her lunch over 5 days.

7 Menzier bought p pears and some bananas.
  a He had three times as many bananas as pears.
    How many bananas did he have?
  b He ate all the bananas except four. How many bananas did he eat?
  *c He ate a quarter of the pears then gave two to his sister. How many pears did he have left?

*8 Rob had £$2x$. Hitesh had £3.
Rob and Hitesh spent all their money on a raffle ticket.
It won three times as much as the ticket had cost.
Write an expression, using brackets, for the amount they won.

*9 Ella, Todd and Debbie each have a number of toffees in a bag.
Ella has 3 more than Todd.
Debbie has 5 times as many as Todd.
Debbie calls the number of toffees she has t.
Write an expression using t for the number of toffees in Ella's bag.

Discussion

A courtyard was built around a tree.
Jesamine wrote this expression for the area of the courtyard.

area of large rectangle = **8** × **12**
area round tree = x × **3**
area of courtyard = **12** × **8** − **3** × x = **96** − **3**x

Is her expression correct? **Discuss.**

Ronan wrote this expression for the area of the courtyard.

area of red rectangle = **5** × **12**
area of blue rectangle = (**12** − x) × **3**
area of courtyard = **5** × **12** + (**12** − x) × **3**
  = **60** + **3**(**12** − x)

Is Ronan's expression correct? **Discuss.**

Are Jesamine's and Ronan's expressions equivalent? **Discuss.**

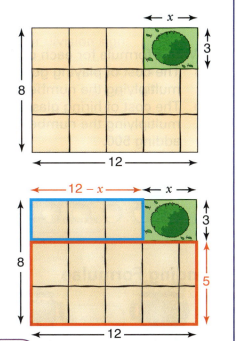

'Equivalent' means they are the same for all values of x.

Writing and finding formulae

A roast chicken takes 40 minutes per kilogram to cook.
2 kilograms would take   40 × 2 minutes to cook.
3 kilograms would take   40 × 3 minutes to cook.
n kilograms would take   40 × n minutes to cook.

We could write a formula for the time taken, T, in minutes to cook n kilograms. $T = 40n$

time (min)   number of kilograms

Exercise 11

1 Write a formula for each of these.
   a The time, T, to drive along a road is the number of kilometres, x, multiplied by 2.
       $T = $ ___ × ___ .
   b The amount of water, w, to be added to an acid is the number of millilitres of acid, a, multiplied by 20. $w = $ ___ × ___ .
   c The number of gallons, g, is found by dividing the number of pints, p, by 8.

10 The formula for density, D, is $D = \frac{m}{V}$ where m is mass and V is volume.
Find
  a D if $m = 28$ and $V = 4$
  b m if $D = 20$ and $V = 12$.

11 Alice is holding n cubes.
  a Two cubes are removed.
     Write an expression for the total number of
     cubes she is holding now.
  b Alice starts again with a row of n cubes.
     Another row the same length is joined on.
     Write an expression for the total number of
     cubes she is holding now.
  c Sam also has some cubes in his hands.
     In one hand there are $3n - 1$ cubes.
     In the other hand there are $3(n - 1)$ cubes.
     Is Sam holding the same number of cubes in each hand?
     Explain your answer.
  d Write an expression for the number of cubes Sam has altogether.

12 Write an expression for the perimeter of each of these shapes. Simplify your
expression as much as possible.
  a 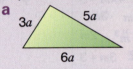 **b** 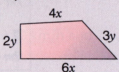 **c** 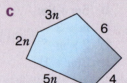 **d**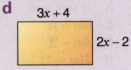

13 Write a formula for this:
The cost, C in pounds, of hiring a dress is found by multiplying the number
of days hired, d, by £18 and adding £20.

8 Sequences and Functions

You need to know

Key vocabulary

consecutive, difference, flow chart, function, input, nth term, output, rule, sequence, term, $T(n)$

Don't tell Fibonaccis

The Fibonacci sequence is generated by this rule.

first terms   1, 1   **rule**   add the two previous terms together

1,   1,   2,   3,   5,   8,   13

1 + 1   1 + 2   2 + 3   3 + 5 ...

1. Write down the first 15 terms of the Fibonacci sequence.

2. Look at every 3rd number.
   What sort of number is it?

3. Look at every 4th number.
   What sort of number is it?

4. What about every 5th number?

5. Fibonacci numbers often occur in nature.
   To find out more you could look at this website.
   www.mcs.surrey.ac.uk/Personal/R.Knott/Fibonacci

2, 8, ... are 3rd numbers.

This daisy has 21 petals.

Algebra

Writing sequences from flow charts

Worked Example

Write down the sequence given by this flow chart.

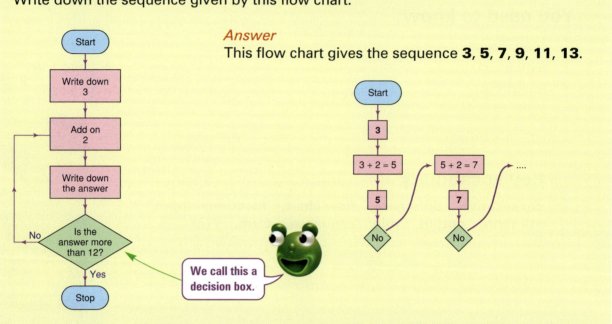

Answer

This flow chart gives the sequence **3, 5, 7, 9, 11, 13**.

We call this a decision box.

Exercise 1

1 Write down the sequences given by these flow charts.

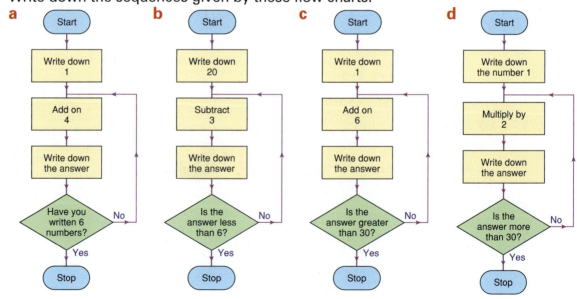

Writing sequences from rules

Term-to-term rules

1, 3, 5, 7, 9, ... is a sequence.

1st term   3rd term

The **first term** is 1 and the **rule** is 'add 2'.

We can write down a sequence if we know the first term and the rule for finding the next term.

Example   **First term**   30   **rule**   subtract 3

gives   30, 27, 24, 21, 18, ...
              −3   −3

Exercise 2

1 Write down the first six terms of these sequences.
Use the number lines to help.

   a Start at 5 and count on in steps of 4.

   b Start at 2 and count on in steps of 0·5.

   c Start at 8 and count back in steps of 3.

   d Start at 10 and count back in steps of 0·2.

2 a The number chain below is part of a **doubling** number chain.   **[SATs Paper 1 Level 4]**
What are the two missing numbers?

     ___ → 40 → 80 → 160 → ___

   b The number chain below is part of a **halving** number chain.
What are the two missing numbers?

     40 → 20 → 10 → ___ → ___

3 Write down the first six terms of these sequences. The first one is done.

	1st term(s)	term-to-term rule
a	6	add 4
b	20	subtract 3
c	1	multiply by 2
d	0	add consecutive numbers starting with 1
e	100 000	divide by 10
f	1, 3	add the two previous terms

6, 10, 14, 18, 22, 26,
   +4   +4   +4   +4   +4

Add 1, 2, 3, 4, ...

Algebra

4 Use a copy of this.
Find a path through the numbers.
The path follows a sequence with the rule 'add 3'.

start

1	4	8	11	14	40	← end
2	7	9	26	32	37	
7	10	13	16	31	34	
4	12	16	15	28	36	
9	17	19	22	25	29	

5 Use the rule to work out the next three numbers in the number chain.

Rule   **Number chain**

a add 5

$1 \xrightarrow{+5} 6 \xrightarrow{+5} 11 \xrightarrow{+5} \_ \xrightarrow{+5} \_ \xrightarrow{+5} \_$

b multiply by 3

$1 \xrightarrow{\times 3} 3 \xrightarrow{\times 3} 9 \xrightarrow{\times 3} \_ \xrightarrow{\times 3} \_ \xrightarrow{\times 3} \_$

*c multiply by 2 then subtract 1

$2 \rightarrow 3 \rightarrow 5 \rightarrow$

*d multiply by 3 then add 1

$1 \rightarrow 4 \rightarrow 13 \rightarrow$

6 Joe makes a sequence by adding three to the previous term.
a Write down the first five terms of a sequence he could make.
b Write down another sequence he could make.

7 Bob wrote down these sequences.
For each one say if it
a goes up or down
b has equal differences between each term.
    i 1, 2, 4, 7, 11, ...
   ii 128, 64, 32, 16, 8, ...
  iii 25, 29, 33, 37, 41, ...

*8 A spider crawls up a wall.
Each minute it crawls 2 cm more than the minute before.
It crawled 1 cm during the first minute.
Give the sequence of the **total** distance crawled after each minute.

*9 Use numbers from the box to make a sequence that has the rule
a add 5
b multiply by 2.

2	11	27	8	1
15	22	6	12	3
16	4	9	17	32
14	7	64	20	30

*10 The next number in a sequence is the sum of the two previous numbers.
What are the missing numbers?
    ☐, ☐, ☐, 1, 1, 2, 3, 5, 8, 13, 21, 34

Sometimes we can **work out the rule** by looking to see what was done to the previous term to get the next term.

Then we can write down the next terms in the sequence.

Example   In the sequence   5, 11, 17, 23, 29, ...   6 has been added to get each next term.   +6 +6 +6 +6

The next three terms are 35, 41, 47, ...   +6 +6

1   Hugo had these sequences in his book.
A cat walked over the book and smudged the next three terms in each sequence.
Write down what the next three terms could be.

a 5, 8, 11, 14, 17 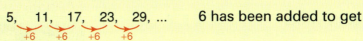   **b** 12, 24, 36, 48

c 25, 21, 17, 13   **d** 8000, 4000, 2000, 1000

e 10, 100, 1000   *** f** 1, 3, 6, 10

2   Work out the rule for each of these number chains.
Write it in words.

a 5 $\longrightarrow$ 11 $\longrightarrow$ 17 $\longrightarrow$ 23 $\longrightarrow$ **b** 400 $\longrightarrow$ 200 $\longrightarrow$ 100 $\longrightarrow$ 50 $\longrightarrow$

c 60 $\longrightarrow$ 56 $\longrightarrow$ 52 $\longrightarrow$ 48 $\longrightarrow$ **d** 8 $\longrightarrow$ 13 $\longrightarrow$ 18 $\longrightarrow$ 23 $\longrightarrow$

e 6 $\longrightarrow$ 10 $\longrightarrow$ 14 $\longrightarrow$ 18 $\longrightarrow$ **f** 23 $\longrightarrow$ 19 $\longrightarrow$ 15 $\longrightarrow$ 11 $\longrightarrow$

g 3 $\longrightarrow$ 9 $\longrightarrow$ 27 $\longrightarrow$ 81 $\longrightarrow$

3   Write down the rule for these sequences.
Write down the next term.

a 4, 10, 16, 22, ...   **b** 2, 10, 50, 250, ...

c 80, 40, 20, 10, ...

4   Mr Patel said to his class,

'The rule for a sequence is add ☐.'

He asked the class to choose a 1st term and a number to go in the box.
What rule and 1st term might these pupils have chosen?

a   Jessie's sequence has all even numbers greater than 1.

b   Barry's sequence has all odd numbers greater than 0.

c   All terms in Amy's sequence end in 0.

Discussion

Mrs James put this sequence on the board.
She asked her class to write down the next three terms.

Penny wrote   1, 2, 3, 4, 5, 6, ...

Blair wrote   1, 2, 3, 5, 8, 13, ...

Who is correct? **Discuss**.

How might the sequence 1, 2, 4, ... continue?
What about 3, 8, 13, ...?

 Puzzle

a I am thinking of a sequence.
The rule is add 3.
The third term is 8.
Write down the first six terms
of the sequence.

b I am thinking of a sequence.
The rule is subtract 2.
The fourth term is 10.
Write down the first six terms
of the sequence.

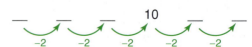

Dry land – a game for one player

You will need 5 counters and
a copy of this board.

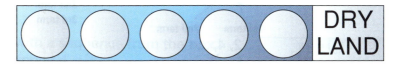

To play • Put one counter on each circle.
• Move all counters to dry land in as few moves as possible.
The rules for moving the counters are
– you can move right or left one space
– you can put one counter on top of another
– if counters are on top of one another, they must be moved as a group
– a group of counters can only move the same number of spaces as there are counters in the group.

Example This group can move 3 spaces left or 3 spaces right.

Sometimes the **rule for the *n*th term** of a sequence is given.

Example $T(n) = n + 3$

this means
the *n*th term

Each term is the term number, *n*, plus 3.

Term number	1	2	3	4	5	6
Working	1 + 3	2 + 3	3 + 3	4 + 3	5 + 3	6 + 3
Term	4	5	6	7	8	9

Worked Example
Write down the first 5 terms for the sequence with the rule $T(n) = 2n + 3$.

Answer

Term number, *n*	1	2	3	4	5
Term (2*n* + 3)	2 × 1 + 3 5	2 × 2 + 3 7	2 × 3 + 3 9	2 × 4 + 3 11	2 × 5 + 3 13

The sequence is **5, 7, 9, 11, 13, ...**

Sequences in practical situations

Discussion

Pia made photo frames from pieces of wood.

size 1
4 pieces

size 2
8 pieces

size 3
12 pieces

What sequence is made from the number of pieces of wood?
What might the next few terms be?
How did you decide this? **Discuss**.

Explain how you would find the number of pieces of wood for size n.
How did you work it out? **Discuss**.

Worked Example

Peter made plant boxes from lengths of wood.

box 1

box 2

box 3

a How many lengths of wood are needed for box 4?
b What sequence is made by the number of lengths needed for box 1, box 2, box 3, ...?
Work out the next few terms.
Use the diagrams to explain how you got this sequence.
c Which of the following is the expression for the number of lengths needed for the
nth box. Explain your answer.
$$2n + 5 \qquad 5n \qquad n + 5 \qquad n - 5$$
Check to see if your answer to **b** is correct.

Answer
a **20**, because each shape needs 5 lengths.
b 5, 10, 15, 20, 25, 30, 35, ...
We add 5 each time because to make the next box, we add one more shape.
Each shape needs 5 lengths.
c **5n**
Each term is $5 \times n$ because for box n there will be n shapes and each shape needs
5 lengths.

Exercise 6

1 Beth made patterns with matchsticks.

 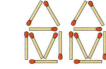 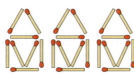

pattern 1   **pattern 2**   **pattern 3**

a What numbers go in the gaps in this table?

Pattern number	1	2	3	4	5
Number of matchsticks	8	16			

b Which of these rules for finding the number of matchsticks is correct?
  A multiply the pattern number by 4
  B add 8 to the pattern number
  C multiply the pattern number by 8

c What goes in the box?

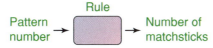

d How many matchsticks are needed for pattern 10?

2 Tim made these patterns with rods.

1 diamond   **2 diamonds**   **3 diamonds**

a What numbers go in the gaps in this table?

Number of diamonds (D)	1	2	3	4	5
Number of rods (R)	4	8			

b Tim wants to make a pattern with 6 diamonds.
How many rods will he need?

c Which of these is the rule for finding the number of rods, R, from the number of diamonds, D?
  A $R = 2 + D$ **B** $R = 4 + D$ **C** $R = 4 \times D$ **D** $R = 2 \times D$

3 Sam made these patterns with squares.

pattern 1   **pattern 2**   **pattern 3**

a Draw pattern 4.
b What numbers go in the gaps in this table?

Pattern number (P)	1	2	3	4	5	6
Number of squares (S)	1	4	7			

c The rule for finding the number of squares is
  '3 times the pattern number then subtract 2'
  Use the rule to find the number of squares in pattern 10.

d Complete the rule for finding the number of squares, S, from the pattern number, P.
  $S =$ _____

Algebra

4 Ellie owned a gift shop.
She designed a logo for her sign.

size 1   **size 2**   **size 3**
4 lines   **8 lines**   **12 lines**

a Draw a size 4 logo.
b What sequence is made by the number of lines for
        size 1, size 2, size 3, size 4, ...?
c Write down the next two terms.
        Use the diagrams to explain how you got these terms.
d How could you work out the number of lines in a size 20 logo?
e Write an expression for the number of lines in a size n logo.
        Use the diagrams to explain how you found this.

5 Jeff makes a sequence of patterns with black and grey
triangular tiles.   [SATs Paper 2 Level 5]

                      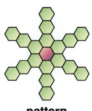

pattern   **pattern**   **pattern**
number   **number**   **number**
1   **2**   **3**

The rule for finding the number of tiles in pattern number N in Jeff's sequence is:

  number of tiles = $1 + 3N$

a The 1 in this rule represents the black tile.
        What does the $3N$ represent?
b Jeff makes pattern number 12 in his sequence.
        How many black tiles and how many grey tiles does he use?
c Jeff uses 61 tiles altogether to make a pattern in his sequence.
        What is the number of the pattern he makes?
d Barbara makes a sequence of patterns with hexagonal tiles.

Each pattern in Barbara's sequence has 1 pink tile in the middle.
Each new pattern has 6 more green tiles than the pattern before.
Write the rule for finding the number of tiles in pattern number N in Barbara's
sequence.

  number of tiles = ... + ...

e Gwenno uses some tiles to make a different sequence of patterns.
The rule for finding the number of tiles in pattern number N in Gwenno's sequence is:

> number of tiles = $1 + 4N$

Draw what you think the first 3 patterns in Gwenno's sequence could be.

6 You can make 'huts' with matches.   [SATs Paper 2 Level 5]

  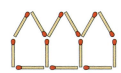   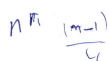

1 hut needs	2 huts need	3 huts need
5 matches	9 matches	13 matches

A rule to find how many matches you need is

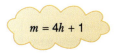

$$m = 4h + 1$$

m stands for the number of matches.
h stands for the number of huts.

a Use the rule to find how many matches you need to make 8 huts.
Show your working.
b I use 81 matches to make some huts.
How many huts do I make?
Show your working.
c Andy makes different 'huts' with matches.

  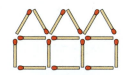 $6h - (h-1)$

1 hut needs	2 huts need	3 huts need
6 matches	11 matches	16 matches

Which rule below shows how many matches he needs?

Remember: m stands for the number of matches.
                 h stands for the number of huts.

$m = h + 5$ $m = 4h + 2$ $m = 4h + 3$

$m = 5h + 1$ $m = 5h + 2$ $m = h + 13$

197

Algebra

Finding the rule for the *n*th term

Term number, *n*	1 +2	2 +2	3 +2	4 +2	...
Term	3	4	5	6	...

For the sequence 3, 4, 5, 6, ...
each term is two more than the term number.

The rule for this sequence can be written as

nth term $= n + 2$ or $T(n) = n + 2$

↑ term number   ↑ term number

To find the **rule for the *n*th term**, we can draw a **difference table**.

Example   3, 5, 7, 9, 11, ...

Term number	1	2	3	4	5
T (*n*)	3	5	7	9	11
Difference		2	2	2	2

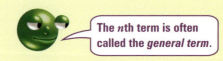

The *n*th term is often called the *general term*.

The difference is **2** so the *n*th term is **2** × *n* + **?**.
The first term is 3.
This is **2** × 1 + **1**.
Check that the other terms are **2** × *n* + **1**.

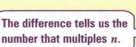

The difference tells us the number that multiples *n*.

The rule for the *n*th term is **2*n* + 1**.

Exercise 7

1 Which of **A**, **B**, **C** gives the expression for the *n*th term?

  a 2, 4, 6, 8, 10, ...
    A $T(n) = n + 2$
    B $T(n) = 2n$
    C $T(n) = 2n + 2$

  b 4, 5, 6, 7, 8, ...
    A $T(n) = n + 3$
    B $T(n) = n + 2$
    C $T(n) = 3n$

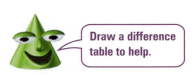

Draw a difference table to help.

  c 4, 9, 14, 19, 24, ...
    A $T(n) = n + 5$
    B $T(n) = 5n$
    C $T(n) = 5n - 1$

  d 90, 80, 70, 60, ...
    A $T(n) = 10n$
    B $T(n) = 100 - 10n$
    C $T(n) = 100 - n$

2 Write an expression for the *n*th term.
  Use the difference table to help.

  a 6, 12, 18, 24, 30, ...

Term number	1	2	3	4	5
T (*n*)	6	12	18	24	30
Difference					

  b 5, 6, 7, 8, 9, ...

Term number	1	2	3	4	5
T (*n*)	5	6	7	8	9
Difference					

*c   1, 4, 7, 10, 13, ...

Term number	1	2	3	4	5
$T(n)$	1	4	7	10	13
Difference					

*3   9, 14, 19, 24, 29, 34, ...
This sequence continues in the same way.
a   Write the next three terms of the sequence.
b   Explain the rule for finding the next term.
c   What is the nth term of the sequence.
d   What is the 20th term of the sequence?

Functions

Remember
We can find the **output** of a function machine if we are given the input.

Example   3, 8, 2, 5 → | multiply by 3 | → | subtract 2 | → 7, 22, 4, 13

We can show the inputs and outputs on a table.

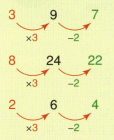

Input	Output
3	7
8	22
2	4
5	13

Exercise 8

T

1   Use a copy of this.
Fill in the input/output table for each function machine.

a x → | multiply by 3 | → | add 2 | → y

Input	Working	Output
5	$5 \times 3 + 2$	
2	$2 \times 3 + 2$	
0	$0 \times 3 + 2$	
12	$12 \times 3 + 2$	

b x → | divide by 2 | → | subtract 1 | → y

Input	Output
4	
10	
24	
68	

199

c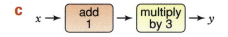

Input	Output
11	
21	
0	
* ⁻5	

d $x \rightarrow$ multiply by 4 $\rightarrow$ subtract 1 $\rightarrow y$

Input	Output
7	
3	
⁻1	
0·5	

*2 Draw input/output tables like the ones in question **1** for these function machines.
The input values are given. Find the output values.

a 4, 7, 10 $\rightarrow$ subtract 2 $\rightarrow$ multiply by 3 $\rightarrow$ ___, ___, ___, ___

b 25, 5, 15 $\rightarrow$ divide by 5 $\rightarrow$ add 2 $\rightarrow$ ___, ___, ___, ___

Sometimes we show the inputs and outputs on a **mapping diagram**.

Example 0, 1, 2, 3, 4 $\rightarrow$ multiply by 2 $\rightarrow$ add 4 $\rightarrow y$

We multiply each number by 2 then add 4.

This is a mapping diagram for the input values 0, 1, 2, 3, 4.

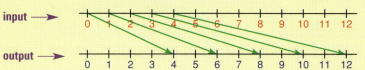

$0 \times 2 + 4 = 0 + 4 = 4$
$1 \times 2 + 4 = 2 + 4 = 6$
$2 \times 2 + 4 = 4 + 4 = 8$
$3 \times 2 + 4 = 6 + 4 = 10$
$4 \times 2 + 4 = 8 + 4 = 12$

This mapping can be written using symbols as $x \rightarrow 2x + 4$ or $y = 2x + 4$.

Exercise 9

T

1 Use a copy of these mapping diagrams.
Fill them in for the function machine and input given.
a is started for you.
Write the function in symbols.

a 0, 1, 2, 3, 4 $\rightarrow$ add 3 $\rightarrow y$

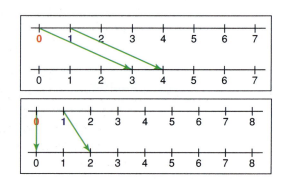

b 0, 1, 2, 3, 4 $\rightarrow$ multiply by 2 $\rightarrow y$

c $0, 1, 2, 3, 4 \rightarrow$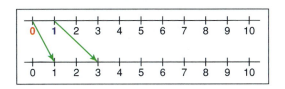

T

2 Use a copy of this mapping diagram.
Fill it in for the function and input given.
$x \rightarrow 2x + 2$ for $x = 0, 1, 2, 3, 4$
e.g. When $x = 0$, $2x + 2 = 2 \times 0 + 2$
$= 0 + 2$
$= 2$

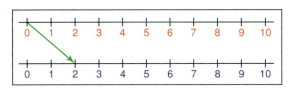

T

*3 **a** Use a copy of these mapping diagrams.
Fill them in for the function given.
Use input values of $x = 1, 2, 3, 4, 5$ and 6.
i $x \rightarrow x + 2$ **ii** $x \rightarrow x + 3$

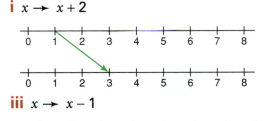   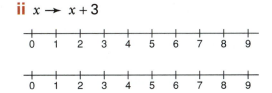

iii $x \rightarrow x - 1$

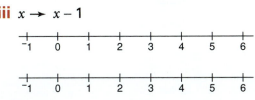

b What do you notice about the lines in each of the mapping diagrams in part **a**?
c What can you say about the lines on a mapping diagram for a function of the
form $x \rightarrow x + c$?

Finding the function given the input and output

   Guess my rule – a game for a class

To play   • Choose a leader.
• The leader stands in front of the class.
He or she thinks of a rule and writes it on a piece of paper.
• The leader chooses one person in the class to say a number.
The leader then says what the output would be.
Example   The leader thinks of

The person says 4.
The leader says the output is 10.
• The person then tries to guess the rule.
If correct, this person becomes the leader.
If not correct, the leader chooses another person to say a
number.

Algebra

Worked Example

Find the rule for these function machines.

a

8, 6, 5, 4, 7 → [?] → 11, 9, 8, 7, 10

b

7, 6, 8, 5, 9 → [multiply by 2] → [?] → 11, 9, 13, 7, 15

Answer

a Put the input and output in order on a table.
The difference is **1**.
This means the rule is $x \rightarrow 1x + $ **?**
When the input (x) is 4, the output is 7.

$$1 \times x + \textbf{?} = 7$$
$$1 \times 4 + \textbf{?} = 7$$
    ↑   ↑
  input output
    so **?** = **3**

The rule is $x \rightarrow 1x + 3$ or $x \rightarrow x + 3$ or $y = x + 3$

Input (x)	4	5	6	7	8
Output (y)	7	8	9	10	11
Difference		1	1	1	1

b Put the input and output in order on a table.
The difference is **2**.
This means the rule is $x \rightarrow 2x + $ **?**
When the input (x) is 5, the output is 7.

$$2 \times x + \textbf{?} = 7$$
$$2 \times 5 + \textbf{?} = 7$$
    ↑   ↑
  input output $10 + {}^-3 = 7$
    so **?** = **⁻3**

The rule is $x \rightarrow 2x - 3$ or $y = 2x - 3$

Input (x)	5	6	7	8	9
Output (y)	7	9	11	13	15
Difference		2	2	2	2

Exercise 10

1 Find the missing operation for these.

a 3, 1, 4 → [?] → 9, 3, 12

b 5, 1, 3 → [?] → 11, 7, 9

c 1, 5, 3, 2, 4 → [?] → 0, 4, 2, 1, 3

d 7, 8, 10, 6, 9 → [?] → 11, 12, 14, 10, 13

Remember to put the input and output in order on a table.

2 What does **?** stand for?

a
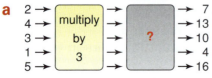

2 →		→ 7
4 →	multiply	→ 13
3 →	by	→ 10
1 →	3	→ 4
5 →		→ 16

with **?** in the output machine

b
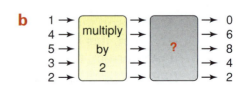

1 →		→ 0
4 →	multiply	→ 6
5 →	by	→ 8
3 →	2	→ 4
2 →		→ 2

with **?** in the output machine

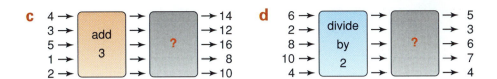

c

4 → | add 3 | → | ? | → 14
3 → | | → | | → 12
5 → | | → | | → 16
1 → | | → | | → 8
2 → | | → | | → 10

d

6 → | divide by 2 | → | ? | → 5
2 → | | → | | → 3
8 → | | → | | → 6
10 → | | → | | → 7
4 → | | → | | → 4

3 Write the functions in question **2** as symbols.

4 John's group were playing a match up game.
Find the cards that would go in the empty boxes.

a

5, 2, 1, 3, 4 → | ? | → | ? | → 9, 3, 1, 5, 7

b

5, 3, 7, 4, 6 → | ? | → | ? | → 13, 7, 19, 10, 16

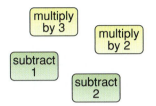

multiply by 3

multiply by 2

subtract 1

subtract 2

Properties of functions

Discussion

● Find the output for each of these function machines.

3, 5, 9, 15 → | add 4 | → | add 3 | → __, __, __, __

3, 5, 9, 15 → | add 7 | → __, __, __, __

What do you notice about the outputs? **Discuss**.

Can two additions, two subtractions or an addition and a subtraction always be replaced with a single addition or subtraction? **Discuss**.
Try these examples.

a

4, 3, 1, 2 → | add 1 | → | add 4 | → __, __, __, __   + 1 + 4 = + 5

4, 3, 1, 2 → | add 5 | → __, __, __, __

b

2, 6, 1, 4 → | add 3 | → | subtract 1 | → __, __, __, __   + 3 − 1 = + 2

2, 6, 1, 4 → | add 2 | → __, __, __, __

Can two multiplications be replaced with a single multiplication? **Discuss**.
Try these examples.

c

4, 3, 1, 2 → | multiply by 2 | → | multiply by 3 | → __, __, __, __

4, 3, 1, 2 → | multiply by 6 | → __, __, __, __   × 2 × 3 = × 6

d

2, 6, 1, 4 → | multiply by 2 | → | multiply by 5 | → __, __, __, __

2, 6, 1, 4 → | multiply by 10 | → __, __, __, __

203

Algebra

1 What single operation could replace the two given?

a $x \rightarrow$ [add 3] $\rightarrow$ [add 5] $\rightarrow y$

b $x \rightarrow$ [subtract 2] $\rightarrow$ [subtract 4] $\rightarrow y$

c $x \rightarrow$ [add 9] $\rightarrow$ [subtract 4] $\rightarrow y$

d $x \rightarrow$ [add 3] $\rightarrow$ [subtract 7] $\rightarrow y$

e $x \rightarrow$ [multiply by 5] $\rightarrow$ [multiply by 3] $\rightarrow y$

f $x \rightarrow$ [multiply by 10] $\rightarrow$ [multiply by 2] $\rightarrow y$

To find the **inverse of a function**, do the inverse operation.

Example

$x \rightarrow$ [add 3] $\rightarrow x + 3$ **function machine**

$x - 3 \leftarrow$ [subtract 3] $\leftarrow x \leftarrow$ start with x and work backwards   **inverse function machine**

The inverse of $x \rightarrow x + 3$ is $x \rightarrow x - 3$.

> The inverse function machine has the arrows the other way round.

To find the inverse when there are two operations, do the inverse operations in the reverse order.

Example

$x \rightarrow$ [add 1] $\xrightarrow{x+1}$ [multiply by 2] $\rightarrow 2 \times (x + 1)$

$\frac{x}{2} - 1 \leftarrow$ [subtract 1] $\xleftarrow{\frac{x}{2}}$ [divide by 2] $\leftarrow x \leftarrow$ start with x and work backwards

The inverse of $x \rightarrow 2 \times (x + 1)$ is $x \rightarrow \frac{x}{2} - 1$.

Worked Examples
Find the input for these.

a $\underline{\quad}, \underline{\quad}, \underline{\quad} \rightarrow$ [add 2] $\rightarrow 6, 8, 4$

b $\underline{\quad}, \underline{\quad}, \underline{\quad} \rightarrow$ [subtract 1] $\rightarrow$ [multiply by 2] $\rightarrow 10, 6, 12$

Answer
Draw inverse function machines first.

a $\underline{4}, \underline{6}, \underline{2} \leftarrow$ [subtract 2] $\leftarrow 6, 8, 4$

$6 - 2 = 4$
$8 - 2 = 6$
$4 - 2 = 2$
The inputs are **4, 6, 2**

b $\underline{6}, \underline{4}, \underline{7} \leftarrow$ [add 1] $\leftarrow$ [divide by 2] $\leftarrow 10, 6, 12$

$10 \div 2 + 1 = 5 + 1 = 6$
$6 \div 2 + 1 = 3 + 1 = 4$
$12 \div 2 + 1 = 6 + 1 = 7$
The inputs are **6, 4, 7**.

---

1 Find the inputs for these.

a $\underline{\quad}, \underline{\quad}, \underline{\quad} \rightarrow$ [multiply by 3] $\rightarrow 9, 6, 15$

$\underline{\quad}, \underline{\quad}, \underline{\quad} \leftarrow$ [ ] $\leftarrow 9, 6, 15$

b $\underline{\quad}, \underline{\quad}, \underline{\quad} \rightarrow$ [add 3] $\rightarrow 6, 4, 5$

$\underline{\quad}, \underline{\quad}, \underline{\quad} \leftarrow$ [ ] $\leftarrow 6, 4, 5$

c ___, ___, ___ → [add 5] → [divide by 2] → 4, 3, 5   **d** ___, ___, ___ → [multiply by 2] → [subtract 5] → 7, 11, 3

___, ___, ___ ← [ ] ← [ ] ← 4, 3, 5   ___, ___, ___ ← [ ] ← [ ] ← 7, 11, 3

2 Use a copy of this.
Fill in the inverse function machine to find the inverse function.
The first one is done.

a x → [multiply by 2] → $2x$

$\frac{x}{2}$ ← [divide by 2] ← x

b x → [add 6] → $x + 6$

← [ ] ← x

c x → [subtract 3] →

← [ ] ← x

d x → [divide by 2] →

← [ ] ← x

e x → [multiply by 3] → $3x$ **function**

___ ← [ ] ← x **inverse function**

f x → [divide by 3] → $\frac{x}{3}$ → [add 2] → $\frac{x}{3} + 2$ **function**

___ ← [ ] ← $x - 2$ ← [subtract 2] ← x **inverse function**

g x → [add 2] → [multiply by 4] → $4(x + 2)$ **function**

___ ← [ ] ← [ ] ← x **inverse function**

Summary of key points

A We can write a sequence from a **flow chart**.
See page 186 for an example.

B We can **write down a sequence if we know the first term and the rule for finding the next term**.

Examples **1st term** 2 **rule** add 3 gives   2, 5, 8, 11, 14, ...
+3 +3 +3 +3

1st term 1 **rule** multiply by 2 gives   1, 2, 4, 8, 16, ...
×2 ×2 ×2 ×2

C We can write a sequence if we know the **rule for the nth term**.
$T(n)$ means the nth term.

Examples $T(n) = n + 3$

Term number	1	2	3	4	5	...
Term ($n + 3$)	1 + 3 4	2 + 3 5	3 + 3 6	4 + 3 7	5 + 3 8	

$T(n) = 3n - 1$

Term number	1	2	3	4	...
Term ($3n - 1$)	3 × 1 − 1 2	3 × 2 − 1 5	3 × 3 − 1 8	3 × 4 − 1 11	

Algebra

D We can find the rule for the *n*th term in a **practical situation**.

Example

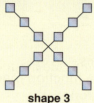

shape 1   shape 2   shape 3
4 squares   8 squares   12 squares

The expression for the number of squares in the *n*th shape is $4n$.

Each time a new shape is drawn 4 new squares are added.

There are *n* lots of 4, where *n* is the shape number.

E We can find the **rule for the *n*th term** using a difference table.

Example For the sequence 3, 5, 7, 9, ...

Term number	1	2	3	4
Term	3	5	7	9
Difference		2	2	2

The difference is **2** so the *n*th term is $2 \times n + ?$

For the 1st term, $2 \times n + ? = 3$

$2 \times 1 + ? = 3$

$2 + ? = 3$ ← **1st term**

So $? = 1$

The rule for the *n*th term is $2n + 1$.

F We can find the **output** of a function machine if we are given the input.

The inputs and outputs can be shown on a table.

Example 3, 6, 1, 4 → [add 1] → [multiply by 2] →

e.g. 3 → [add 1] → 4 → [multiply by 2] → 8

Input	Output
3	8
6	14
1	4
4	10

G We can show inputs and outputs on a **mapping diagram**.

Example 0, 1, 2, 3, 4 → [add 3] → 3, 4, 5, 6, 7

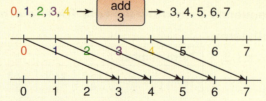

This mapping can be written as $x \rightarrow x + 3$ or $y = x + 3$.

If we are given the input and output we can **find the rule** for the function machine.

Example   5, 1, 4, 2, 3 → | multiply by 3 | → | ? | → 13, 1, 10, 4, 7

Put the input and output in order on a table.

Input (x)	1	2	3	4	5
Output (y)	1	4	7	10	13
Difference		3	3	3	3

The difference is **3**.

The rule is x → **3**x + **?**

When the input is 1, the output is 1.

$$3x + ? = 1$$
$$3 \times 1 + ? = 1$$
$$3 + ? = 1$$
$$? = {}^-2$$

The rule is x → **3**x − **2** or y = **3**x − **2**

Sometimes two operations can be combined.

Example   → | add 6 | → | add 1 | →   can be written   → | add 7 | →

 The **inverse of a function** is found by doing the inverse operation in the reverse order.

We can use an inverse function machine to find the inverse of a function.

Example

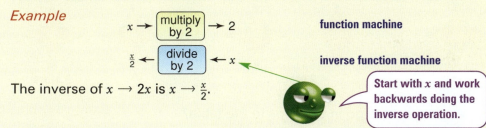

x → | multiply by 2 | → 2   **function machine**

$\frac{x}{2}$ ← | divide by 2 | ← x **inverse function machine**

The inverse of x → $2x$ is x → $\frac{x}{2}$.

Start with x and work backwards doing the inverse operation.

Test yourself

1 Write down the sequence given by this flow chart.

2 Write down the first five terms of these sequences.

	1st term(s)	**term-to-term rule**
a	5	add 10
b	1	multiply by 3
c	1, 2	add the two previous terms.

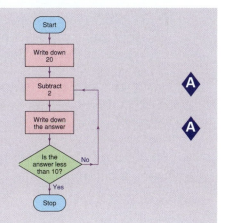

A

A

Algebra

3 a Do these sequences go up or down?
  b Are the differences between each term equal?
    i 25, 36, 49, 64, ...
    ii 40, 38, 34, 28, 20, ...
    iii 1, 3, 5, 7, 9, ...

4 Write down the next three terms in these.
  a 6, 15, 24, 33, 42, ...   **b** 64, 60, 56, 52, 48, ...   **c** 320, 160, 80, 40, ...

5 Emily went on a running programme.
Each day she ran 1 km more than the day before.
On the first day she ran 3 km.
On the second day she ran 4 km.
  a Write down how far she ran each day for the first 6 days.
  *b Write down the sequence of the **total** distance run after each day.

6 Write down the first 5 terms of these.
The first one has been started.
  a $T(n) = 3n$ **b** $T(n) = n + 4$ **c** $T(n) = 2n - 1$
    1st term $T(\mathbf{1}) = 3 \times \mathbf{1} =$
    2nd term $T(\mathbf{2}) = 3 \times \mathbf{2} =$
    3rd term $T(\mathbf{3}) = 3 \times \mathbf{3} =$

7 Andrew was making Christmas decorations.

    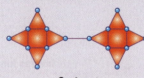   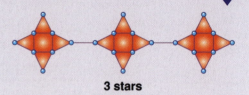

  1 star   2 stars   3 stars
  8 dots   16 dots

  a What numbers go in the gaps in the table?

Star number (S)	1	2	3	4	5
Number of dots (D)	8	16			

  b Andrew wants to make a decoration with 6 stars.
    How many dots will he use?
  c Which of these is the rule for finding the number of dots, D, from the number
    of stars, S?
    A $D = 4S + 4$ **B** $D = 8 \times S$ **C** $D = 8 + S$ **D** $D = 4 \times S$
  d Use the diagrams to explain your answer to **c**.

8 Find the rule for the nth term of the sequence.
Draw a difference table to help.

  a 5, 7, 9, 11, 13, ...   **b** 2, 5, 8, 11, 14, ...

9 Use a copy of this.
Fill in the input/output table for the function machine.

$x \rightarrow$ [multiply by 4] $\rightarrow$ [subtract 2] $\rightarrow y$

Input	Output
2	
10	
4	
0	

10 Use a copy of these mapping diagrams.
Fill them in for the function machine and input given.

a $0, 1, 2, 3, 4 \rightarrow$ [add 2] $\rightarrow$

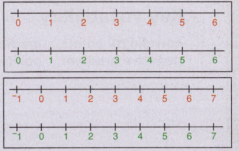

b $0, 1, 2, 3, 4 \rightarrow$ [multiply by 2] $\rightarrow$ [subtract 1] $\rightarrow$

11 Write the functions in question **9** in symbols.

12 Write down what goes in the empty box.

a $5, 1, 3, 4, 2 \rightarrow$ [multiply by 2] $\rightarrow$ [ ] $\rightarrow 14, 6, 10, 12, 8$

b $3, 1, 2, 5, 4 \rightarrow$ [add 1] $\rightarrow$ [ ] $\rightarrow 8, 4, 6, 12, 10$

13 What single operation could replace the two given?

a $x \rightarrow$ [add 6] $\rightarrow$ [subtract 5] $\rightarrow y$

b $x \rightarrow$ [multiply by 2] $\rightarrow$ [multiply by 6] $\rightarrow y$

14 Find the inputs for these.

a
$\underline{} \rightarrow$ [subtract 2] $\rightarrow$ 8
$\rightarrow$ 3
$\rightarrow$ 12

b
$\underline{} \rightarrow$ [add 2] $\rightarrow$ [divide by 3] $\rightarrow$ 1
$\rightarrow$ 5
$\rightarrow$ 3

15 Use a copy of this.
Fill in the inverse function machine to find the inverse function.

a $x \rightarrow$ [add 4] $\rightarrow x + 4$

$\leftarrow$ [ ] $\leftarrow x$

b $x \rightarrow$ [divide by 5] $\rightarrow \frac{x}{5}$

[ ] $\leftarrow x$

9 Graphs of Functions

You need to know

✓ graphs
  — graphs of real-life situations page 138

Key vocabulary

coordinate points,   gradient,   intercept,   linear function,
negative gradient,   positive gradient

 Dealing in Dollars

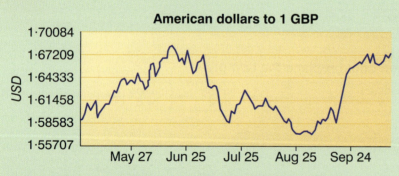

American dollars to 1 GBP

This graph of how many American dollars you got for £1 versus time, came from the Internet.
Find some other real-life graphs in newspapers, magazines or the Internet.
Make a poster of your graphs explaining what each is about.
Are any of them misleading?

Graphing functions

$y = 2x - 3$ is a **linear function**.
It gives the rule for finding y when we know x.
We can find coordinate pairs that follow this rule.

Example   For $y = 2x - 3$

x	Working $2 \times x - 3$	y
3	$6 - 3 = 3$	3
2	$4 - 3 = 1$	1
0	$0 - 3 = {}^-3$	${}^-3$
${}^-1$	${}^-2 - 3 = {}^-5$	${}^-5$

We can plot the coordinate points from the table, $(3, 3)$, $(2, 1)$, $(0, {}^-3)$, $({}^-1, {}^-5)$, on a grid.

We label the line with the
equation $y = 2x - 3$.

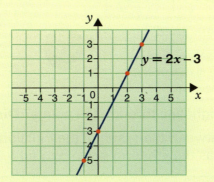

Worked Example

a Copy and complete the table of values for $y = 2x + 1$.
b Plot the points.
   Draw a line through the points.
c Does $y = 2x + 1$ go through the point $(4, 9)$?

x	Working $2 \times x + 1$	y
2		
1		
0		
${}^-1$		

Answer

a Substitute the given x-values into $y = 2x + 1$ to find y.

x	Working $2 \times x + 1$	y
2	$4 + 1 = 5$	5
1	$2 + 1 = 3$	3
0	$0 + 1 = 1$	1
${}^-1$	${}^-2 + 1 = {}^-1$	${}^-1$

b

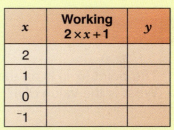

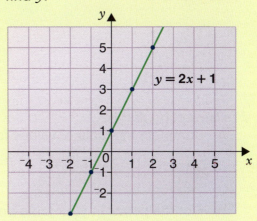

Always label your
line.

Algebra

c We can check if (4, 9) satisfies $y = 2x + 1$.   **The x-coordinate is 4.**

$$y = \mathbf{4} \times 2 + 1$$
$$= 8 + 1$$
$$= 9 \checkmark \quad \text{9 is the } y\text{-coordinate}$$

(4, 9) does satisfy $y = 2x + 1$ so (4, 9) does lie on the line.

A graph with an equation of the form $y = mx + c$ is a **straight line**.
x and y have a **linear relationship**.

Discussion

Isaac plotted this graph of the linear sequence given in the table below.

Term number	1	2	3	4	5
T(n)	1	3	5	7	9

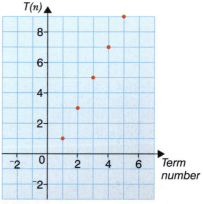

Do the points lie on an 'imagined' straight line?

Is it possible to have a term number of $1\frac{1}{2}$?

$T(n)$ is the notation for the nth term of a sequence.

Should Isaac draw a straight line through the points? **Discuss.**

What is the difference between the graph of the function $y = 2x - 1$ and the graph of the sequence which has the nth term given by $T(n) = 2n - 1$?

Exercise 1

1 The point (6, n) is on the line marked m. What is the value of n?

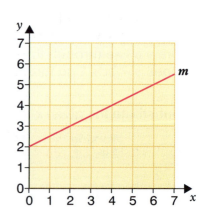

T

2 Use a copy of this.
 a Complete the table of values for $y = x - 1$.

x	3	1	⁻1	⁻3
y				

 b Plot the points in your table on the grid. Join with a straight line.
 c The point (2, a) is on the line. What is the value of a?

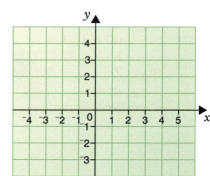

3 Use a copy of this.

   a Complete the table of values for $y = x + 2$.

x	5	2	⁻1	⁻4
y				

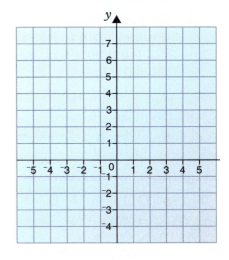

   b Plot the points in the table on the grid.
Draw a straight line through these points.

   c What is the value of y when $x = 1$?

   d The point $(b, {}^-1)$ lies on the line.
What is the value of b?

4 Use a copy of this.

   a Complete the table of values for $y = 2x$.

x	2	1	0	⁻1	⁻2
y					

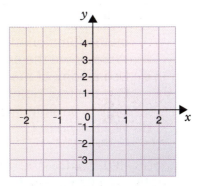

   b On the grid, draw the graph of $y = 2x$ by plotting
the points given in your table and joining them
with a straight line.

   c What is the value of y when $x = 1{\cdot}5$?

5 Use a copy of this.

   a Complete the table of values for $y = 2x - 1$.

x	3	2	0	⁻1	⁻3
y					

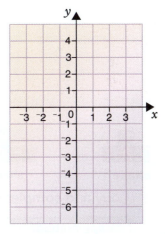

   b On the grid, draw the graph of $y = 2x - 1$.

   c The line goes through the point $(a, 7)$.
What is the value of a?

6 Copy and complete these coordinate pairs so they satisfy
the given function.

   a $y = 2x$ (2, ___), (1, ___), (0, ___), (⁻1, ___), (⁻2, ___)

   b $y = 2x + 5$ (2, ___), (1, ___), (0, ___), (⁻1, ___), (⁻2, ___)

   c $y = 3x - 3$ (2, ___), (1, ___), (0, ___), (⁻1, ___), (⁻2, ___)

   d $y = 10 - x$ (3, ___), (2, ___), (1, ___), (0, ___), (⁻1, ___)

> Work out the positive
> values first. We can
> sometimes use the
> pattern to help find the
> negative values.

Algebra

7 For each of the functions given in question **6** draw a set of axes with x- and y-values from $^-10$ to 12.
Plot the points and draw a straight line through them.
Label the line.

8 Do these coordinate points lie on the line $y = x - 2$?   Write yes or no.
    a (3, 1)   **b** (10, 6)   **c** (5, 3)   **d** ($^-1$, 3)   **e** (0, $^-2$)   **f** $(2\frac{1}{2}, \frac{1}{2})$

9 Do these coordinate pairs lie on the line $y = 2x - 3$?   Write yes or no.
    a (2, 1)   **b** (4, 1)   **c** (10, 23)   **d** (6, 9)   ＊**e** $(\frac{1}{2}, {}^-2)$

⊤ **10** $x \rightarrow$ [multiply by 3] $\rightarrow$ [subtract 4] $\rightarrow y$

x	0	1	2	3
y				

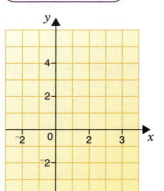

    a Copy and complete this table for the rule given by the function machine.

> There is more about function machines on page 137.

    b Write down the coordinate pairs from the table.
    c Use a copy of the grid.
        Plot the four points.
        Draw a straight line through them.
    d Will the point $(\frac{1}{2}, {}^-2\frac{1}{2})$ lie on the line?

⊤ **11**

 shape 1   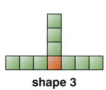 shape 2   shape 3

    a Copy and complete this table.

Shape number	1	2	3	4
Number of squares	4	7		

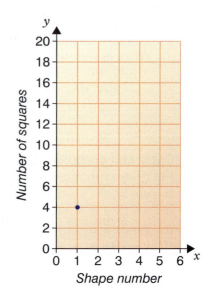

    b Use a copy of this grid.
        On your grid, plot the coordinate pairs from the table.
        The first pair is plotted for you.
        Do the points lie in a straight line?
    c Use your graph to find the number of squares in shape 6.

Equations of straight-line graphs

Investigation

Features of graphs

You will need graph paper, a graph plotting software package and a graphical calculator.

A 1 For each of the following, draw up a table of values for x and y, like this one. Choose x-values between $^-3$ and 3.

x	3	2	0	$^-1$	$^-3$
y					

You could choose different values for x.

Plot the points on a grid.
Draw and label the line for each.

$y = 2x$, $\quad\quad$ $y = 2x + 1$, $\quad$ $y = 2x + 5$, $\quad$ $y = 2x - 3$, $\quad$ $y = 2x - 1$

Describe the similarities and differences.

What do you notice about the equations of the lines and
a the slopes of the lines $\quad\quad$ **b** where the lines cut the y-axis?

∗2 The graph of a straight line is given by $y = mx + c$.
What does m represent?
What does c represent?

You could check using a graph plotter or graphical calculator.

What do you think the graphs of these would look like?
$y = x$ $\quad$ $y = 3x - 2$ $\quad$ $y = 4x + 2$ $\quad$ $y = x - 4$ $\quad$ $y = 3x + 3$ $\quad$ $y = 2x - 1$

Example $\quad$ Josh thinks that $y = 3x - 2$ is a straight line with a gradient steeper than $y = x$. He thinks it cuts the y-axis at $^-2$.

B 1 Use a graphical calculator or graph plotter to **investigate** these families of straight-line graphs.

i $\quad$ $y = x, y = x + 1, y = x + 2, y = x + 3, y = x - 1, y = x - 2, y = x - 3$

ii $\quad$ $y = 3x, y = 3x + 1, y = 3x + 2, y = 3x + 3, y = 3x - 1, y = 3x - 2, y = 3x - 3$

iii $\quad$ $y = ^-x, y = 1 - x, y = 2 - x, y = 3 - x, y = 4 - x$

iv $\quad$ $y = x + 2, y = 2x + 2, y = 3x + 2, y = 4x + 2$

2 For each family of graphs above, describe the similarities and differences.

Algebra

$y = mx + c$ is the equation of a straight line.

m gives the **gradient** or slope.

The greater the value of m the steeper the line.

Lines with the same value of m have the same slope.

$y = x$ is the same as $y = 1 \times x$.

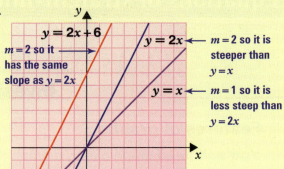

Examples $y = \mathbf{2}x + 3$ and $y = \mathbf{2}x - 4$ have the same slope.

$y = \mathbf{4}x + 2$ has a steeper slope than $y = \mathbf{2}x + 4$.

If m is positive, the gradient is positive. (/)

If m is negative, the gradient is Negative. (\\)

Examples $y = 3x + 2$ has a **positive** gradient.

$y = 5 - 2x$ has a **negative** gradient.

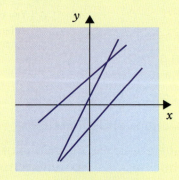

These lines have **positive gradients** (m is positive).

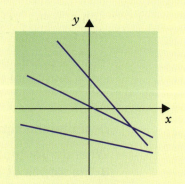

These lines have **negative gradients** (m is negative).

Exercise 2

1 Eight equations of straight-line graphs are given in the box. If each graph was drawn

   a which would have the steepest slope

   b which would have a positive slope

   c would $y = 2x + 4$ or $y = 3x - 5$ have a steeper slope

   d which would have a negative slope?

> $y = 2x + 4$
> $y = 3x - 5$
> $y = \frac{1}{2}x - 2$
> $y = 5 - 2x$
> $y = x + 2$
> $y = x - 5$
> $y = {}^-x$

2 Which two of these lines have the same slope?

   $y = 2x + 3$ $y = 3x + 2$ $y = 2$ $y = 2x - 3$

*3 Which number in these equations represents the gradient?

   a $y = 2x$ **b** $y = 3x$ **c** $y = 3x + 2$ **d** $y = \frac{1}{2}x - 4$

   e $y = \frac{1}{2}x$ **f** $y = \frac{1}{3}x + 2$ **g** $y = 4 - 2x$

In $y = mc + c$, c tells us where the straight line crosses the y-axis.
This is often called the **y-intercept**.

Examples $y = 4x - \mathbf{3}$ crosses the y-axis at $^-\mathbf{3}$.
                  $y = \mathbf{10} - 2x$ crosses the y-axis at $\mathbf{10}$.

c is often called the constant.

Exercise 3

1   Where do these lines cross the y-axis?

    a $y = 2x + 2$ **b** $y = 3x + 1$ **c** $y = 3x + 2$ **d** $y = \frac{1}{2}x - 4$

    e $y = \frac{1}{2}x$ **f** $y = \frac{1}{3}x + 2$ **g** $y = 4 - 2x$ **h** $y = 3 - 5x$

2   **a**   Which of these could be the equation of l_1?
      b   Which could be the equation of l_2?

          A $y = x + 2$
          B $y = x - 2$
          C $y = -x$
          D $y = 2x + 1$
          E $y = x$

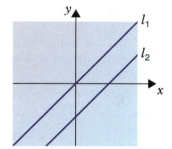

 *Practical

    You will need a graphical calculator.

    Create these displays using your graphical calculator.

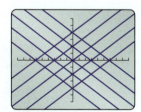   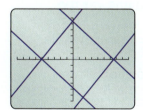 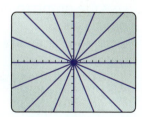

Reading and plotting real-life graphs

Sometimes we are given data in a table and when we plot it on a grid, it gives a straight line.
We can estimate values from the graph.

Algebra

Worked Example

The table shows the cost of hiring different numbers of videos.
A joining fee is charged.

a Plot these pairs of values on the grid.

Number of ideas	1	2	4	6	8
Cost (£)	5	6	8	10	12

b What is the cost of hiring three videos?
c How many videos can be hired for £9?
d How much is the joining fee?

Answer

a We plot the points (1, 5), (2, 6), (4, 8), (6, 10), (8, 12).
These are shown by the dots on the grid.
We then join the dots with a straight line.

b We draw the red dashed line up from 3.
Then draw across to the cost axis.
The cost is about £7.

c We draw the blue dashed line across from £9.
Then draw down to the number of videos axis.
The number of videos is about £44.

d £4 because the line crosses the cost axis at £4.

> Always give the graph a title and label its axes.

> Each two squares on the cost axis is £1.

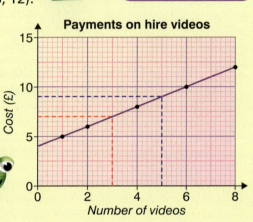

Payments on hire videos

Exercise 4

1 This conversion graph can be used to convert between miles and kilometres. Use the graph to help you answer the questions.
Give your answers to the nearest whole number.

  a Convert these miles to kilometres.
    i 20 miles
    ii 27 miles
    iii 33 miles
    iv 39 miles

  b Convert these kilometres to miles.
    i 30 km
    ii 50 km
    iii 35 km
    iv 56 km

  c Explain how to use the graph to convert 10 000 km to miles.

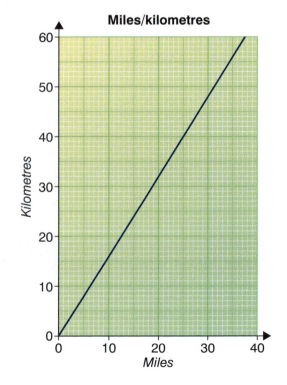

Miles/kilometres

2

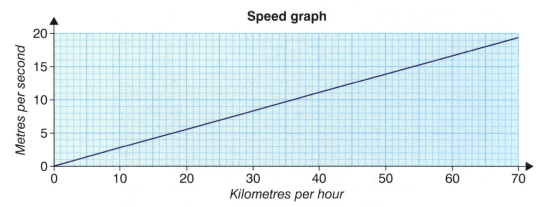

Speed graph

a   The speed limit in a town is 50 km/h.
    A cyclist is travelling at a speed of 15 m/sec.
    Is this cyclist going faster than the speed limit?
b   This table shows the top speeds (in metres per second)
    of some animals.
    Estimate their top speeds in kilometres per hour.
    Give the answers to the nearest km/h.

racehorse	18
antelope	15·5
deer	12·5

3   Kim went to a rock concert in London.
    She drove for 5 hours.
    The table shows the amount of petrol used by Kim's car.

Number of hours travelled	1	2	3	4	5
Amount of petrol used	5	10	15	20	25

a   Use a copy of this grid.
    On the grid, plot the pairs of points given in the table.
    Draw a straight line through the points.

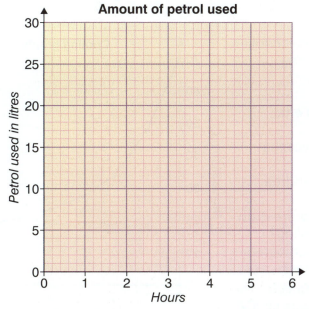

Amount of petrol used

b   Use your graph to find the amount of petrol Kim had used after 2·5 hours.
c   How long does it take for Kim to use 18 litres of petrol?

4 Fudge bars can be cut into different lengths.
This graph shows the mass of different lengths of two types of fudge, A and B.

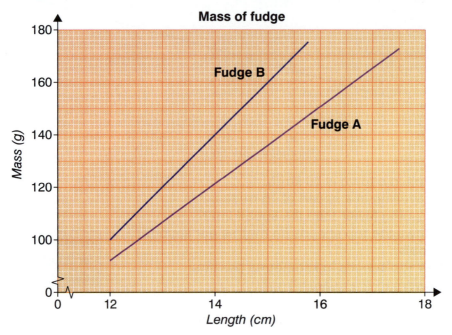

Mass of fudge

a What is the mass of a 14 cm length of fudge A?
b Estimate the length of fudge B if it has a mass of 120 g.
c Estimate the mass of these lengths of fudge B.
   i 14 cm   **ii** 15 cm   **iii** 13·5 cm
d Fudge A and Fudge B are both 13 cm long.
Use the graphs to estimate the difference in their masses.
e Fudge A and Fudge B both weigh 150 g.
Use the graphs to estimate the difference in their lengths.
∗f Fudge A costs £1·25 per 100 g. Ben buys 16 cm of Fudge A.
How much does this cost?

5 Andrea heated a liquid, then let it cool.
While it was cooling, she took the
temperature every minute.
Andrea drew this graph.
   a What was the temperature after
      1 minute?
   b After how many minutes was the
      temperature 24 °C?
   c To what temperature did Andrea
      heat the liquid?
   d How much did the temperature drop
      in the first 5 minutes of cooling?
   e What do you estimate the temperature
      to have been after $2\frac{1}{2}$ minutes?
   ∗f After how many minutes do you
      estimate the temperature was 50 °C?
   ∗g What does the shape of this graph tell
      you about the cooling of the liquid?

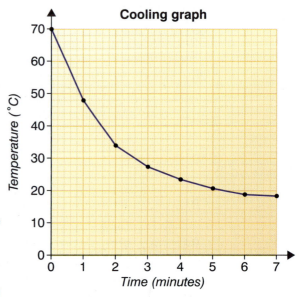

Cooling graph

T

6 The graph shows the average heights of 1- to 4-year-olds.

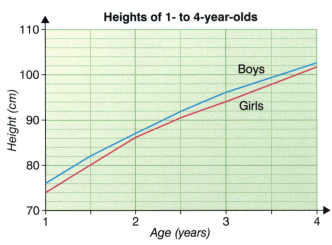

a Estimate the average height of $2\frac{1}{2}$-year-old girls.

b Estimate the average height of 3-year-old boys.

c Olivia is of average height.
Her height is 98 cm.
Use the graph to find out about how old Olivia is.

d Use a copy of this table.
Use the graph to fill it in.

Age of girl (in years)	Height in cm at start of year (approximate)	Height in cm at end of year (approximate)	Approximate growth in cm
1–2	74	86	12
2–3	86		
3–4			

e About how much taller, on average, are
$2\frac{1}{2}$-year-old boys than
$2\frac{1}{2}$-year-old girls?

f What does the shape of this graph tell you about growth of 1–4-year-olds?

*7 $d = 4 + 0{\cdot}5s$ gives the stopping distance d, in metres, for a car travelling at a speed s, in km/h, on a road.

a Copy and complete this table.

s	10	50	100
d		29	

Use the formula to work out the other values for d.

b Copy and complete these coordinate pairs (10, ___), (50, 29), (100, ___).

c Draw the graph of stopping distance against speed.
Have s on the horizontal axis.

d Use your graph to estimate the distance a car, travelling at 70 km/h, takes to stop.

e It took 40 metres for Jenny to stop from the time she put her foot on the brake.
Use your graph to estimate the speed at which Jenny was travelling.

You will need to choose suitable scales for the axes.

Algebra

T

*8 You pay £2·60 each time you go to an aerobics class.

Number of classes	0	10	20	30
Total cost (£)	0	26		

  a Copy and complete this table.
  b Use a copy of this grid.
     Show the information in the table on the grid.
     Join the points with a straight line.
  c A different way of paying is to pay a yearly fee of £24.
     Then you pay £1·60 for each class.
     Copy and complete this table.

Number of classes	0	10	20	30
Total cost (£)	24	40		

  d Show the information in this table on the same graph.
     Join these points with a straight line.
  e For how many classes does the graph show that the cost is the same for both ways of paying?
  f Janita wants to go to aerobics classes once every fortnight for a year.
     Which way is cheaper for her to pay?
     Use the graph to estimate by how much.
     Show how you did this.

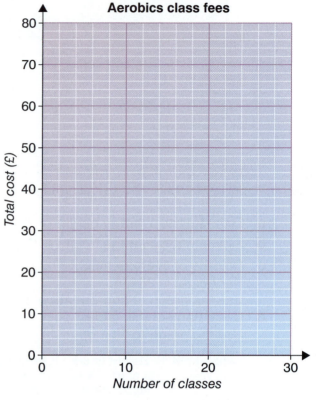

Aerobics class fees

Discussion

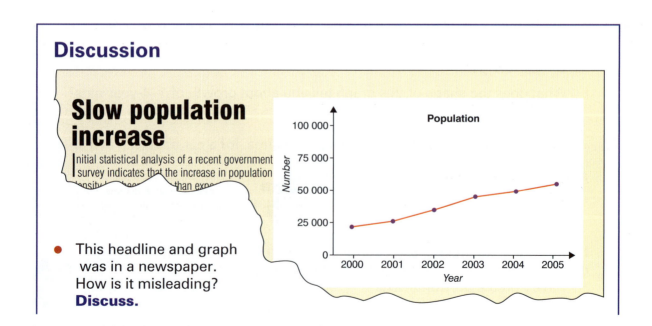

Slow population increase

Initial statistical analysis of a recent government survey indicates that the increase in population ~~density~~ ~~than exp~~

- This headline and graph was in a newspaper. How is it misleading?
  Discuss.

Population

● Explain why this graph is misleading.

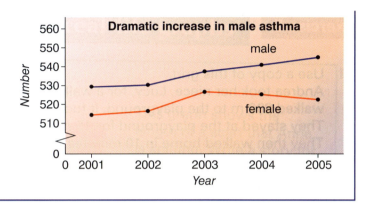

Distance/time graphs

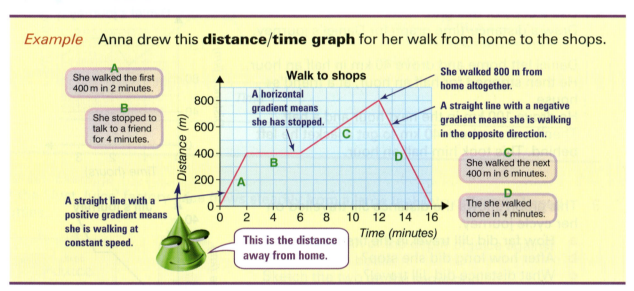

Example   Anna drew this **distance/time graph** for her walk from home to the shops.

A
She walked the first 400 m in 2 minutes.

B
She stopped to talk to a friend for 4 minutes.

A straight line with a positive gradient means she is walking at constant speed.

A horizontal gradient means she has stopped.

This is the distance away from home.

She walked 800 m from home altogether.

A straight line with a negative gradient means she is walking in the opposite direction.

C
She walked the next 400 m in 6 minutes.

D
The she walked home in 4 minutes.

Discussion

Discuss how to draw the distance/time graph for Peter's journey.

Peter left home and drove the 5 km to his friend's house in 5 minutes. He stayed at his friend's house for 8 minutes.
He then drove the 5 km home in 8 minutes.

Going home is like going backwards.

You may like to use this grid.

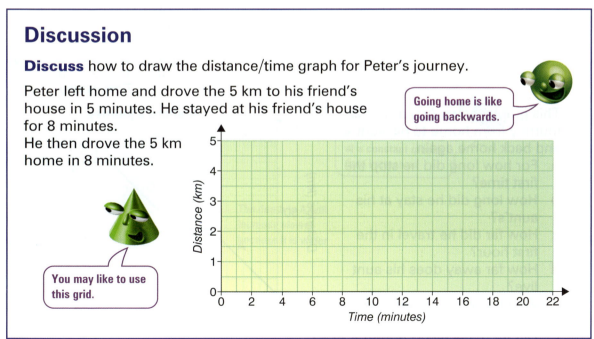

Algebra

T

8 Use a copy of this grid.
Mahmud went on a training run.
He ran 12 km in the first hour.
He rested for $\frac{1}{2}$ an hour.
He then jogged the 12 km home in $1\frac{1}{2}$ hours.

Draw a distance/time graph for Mahmud's run.

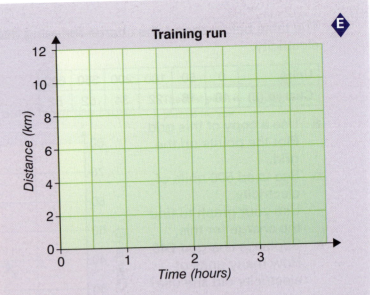

9 Richard sketched this graph to show the depth of snow on a ski field each night.
  a One day it snowed heavily. Which day do you think this was?
  b On sunny, warm days some snow melted. Which days do you think were sunny and warm?

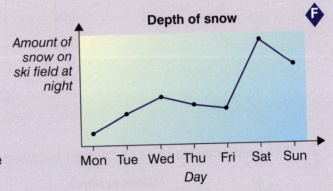

10 Water is flowing steadily into two bowls from Tap A and Tap B.
Tap A has a bigger opening than Tap B.

Which of the lines, l_1 or l_2, shows A filling and which shows B filling?

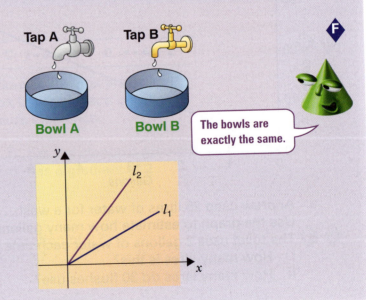

The bowls are exactly the same.

Shape, Space and Measures Support

Lines

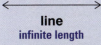
line
infinite length

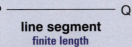
line segment
finite length

Two straight lines must **intersect** or be **parallel**.

Parallel lines never meet.
We show parallel lines with arrows.
SR is parallel to UT.

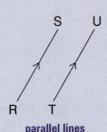
parallel lines

We show lines are **perpendicular** using

the symbol ⌐ .

AB is perpendicular to CD.

Parallel and perpendicular lines can be drawn
using a **ruler and set square**.

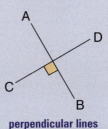
perpendicular lines

Practice Question 1

Angles

Angle is a measure of turn.
We measure angles in degrees.

One complete turn or 360°

Right angle or 90°

Straight angle or 180°

Acute angle less than 90°

Obtuse angle between 90° and 180°

Reflex angle between 180° and 360°

We **name angles** using the letter at the vertex **or**
using three letters, the middle letter being the vertex.

Example   This angle is named as angle T or ∠T
or ∠STR or ∠RTS

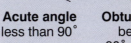

If there is more than one
angle at a vertex, always
use three letters.

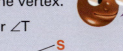

235

Measuring and drawing angles

We use a **protractor** or **angle measurer** to measure and draw angles.

It is best to measure and draw reflex angles using a 360° protractor.

To measure a reflex angle with a 180° protractor measure the acute or obtuse angle and then subtract it from 360°.

Example   Obtuse angle is 130°
Reflex angle = 360° − 130°
= 230°

measure this as 130°

When measuring angles always **estimate the size** of the angle first.
Do this by comparing the acute or obtuse angle to a right angle.
Angle A is about $\frac{1}{3}$ of a right angle.
It is about 30°.

A

The corner of a page is a right angle.

Practice Questions 3, 4, 7, 16

Calculating angles

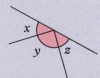

a = b
Vertically opposite angles are equal

x + y + z = **180°**
Angles on a straight line add to 180°

c + d + e = **360°**
Angles at a point add to 360°

Example   x + 90° + 30° + 85° = 360°
x + 205° = 360°
x = 360° − 205°
= **155°**

30°

85°

x

Angles in a triangle add to 180°.
Example   m + 95° + 35° = 180°
m + 130° = 180°
m = 180° − 130°
= **50°**

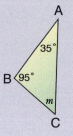

A

35°

B 95°

m

C

This triangle is named ABC.

Practice Questions 8, 23

2-D shapes

A polygon is a closed 2-D shape.

a polygon

2-D is short for two-dimensional.

Properties of triangles

A **triangle** is a 3-sided polygon.

right-angled
one angle is a
right angle

isosceles
2 equal sides
2 base angles equal

equilateral
3 equal sides
3 equal angles

scalene
no 2 sides are equal
no 2 angles are equal

Properties of quadrilaterals

A **quadrilateral** is a 4-sided polygon.

These are the **special quadrilaterals**.

square
4 equal sides
4 right angles

rectangle
2 pairs of opposite
sides equal
4 right angles

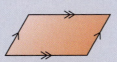

parallelogram
opposite sides equal
and parallel

rhombus
a parallelogram with
4 equal sides
opposite angles equal

trapezium
1 pair of opposite
sides parallel

kite
2 pairs of adjacent
sides equal
1 pair of equal angles

arrowhead or **delta**
2 pairs of adjacent
sides equal
1 reflex angle
1 pair of equal angles

Adjacent sides are
next to each other.

Polygons

A 3-sided polygon is a triangle.
A 5-sided polygon is a pentagon.
A 7-sided polygon is a heptagon.

A 4-sided polygon is a quadrilateral.
A 6-sided polygon is a hexagon.
An 8-sided polygon is an octagon.

A **regular polygon** has all its sides equal and all its angles equal.

Shape, Space and Measures

Tessellations

If shapes fit together leaving no gaps we say they **tessellate**.

Example will tessellate to give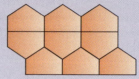

Congruent shapes are exactly the same shape and size.

Example and are congruent.

Practice Questions 9, 11, 12, 18, 32, 33

Constructing triangles

We can **construct triangles** using a ruler and protractor.

Example   To construct triangle JKL,
draw JL 3 cm long
draw an angle of 80° at L
draw LK 3·5 cm long
join J to K.

J — 3 cm, 3·5 cm, 80° — K, L

Practice Question 40

3-D shapes

3-D shapes can be drawn on **isometric** (triangle dotty) paper.

3-D stands for three-dimensional.

We often construct 3-D shapes by first drawing a net.

A 2-D shape that can be folded to make a 3-D shape is called a **net**.

Example   The net below folds to make this cuboid.

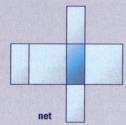

net

base
cuboid

Practice Questions 10, 17, 31, 41

Coordinates

We use **coordinates** to give the position of a point on a grid.
The coordinates of A are (⁻2, ⁻3).
We always give the x-coordinate first.
The coordinate of the **origin** are (0, 0).

Practice Questions 13, 21

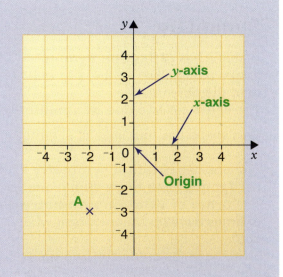

Transformations

Reflection

PQR has been reflected in the mirror line to give the image P′Q′R′.
Each image point is the same perpendicular distance behind the mirror as the original point is in front.
Q is on the mirror line so it does not move.

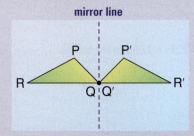

Rotation

To rotate a shape we need to know the **centre of rotation** and the **angle of rotation**.

Example   PQRS has been rotated 90° to P′Q′R′S′.

When only an angle of rotation is given we always rotate anticlockwise.

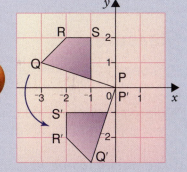

Translation

To **translate** a shape we slide it without turning.

Example   The blue shape has been **translated** 3 units to the right and 1 down to get the green shape.

The **inverse translation** is 3 units left and 1 unit up.

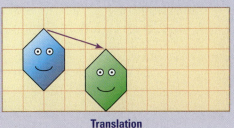

Translation

Practice Questions 22, 24, 25, 26, 36

Symmetry

A shape has **reflection symmetry** if it can be folded along a line so that one half fits exactly onto the other half.
The line is called a **line of symmetry**.

This shape has four lines of symmetry.

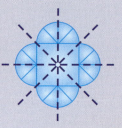

A shape has **rotation symmetry** if it fits onto itself **more than once** during one full turn.
The **order of rotation symmetry** is the number of times a shape fits onto itself when it is rotated 360°.
This shape fits onto itself four times in a 360° turn.
It has order of rotation symmetry 4.

Practice Questions 5, 6, 14, 30, 44

Enlargement

The blue L has been **enlarged** to give the green L.
Each length on the green L is two times longer than on the blue L.
The **scale factor** of the enlargement is 2.

Practice Question 20

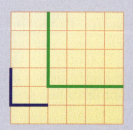

Measures

You need to know these **metric conversions**.

length
1 kilometre (km) = 1000 metres
1 metre (m) = 1000 millimetres (mm)
1 metre (m) = 100 centimetres (cm)
1 centimetre (cm) = 10 millimetres (mm)

capacity
1 litre (ℓ) = 1000 millilitres (mℓ)
1 litre (ℓ) = 100 centilitres (cℓ)
1 centilitre (cℓ) = 10 millilitres (mℓ)

mass
1 kilogram (kg) = 1000 grams (g)

Examples   0·45 kg = 0·45 × 1000 g
                  = 450 g

24 cm = (24 ÷ 100) m
         = 0·24 m

3200 mℓ = (3200 ÷ 1000) ℓ
              = 3·2 ℓ

time
1 minute = 60 seconds
1 hour = 60 minutes
1 day = 24 hours
1 week = 7 days
1 year = 365 days (a leap year has 366 days)

You need to know these **metric and imperial equivalents**.

length	mass	capacity
8 km ≈ 5 miles	1 kg ≈ 2·2 lb	1 gallon = 4·5 ℓ
		1 pint is just over $\frac{1}{2}\ell$

≈ means is approximately equal to.

When **reading scales** you need to work out the value of each small division.

Example   There are five small gaps between 0 and 1.
Each small gap is $1 \div 5 = \frac{1}{5}$ or 0·2 kg.
The pointer is at 0·8 kg.

When we **estimate** a measurement it is a good idea to give a **range** for the estimate.

Example   12 cm < length of calculator < 20 cm

Practice Questions 2, 15, 19, 29, 34, 35, 37, 43

Perimeter, area and surface area

The distance right around the outside of a shape is called the **perimeter**.
Perimeter is measured in mm, cm, m or km.

The amount of surface a shape covers is called the **area**.
Area is measured in mm², cm², m², km².

Perimeter of a rectangle = 2 × length + 2 × width or 2 × (length + width)
  = **2l + 2w** or **2(l + w)**

Area of a rectangle = length × width
  = **lw**

Example   Perimeter of rectangle = 2 × (8 + 3)
  = 22 cm

Area of rectangle = 8 × 3
  = 24 cm²

Area of a right-angled triangle = $\frac{1}{2}$ area of rectangle
  = $\frac{1}{2}$ × base length × height
  = **$\frac{1}{2}bh$**

Example   Area of triangle ABC = $\frac{1}{2}$ × 3 × 4
  = $\frac{1}{2}$ × 12
  = **6 m²**

Shape, Space and Measures

The **surface area** of a solid is the total area of all of the faces.

The **surface area of a cuboid**
= 2(length × width) + 2(length × height) + 2(height × width)
= $2lw + 2lh + 2hw$

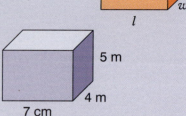

Example   Surface area = $2lw + 2lh + 2hw$
= 2(7 × 4) + 2(7 × 5) + 2(5 × 4)
= 56 + 70 + 40
= **166 cm²**

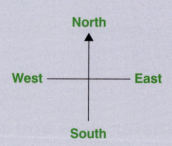

Practice Questions 27, 28, 39, 42

Compass directions

This diagram shows the directions on a compass.

Practice Question 38

```
            North
              ↑
              |
West ---------+--------- East
              |
              |
            South
```

Practice Questions

1 a Estimate the lengths of AB and GH to the nearest centimetre.
Then measure them accurately.

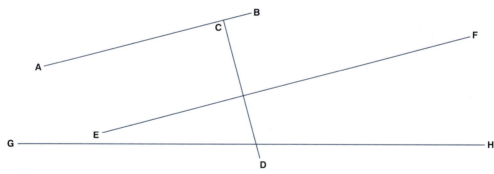

b Which of the lines are parallel?
c Name two lines that are perpendicular.

2 a Paul's birthday is on 1 March.
Ray's is on 3 April.
How many days are **between** their birthdays?
b Oscar's birthday is on 15 February.
His sister's is on 15 March.
How many days are there **between** their birthdays in 2008?

Don't count the birthdays.

3 Write down the colour of the
   a acute   **b** reflex
   c obtuse   **d** right angles.

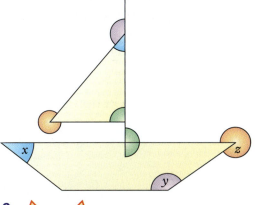

4 In the diagram in question **3**, measure the size of angles x, y and z.

5 Do these shapes have rotational symmetry?
   a    **b**   **c**

6 Use a copy of these shapes?
   Draw on all the lines of symmetry.
   a 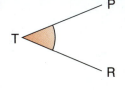   **b** 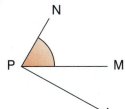   **c**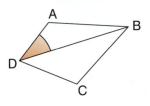

7 Name the shaded angles.
   a   P   **b**   N   **c**   A
       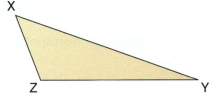

8 Name the triangles.
   a   S   **b**   X

9 Use a copy of this.
   Draw some more tiles to show how they tessellate.

243

10 These three cubes are put together in a line with the red cube in the middle.

   a How many blue faces are showing?
   b How many red faces are showing?

11 Name all the congruent shapes in this diagram.

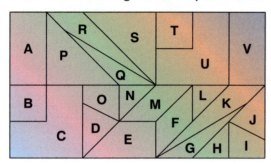

12

> **A** trapezium   **B** parallelogram   **C** rhombus   **D** arrowhead (delta)
> **E** kite   **F** equilateral triangle   **G** scalene triangle
> **H** isosceles triangle   **I** rectangle   **J** right-angled triangle

Choose the best name from the box for each of these.

a 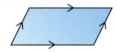   **b**   **c**

d 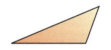   **e**    **f**

g   **h**    **i**

13 Write down the letters at the following coordinates. What does the message say?

$(5, 6)$ $(^-2, 1)$ $(^-4, ^-3)$ $(4, ^-2)$
$(^-3, ^-2)$ $(6, 2)$
$(1, ^-1)$ $(^-4, 3)$ $(4, ^-2)$
$(1, ^-1)$ $(0, 5)$
$(2, ^-3)$ $(^-1, 1)$ $(4, ^-2)$ $(3, ^-4)$ $(4, ^-2)$
$(3, 3)$ $(1, ^-1)$ $(^-3, ^-2)$ $(4, ^-3)$ $(3, ^-4)$ $(4, ^-2)$ $(^-2, ^-6)$ $(0, 5)$

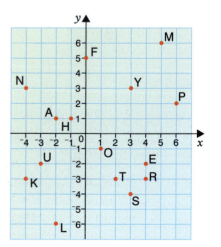

14 What is the order of rotation symmetry of each of these?

a   b   c   d

15 Find the missing numbers.
a   4 cm = ____ mm   b   3 km = ____ m   c   5 ℓ = ____ mℓ
d   6000 g = ____ kg   e   800 cm = ____ m   f   4200 m = ____ km
g   8·2 kg = ____ g   h   8400 g = ____ kg   i   300 mℓ = ____ ℓ
j   52 mm = ____ cm   k   600 cℓ = ____ ℓ   l   4·2 ℓ = ____ cℓ

16 Draw these angles.
a   80°   b   165°   c   290°

T

17 Use a copy of this.
Draw these shapes on the isometric dotty paper.

a 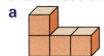   b   c

18 Write true or false for these.
a   An arrowhead has one line of symmetry.
b   A rectangle has 4 lines of symmetry.
c   A parallelogram has opposite sides equal and parallel.
d   An isosceles triangle has 2 lines of symmetry.
e   A trapezium may have 2 equal sides and 2 pairs of equal angles.

T

19

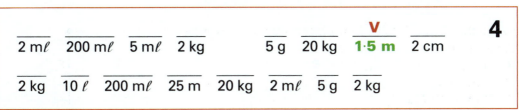

4

____	____	____	____		____	____	V ____	____
2 mℓ	200 mℓ	5 mℓ	2 kg		5 g	20 kg	**1·5 m**	2 cm

____	____	____	____	____	____	____	____
2 kg	10 ℓ	200 mℓ	25 m	20 kg	2 mℓ	5 g	2 kg

Use a copy of this box.
Put the letter beside each measurement above its estimate in the box.
V length of a table ≈ **1·5 m**   **H** mass of a pencil
S mass of a small puppy   **O** capacity of a glass
A mass of a full suitcase   **T** capacity of a bucket
M length of a driveway   **W** capacity of a teaspoon
E length of a pencil sharpener   **C** volume of an eye dropper

Shape, Space and Measures

20 What is the scale factor of each of these enlargements?

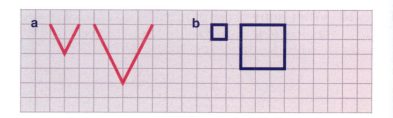

T **21** Use a copy of this grid.
   a Plot the points (1, 0), (1, 3), (4, 3), (4, 0).
      Join the points in order.
   b How many lines of symmetry does the shape have?

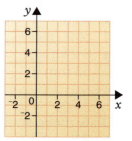

   c The points ($^{-}$1, 4) and (5, 4) are two corners of a rectangle.
      What might the coordinates of the other two corners be?

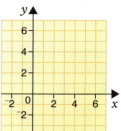

T **22** Use a copy of these.
Draw the reflections in the dotted mirror lines.
   a    **b**

23 Find the angles marked with letters.

   a   **b** 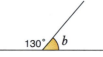   **c**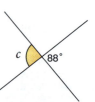

   d 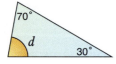   **e** 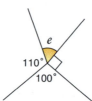   **f**

   g 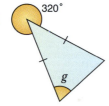   **h** 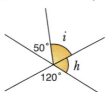   ***i**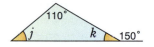

T

24 Use a copy of this.
Draw the image after the given translation.

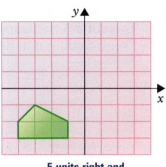

5 units right and
3 units up

*25 What translation is the inverse of the translation
given in question **24**?

26 Which of the diagrams below shows this shape after
 a a rotation of 270° about (0, 0)
 b a rotation of 180° about (0, 0)
 c a translation of 2 units left and 3 units down?

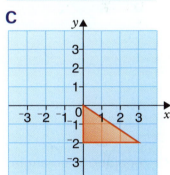

A

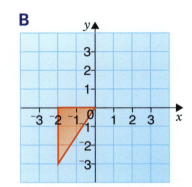

B

C

27 Find the perimeter of each of these.
 a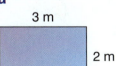
 3 m
 2 m

 b
 8 cm
 6 cm

 c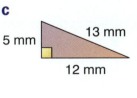
 5 mm   13 mm
 12 mm

 d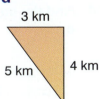
 3 km
 5 km   4 km

28 Find the area of each of the shapes in question **27**.

29 What goes in the gaps?
 a 70 mm + 20 cm = ___ cm   **b** 1·5 ℓ + 450 mℓ = ___ mℓ
 c 250 g + 1·6 kg = ___ kg

Shape, Space and Measures

30 I have a square made of paper.
The square measures 24 cm by 24 cm.
I keep folding it in half until I have a square that is 12 cm by 12 cm.

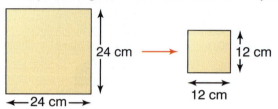

How many times do I fold it?

31 A   B    C

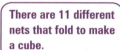

There are 11 different nets that fold to make a cube.

  a Which of these nets will fold to make a cube?
  b Draw another two nets that fold to make a cube.

32 Find the angles named with letters.
Use the properties of shapes.

  a 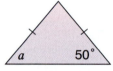   **b** 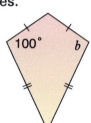   **c**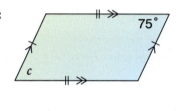

33 Imagine a rectangle.
Join the middle of one side to the middle of the next side.
Do this all the way round.
What shape is made by the new lines?

34 Find the measurements given by pointers **A**, **B** and **C**.

  a 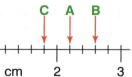   **b** 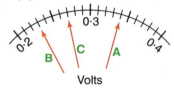   **c**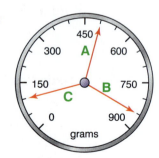

35 Choose the best range for the length of a bike.
  A $0{\cdot}5\ \text{m} \leqslant \text{length of a bike} \leqslant 1\ \text{m}$
  B $1\ \text{m} \leqslant \text{length of a bike} \leqslant 2\ \text{m}$
  C $1{\cdot}5\ \text{m} \leqslant \text{length of a bike} \leqslant 2{\cdot}5\ \text{m}$

36 Use a copy of these shapes. The centre of rotation and angle of rotation are given. Draw the image shapes.

a

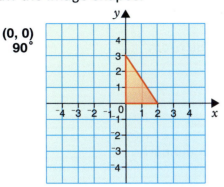

(0, 0)
90°

b

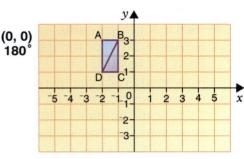

(0, 0)
180°

37 a Give the rough metric equivalent for each of these.
    i   5 miles   **ii** 2·2 lb   **iii** 1 gallon   **iv** 1 pint
  b Give the rough imperial equivalent for each of these.
    i   2 kg   **ii** 4·5 ℓ **iii** 16 km   **iv** 1 ℓ

38 Rebecca was facing North. She turned
    $\frac{1}{4}$ turn clockwise then
    $\frac{1}{2}$ turn anticlockwise.
What direction is she facing now?

39 Tom has a triangular shaped piece of stained glass.
  a What is its perimeter?
  b What is its area?

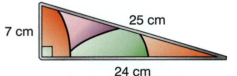

7 cm   25 cm   24 cm

40 Use compasses and a ruler to accurately construct this triangle.
On *your* drawing, measure the size of ∠RPQ.
Give your answer to the nearest degree.

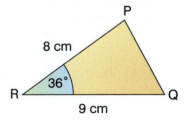

P
8 cm
36°
R   9 cm   Q

41 Mary made this box for a gift.

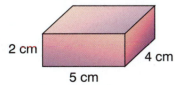

2 cm   4 cm   5 cm

Use a copy of this.
Complete the net for the box.

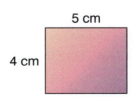

5 cm
4 cm

Shape, Space and Measures

42 Find the surface area of the cuboid in question **41**.

43 Michael is having a party.
  a He needs 30 m of red paper to decorate the room.
   Each roll has 500 cm.
   How many rolls will he need?
  b He decides he needs 50 ℓ of fizzy drink.
   Each bottle contains 250 cℓ.
   How many bottles will he need?
  c His best friend travels 50 miles to come to the party.
   About how many kilometres is this?
  d He orders 10 kg of sausage rolls and savouries.
   About how many pounds is this?

44 This is a right-angled isosceles triangle.
   It is folded along the line of symmetry.
   What angles will be in the folded shape?
   Explain.

You need to know

✓ lines and angles   page 235

Key vocabulary

alternate angles, complementary angles, corresponding angles, equilateral triangle, exterior angles, interior angles, intersecting lines, isosceles triangle, parallel lines, supplementary angles

Snap it!

Photos often contain interesting lines and angles. Find some photos. You could look at home, in magazines, newspapers, on the Internet or on a CD-ROM.
Make a poster or collage of your photos.

Angles made with intersecting and parallel lines

Practical

T

You will need a copy of this and a dynamic geometry software package **or** some sheets of acetate.

Draw three lines that **intersect** at the same point.

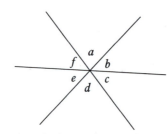

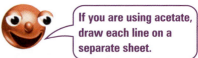

If you are using acetate, draw each line on a separate sheet.

Name the angles as shown.

Which angles are equal?

____ , ____ and ____ ;   ____ , ____ and ____ ;   ____ , ____ and ____

Which angles add to 180°?

____ , ____ and ____ ;   ____ , ____ and ____ ;   ____ , ____ and ____ ;

____ , ____ and ____ ;   ____ , ____ and ____ ;   ____ , ____ and ____

Rotate one of the lines.
Make sure all the lines still intersect at one point.
Which angles have stayed equal to each other?
Which angles add to 180°? Explain why.

Intersecting lines

Remember

$a = 80°$
Vertically opposite angles are equal.

$a + b + c + 90° = 360°$
Angles at a point add to 360°.

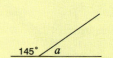

$a + 145° = 180°$
Angles on a straight line add to 180°.

Exercise 1

T

1 Use a copy of this.

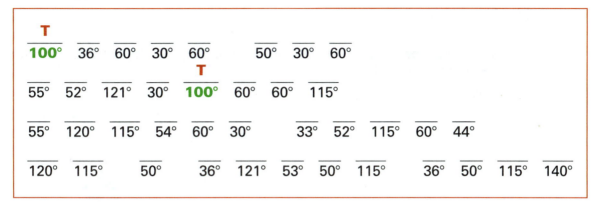

T								
100°	36°	60°	30°	60°		50°	30°	60°

T

55° 52° 121° 30° **100°** 60° 60° 115°

55° 120° 115° 54° 60° 30°   33° 52° 115° 60° 44°

120° 115°   50°   36° 121° 53° 50° 115°   36° 50° 115° 140°

Calculate the shaded angles.
Put the letter that is beside each diagram above its answer in the box.

T 100° 80°

G 54°

I 80° 160°

S 136°

A 45° 85°

R 50° 80°

D 47° 83°

U 121° 39°

O 52° 38°

H 88° 92° 52°

B 57°

N 155°

E 31° 65° 24°

M 52° 85° 170°

F 93° 32° 118°

*2 AEC is a straight line.
Angle BED is a right angle.
$x + y + z = 242°$
Find the size of angles x, y and z.

Hint: $x + y + z = 242°$
$y + z = 90°$

so what does x equal?

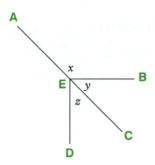

Remember
Parallel lines are always the same distance apart.
We show them with arrows.

Practical

You will need some sheets of acetate.

Draw two intersecting lines.
Draw a third line which is parallel to one of them.

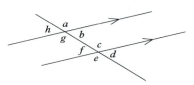

If you use acetate,
draw the third line
on a separate sheet.

Name the angles with letters.

Which angles are equal?

Move the third line you drew so that it always stays parallel.

What happens to the sizes of a, b, c, d, e, f, g and h?
Which angles stay equal to each other?
Explain why.

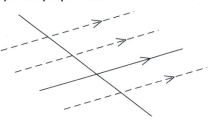

Angles made with parallel lines

In each of these diagrams

angle a = angle b.

Think of the letter **F** .

Angle a and angle b are called **corresponding angles**.

In each of these diagrams

angle x = angle y.

Think of the letter **Z** .

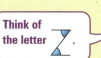

Angle x and angle y are called **alternate angles**.

Worked Example
Find the value of the angles marked with letters.
Give reasons.

a 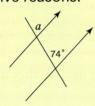   **b**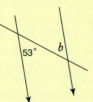

Answer
a $a = $ **74°** (corresponding angles)
b $b = $ **53°** (alternate angles)

Exercise 2

1 i Name the angle that is alternate to the shaded angle.
ii Name the angle that is corresponding to the shaded angle.

a   **b** 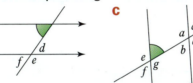   **c** 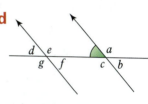   **d**

2 Name all the pairs of corresponding angles in each of the following diagrams.
The diagrams show some trellis and a roof.

a 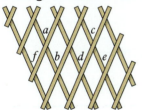   **b**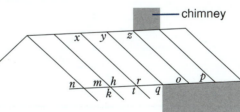
chimney

3 Find the size of each of the shaded angles. Give a reason.

a 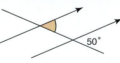   **b** 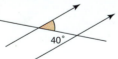   **c**

d 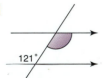   **e**

4 Find the size of each of the shaded angles. Give a reason.

a 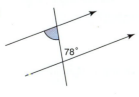   **b** 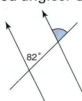   **c**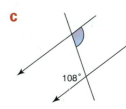

Shape, Space and Measures

5 Name all the angles that are equal to each of these.
  a *a*
  b *b*
  c *h*
  d *d*

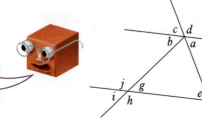

You may need to know about other angle properties to do this. See page 252.

6 For each of the following diagrams, write down the letters of all the angles that are equal to angle *a*.

a 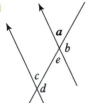 **b** 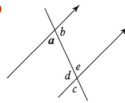 **c**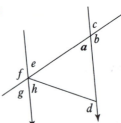

A pair of angles that add to 90° are called **complementary angles**.

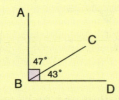

47° and 43° are complementary.

A pair of angles that add to 180° are called **supplementary angles**.

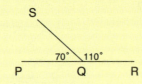

70° and 110° are supplementary.

Exercise 3

1 What goes in the gaps?
  40° and ____ are complementary angles.
  120° and ____ are supplementary angles.

2 Look at this diagram.
  e and *d* are supplementary because they add to 180°.
  Name two other pairs of supplementary angles.

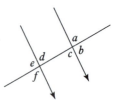

3 a Which of these pairs of angles are complementary angles?
  i 53° and 37° **ii** 27° and 63° **iii** 114° and 66°
  iv 46° and 87° **v** 57° and 123°
  b Which of the pairs of angles given in **a** are supplementary angles?

Angles in triangles

Practical

You will need a dynamic geometry software package.

Ask your teacher for a copy of the **Angles in Triangles** ICT worksheet.

Discussion

● Robina **showed** that the angles of a triangle add to 180° like this.

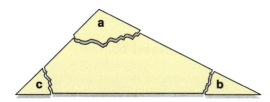

She ripped off the corners of the triangle.
She put them along a straight line.

Does this work for all triangles? **Discuss.**

● Robina wanted to **prove** that the three angles of a triangle add to 180°.
She wrote this:

$x = a$	(alternate angles on parallel lines are equal)	
$z = b$	(alternate angles on parallel lines are equal)	
$x + c + z = 180°$	(angles on a straight line add to 180°)	
$a + c + b = 180°$	(substituting a for x and b for z)	

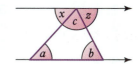

So the 3 angles in a triangle add to 180°.

Is this true for all triangles? **Discuss.**

Remember

The angles a, b and c are called **interior angles** of a triangle. They are the angles inside the triangle.

The sum of the interior angles of a triangle is 180°.

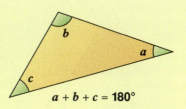

$a + b + c = 180°$

Example   The third angle in the triangle = 360° − 310°
                                     = 50°   (angles at a point add to 360°)

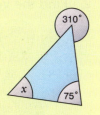

$x = 180° − 50° − 75°$ (angles in a triangle add to 180°)
$x = \mathbf{55°}$

Angle a is called an **exterior** angle. It is outside the triangle.

The exterior angle of a triangle is equal to the sum of the two opposite interior angles.

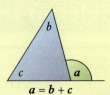

$$a = b + c$$

Examples   **a** $y = 57° + 59°$
                       $= \mathbf{116°}$

> The two opposite angles are the two angles that are not next to the exterior angle.

b This is an isosceles triangle.
      The other base angle is equal to y.
      $2y = 146°$ (exterior angle = sum of two opposite interior angles)
      $y = \mathbf{73°}$

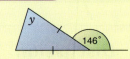

Exercise 4   **Only use a calculator if you need to.**

T

1 Use a copy of this crossnumber.
      Calculate the value of each unknown.
      Using these values, fill in the crossnumber.

Across	**Down**
1. a	**1.** f
2. b	**3.** g
5. c	**4.** h
8. d	**6.** i
9. e	**7.** j

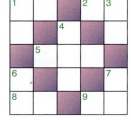

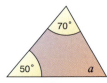 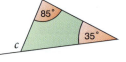 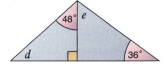

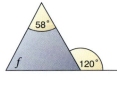 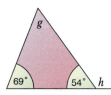 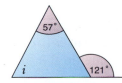 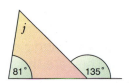

2 Find the value of x.

   a

   b
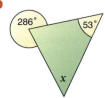

   c

3 Find the value of x.

a   b   c  d   e

4 Find the value of x and y. Give reasons.

a 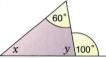 b   c   d

5 Use a copy of this diagram.
  a The angle between Caley Drive and Raven Road is 105°.
  Which of the two angles between Caley Drive and
  Raven Road will this be?
  Write this angle on your diagram.
  The angle between Caley Drive and Gresham Avenue is 60°.
  Which angle is this?
  Write this angle on your diagram.
  b Calculate the angles at the three corners of Bell Park.

6 Find the values of x and y.

a   b   c 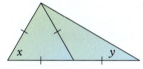 d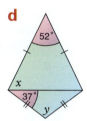

*7 Explain why a triangle can never have a reflex angle within it.

 Puzzle

  1 Draw two equilateral triangles by drawing just five lines.
  2 Draw two equilateral triangles by drawing just four lines.

Using geometrical reasoning to find angles

Sometimes we use **geometrical reasoning**.
It helps to do these.

  1 Write down what you know.
  2 Work out any angles that it looks like you might need.
  3 Write down the steps needed to find each angle.
  Always give reasons.

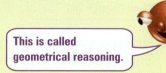

This is called
geometrical reasoning.

Shape, Space and Measures

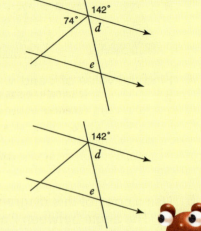

Worked Example

Prove that $e = 38°$.

Show your working clearly and give reasons.

Answer

e and d are alternate angles.

We can find d using angles on a straight line.

$d + 142° = 180°$ (angles on a straight line add to 180°)
$d = 180° - 142°$
$= 38°$

$e = \mathbf{38°}$ (alternate angles on parallel lines are equal)

There is often more than one way to find the answer.

> When you are asked to prove something you must show each step one by one and give reasons.

Exercise 5   **Only use a calculator if you need to.**

1 For each of these, prove that x has the value shown.
Show your working clearly and give reasons.
The first two have been started.

a
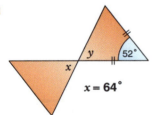
$x = 64°$

b
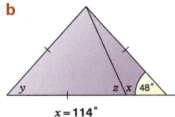
$x = 114°$

c
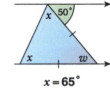
$x = 65°$

a $y + y + 52° = 180°$ (angles in a triangle add to 180°)
$2y = 180° - 52°$
$= \underline{\quad}$
$y = \underline{\quad}$

b $y = 48°$ (base angle in isosceles triangle)
$z = \underline{\quad}$ (angles in a triangle add to 180°)
$x = \underline{\quad}$

d

$x = 25°$

> You'll need to use some of the angle properties given on page 252.

e

$x = 108°$

f

$x = 29°$

2 Find the size of the angles marked a, b and c.
Show your working clearly and give reasons.
It has been started for you.

$a = \underline{\quad}$ (alternate angles on parallel lines are equal)
$b + 40° = \underline{\quad}$ (vertically opposite angles)

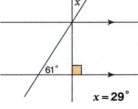 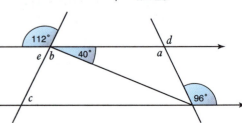

3 Vanessa drew this triangle diagram.
The lines BF and CE are parallel.
AE and BD are parallel and so are DF and CA.
Find the size of x and y.
Show your reasoning clearly.
It has been started for you.

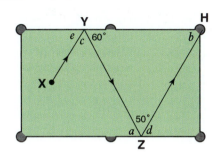

$a =$ _____ (alternate angles on parallel lines CA and DF are equal)

$x =$ _____ (alternate angles on parallel lines BD and FE are equal)

$y =$ _____ (corresponding angles on parallel lines _____ and _____ are equal)

4 The arrows show the path of a billiard ball from X to H.
XY and ZH are parallel paths.
Find the values of a, b, c, d and e.
Show your working and give reasons.

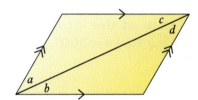

*5 Maria used alternate angles to prove that the opposite angles of a parallelogram are equal.
Show how she might have done this.

*6 Calculate the value of x.
You will need to write and solve an equation.
Show your working clearly and give reasons.
The first two have been started.

a

$x + 20°$ /68°

b

$x + 20°$ 68°

$x + 20° = 68°$

c

168° / $x + 20°$

$x + 50°$

$x + 20° + 68° = 180°$ (angles on a straight line add to 180°)

$x +$ _____ $= 180°$

$x = 180° -$ _____

$=$ _____

There is more about solving equations on page 148.

*7

This tile is made from eight identical triangles.
a Find the size of angle x. Show your reasoning clearly.
b Find the size of angle y. Show your reasoning clearly.

Angles in quadrilaterals

⚪⚪⚪⚪⚪⚪⚪⚪⚪⚪⚪⚪⚪⚪⚪⚪⚪⚪

Investigation

Finding the angle sum

Martina drew this quadrilateral.
She divided it into two triangles as shown.

How could she use this diagram to prove the sum of
the interior angles of any quadrilateral is 360°? **Investigate.**

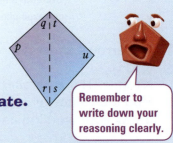

Remember to
write down your
reasoning clearly.

The sum of the interior angles of any quadrilateral is 360°.
$p + q + r + s = \mathbf{360°}$

Worked Example
Find the value of n.

Answer

$n + 68° + 74° + 52° = 360°$ (angles in a quadrilateral add to 360°)
$n + 194° = 360°$
$n = 360° - 194°$
$= \mathbf{166°}$

We have used
geometrical reasoning
to find the answer.

Exercise 6   Only use a calculator if you need to.

1   Find the value of y.
The first two have been started.

a

$y + 134° + 42° + 81° = 360°$
(angles in a quadrilateral add to 360°)

b

$y + 69° + 80° + 74° = 360°$
(angles in a quadrilateral add to 360°)

c

d

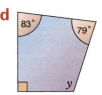

e

f

g

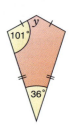

2 What is the sum of the interior angles of these?
  a a square **b** a parallelogram **c** a rhombus **d** a kite

3 Find the values of the angles marked with letters.
Show your working and give reasons.
The first two have been started.

a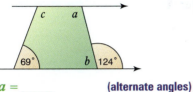

$a =$ ____ (alternate angles)
$b +$ ____ = ____ (angles on a straight line add to 180°)

b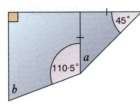

$a =$ ____ (base angle in isosceles triangle)

*c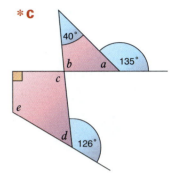

*4 Prove that x has the value given in green.

a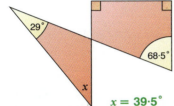

$x = 39.5°$

b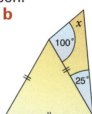

$x = 55°$

Summary of key points

$a = b$
Vertically opposite angles are equal.

$p + q + r = 360°$
Angles at a point add to 360°.

$x + y = 180°$
Angles on a straight line add to 180°.

 Angles made with parallel lines

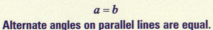
$a = b$
Alternate angles on parallel lines are equal.

$c = d$
Corresponding angles on parallel lines are equal.

 Two angles that add to 90° are **complementary angles**.
Two angles that add to 180° are **supplementary angles**.

Shape, Space and Measures

D   *a* is an **exterior angle** of the triangle.

b, *c* and *d* are all **interior angles** of the triangle.

Interior angles of a triangle add to 180°.

The exterior angle of a triangle equals the sum of the two interior opposite angles.

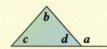

Example $e = 56° + 42°$

          $= 98°$

E   When finding angles we use **geometrical reasoning**.

Each step has an equals sign and a reason.

Example $a = 62°$ (alternate angles on parallel lines are equal)

     $a + b = 180°$ (angles on a straight line add to 180)°

   $62° + b = 180°$

        $b = 180° - 62°$

           $= 118°$

F   The **interior angles of a quadrilateral add to 360°.**

$a + b + c + d = 360°$

Example $x + 90° + 75° + 110° = 360°$ (angles in a quadrilateral add to 360°)

            $x + 275° = 360°$

                 $x = 360° - 275°$

                 $x = \mathbf{85°}$

Test yourself

1   Find the size of *y*.

   a   **b**   **c**

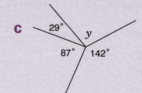

2   Find the angles marked with letters.

Give reasons.

   a   **b**   **c**

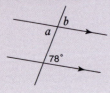 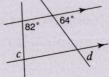 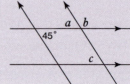

3 a Are 121° and 59° complementary or supplementary angles?

b Are 32° and 58° complementary or supplementary angles?

C

4 Find the angles marked with letters.
Write down a reason for each line.

D **E**

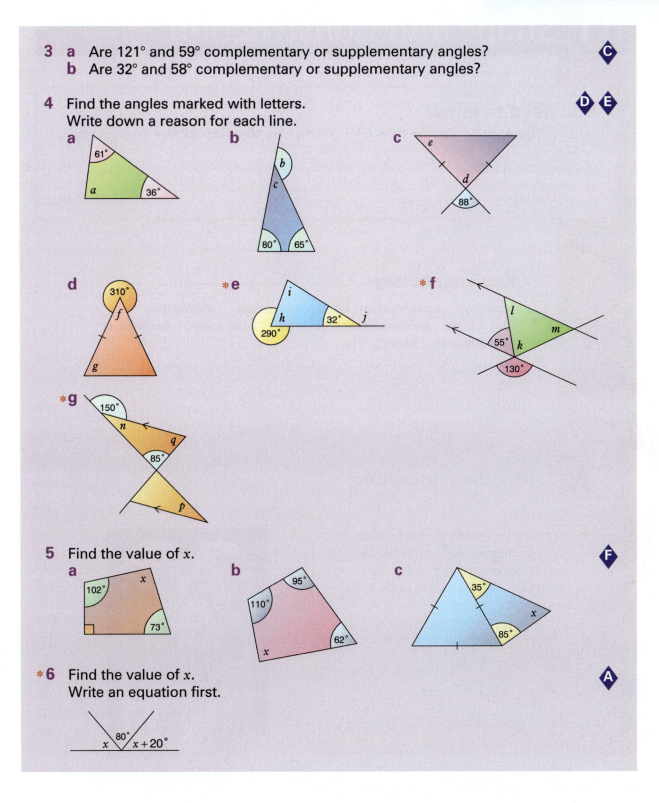

a

b

c

d

∗e

∗f

∗g

5 Find the value of x.

F

a

b

c

∗6 Find the value of x.
Write an equation first.

A

11 Shape and Construction

Key vocabulary

bisector, construction lines, elevation, isometric, mid-point, perpendicular bisector, plan view, tessellate, tessellation, view

Mission impossible

- Maurits Escher drew this picture.
  Look closely at it to see if you can see anything 'impossible'.
  Find other M.C. Escher 'impossible' drawings.

- This rectangle is an 'impossible rectangle'.

Try to draw an 'impossible triangle'.

Visualising and sketching 2-D shapes

Discussion

Imagine a rectangle, cut along the diagonal to make two triangles. Place the diagonals together in a different way. What shape is formed?

Dylan

What might Dylan's answer be? **Discuss**

Exercise 1

1 Imagine two identical parallelograms put together along sides of equal length.
What shape is made?
Is this the only possible shape?

2 Imagine two identical rhombuses put together along one side.
What shape is made?
Is this the only possible shape?

3 Imagine two identical isosceles triangles put together along sides of equal length.
What shape is made?
Is this the only possible shape?

4 Imagine a quarter of a shape is
   a a right-angled isosceles triangle   **b** a parallelogram
   What is the shape?
   Is this the only possible shape?
   One possible answer for **a** is a rectangle.

   I indicates that all these lengths must be identical

5 Imagine two identical equilateral triangles put together along one edge.
   a What shape is made? Explain why.
   b If you add a third equilateral triangle, what shape is formed?
      Sketch this shape.
      Is this the only possible shape?
   c Add a fourth equilateral triangle.
      What shape is formed?
      Is this the only possible shape?
      Make some sketches.

Shape, Space and Measures

Practical

1 **You will need** some pictures or posters with geometrical patterns.

Examples   Tiling patterns, Escher drawings, wallpaper, posters with patterns, ...

Describe the patterns.
Give as much detail as possible.
You could work by yourself, in pairs or in groups.
You could make a poster or booklet.
You could use a computer.

2 **You will need** a topic such as bridges or quilts or fences or seeds or cones or buildings or windows or churches or ...
Describe the shapes you see.
Give a reason why these objects are the shape they are.

Example   Roofs are often a triangular shape because a triangle is a very strong shape.

Investigation

Rectangles and squares

Joshua drew a 4 by 2 rectangle on squared paper. He found he could **cut** it into squares in three different ways.

He was only allowed to cut along grid lines.

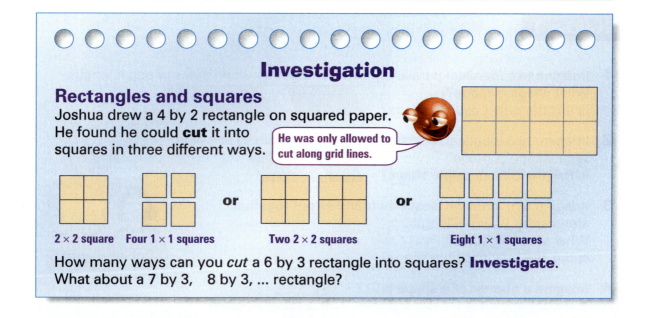

2 × 2 square   Four 1 × 1 squares   Two 2 × 2 squares   Eight 1 × 1 squares

How many ways can you *cut* a 6 by 3 rectangle into squares? **Investigate**.
What about a 7 by 3,   8 by 3, ... rectangle?

Properties of triangles, quadrilaterals and polygons

Investigation

Properties

You will need a dynamic geometry software package.

Ask your teacher for a copy of the **Different Parallelograms** ICT worksheet.

Investigation

Properties of shapes

You will need two large copies of each of these shapes and a copy of the tables.

An isosceles trapezium has two equal sides.

equilateral triangle

isosceles triangle

square

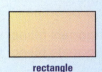

rectangle

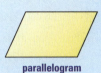

parallelogram

rhombus

kite

isosceles trapezium

trapezium

arrowhead

A

Shape	Number of lines of symmetry	Order of rotation symmetry
Equilateral △		
Isosceles △		
Square		4
Rectangle		
Parallelogram	0	2
Rhombus	2	
Kite		
Isosceles trapezium		1
Trapezium		
Arrowhead		

1   How many lines of symmetry has each shape got?
    Check by folding.
    Fill in the answers on your table.

2   A parallelogram has rotation symmetry of order 2.
    As you turn it one complete turn it fits onto itself in
    two different positions.
    Turn each of your other shapes.
    What is the order of rotation symmetry of each?
    Fill in the answers in your table.

*3 Jake drew the diagonals on his rhombus.
A rhombus has 2 lines of symmetry.
All 4 triangles Jake made are the same.
He wrote:

The angles with ticks are all equal.
They add to 360°. **(angles at a point add to 360°)**
So each one must be 90°.
So the diagonals cross at right angles.

Show that these are true for a rhombus using symmetry properties.

> **Remember rotation symmetry of order 2 means it fits onto itself twice as you turn it 360°.**

    the diagonals bisect each other
    the diagonals bisect the angles

Use the symmetry properties of the other shapes to fill in the gaps in this table.

Quadrilateral	Diagonals equal	Diagonals cross at right angles	Diagonals bisect each other	Diagonals bisect the angles	Sides	Angles
Square				✓		4 right angles
Rectangle			✓	✗		
Parallelogram	✗	✗	✓	✗	opposite sides equal	opposite angles equal
Rhombus	✗	✓	✓	✓	4 equal	
Kite		✓	✗	✗	2 pairs of adjacent sides equal	
Isosceles trapezium	✓	✗	✗	✗		2 pairs of equal angles
Trapezium	✗	✗	✗	✗		
Arrowhead		✓ (outside the shape)	✗	✗		1 pair of equal angles, 1 reflex angle

Once you've filled in the table, you need to know the information on it.

B You will need a 3 × 3 pinboard and some square dotty paper.

1 There are 16 different quadrilaterals you can make on a 3 × 3 pinboard.
Make all 16 and draw them on square dotty paper.
Put them into groups with the same properties.
Some quadrilaterals will be in more than one group.

Example   Tessa made this trapezium.
She put it into the following groups.
    quadrilaterals with a right angle
    quadrilaterals with one pair of parallel sides
    quadrilaterals with no lines of symmetry

> **This is the same as Tessa's shape.**

2 Decide if it is possible to make these on the pinboard.
If not, explain why not.
- a triangle with a reflex angle
- a trapezium with one line of symmetry
- a rhombus which is not a square
- an equilateral triangle

> **A reflex angle is bigger than 180°.**

Once you have decided, check if you are correct by trying.

Properties of triangles

right-angled
1 right angle

isosceles
2 equal sides
base angles equal
1 line of symmetry

equilateral
3 equal sides
3 equal angles
3 lines of symmetry

scalene
no 2 sides are equal
no 2 angles are equal

Properties of quadrilaterals

A quadrilateral has four sides.

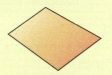
quadrilateral

These are the **special quadrilaterals**.

square
4 equal sides
4 right angles
4 lines of symmetry

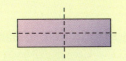
rectangle
2 pairs of opposite sides equal
4 right angles
2 lines of symmetry

parallelogram
opposite sides equal
and parallel

rhombus
a parallelogram with 4 equal sides
2 lines of symmetry
opposite angles equal

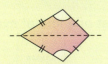
trapezium
1 pair of opposite
sides parallel

kite
2 pairs of adjacent sides equal
1 line of symmetry
1 pair of equal angles

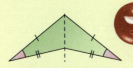
arrowhead or delta
2 pairs of adjacent sides equal
1 reflex angle
1 line of symmetry
1 pair of equal angles

Adjacent sides are next to each other.

Quadrilateral	Diagonals equal	Diagonals cross at right angles	Diagonals bisect each other	Diagonals bisect the angles	Sides	Angles
Square	✓	✓	✓	✓	4 equal	4 right angles
Rectangle	✓	✗	✓	✗	opposite sides equal	4 right angles
Parallelogram	✗	✗	✓	✗	opposite sides equal	opposite angles equal
Rhombus	✗	✓	✓	✓	4 equal	opposite angles equal
Kite	✗	✓	✗	✗	2 pairs of adjacent sides equal	1 pair of equal angles
Isosceles trapezium	✓	✗	✗	✗	1 pair equal 1 pair parallel	2 pairs of equal angles
Trapezium	✗	✗	✗	✗	1 pair parallel	no equal angles
Arrowhead	✗	✓ (outside the shape)	✗	✗	2 pairs of adjacent sides equal	1 pair of equal angles, 1 reflex angle

A shape which has no reflex angles is a **convex shape**.
A shape which has one or more reflex angles is a **concave** shape.

convex shape

concave shape

Shape, Space and Measures

Exercise 2 Look at the table on page 271 to help you answer these

1 Write true or false for each of these.
  a A parallelogram has opposite sides equal and parallel.
  b A kite has two pairs of opposite sides parallel
  c A trapezium has two pairs of parallel sides.
  d A rhombus has no equal angles.
  e An arrowhead has two pairs of adjacent sides equal.

2 What shape am I?
Choose from the box.

| isosceles triangle | equilateral triangle | rectangle | rhombus |

  a I am a quadrilateral.
    I have only two lines of symmetry.
    I have four right angles

  b I have three sides.
    I have three lines of symmetry.

  c I am a quadrilateral.
    I have four equal sides.
    I have only two lines of symmetry.

  d I am a triangle.
    I have only one line of symmetry.

3 a Which of these is a quadrilateral with just one pair of parallel sides?
    A  B   C  D

  b Which of these is a quadrilateral with just one line of symmetry?
    A 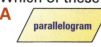 B   C   D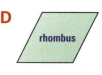

  c Which of these is a quadrilateral with equal diagonals that are also perpendicular?
    A   B   C   D

  d Which of these is a quadrilateral with parallel sides and equal diagonals?
    A   B   C   D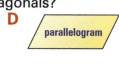

4 a Mary's birthday cake is a quadrilateral with 2 lines of symmetry and rotation symmetry of order 2.
    What shape could it be?
  b Tom's logo on his science book is a quadrilateral with no lines of symmetry and rotation symmetry of order 2.
    What shape could it be?

5 Which of these statements are true?
Explain why or why not.
   a All equilateral triangles are isosceles triangles.
   b All rectangles are squares.
   c All squares are rhombuses.
   d All quadrilaterals are parallelograms.

∗6 Explain why a triangle can never have a reflex angle but a quadrilateral can.

∗7 Explain why it is always possible to make an isosceles triangle from two identical right-angled triangles.

∗8 Use a larger copy of this table.
This trapezium is symmetrical.
It has two pairs of equal angles
and one pair of parallel sides.
It is filled in on the table as shown.
   i Fill these in on the table.
     a square
     b rectangle
     c kite
     d rhombus
     e arrowhead (delta)
     f non-symmetrical trapezium

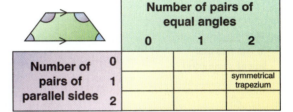

		Number of pairs of equal angles		
		0	1	2
Number of pairs of parallel sides	0			
	1			symmetrical trapezium
	2			

A box could have more than one shape in it.

   ii Some of the boxes are still empty. Is it possible
to draw quadrilaterals for these? If not, explain why not.
Use the properties you know about quadrilaterals in your explanation.

We can use the **properties of 2-D shapes to find unknown angles**.

Worked Example
Find *a* and *b*.

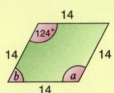

Answer
This shape is a rhombus.
The diagonals are lines of symmetry.
By symmetry *a* = **124°**

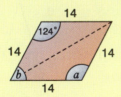

This diagonal cuts angle *a* and the 124° angle in half.
The blue triangle is isosceles.
$62° + 62° + b = 180°$
$\qquad 124° + b = 180°$
$\qquad\qquad b = 180° - 124°$
$\qquad\qquad\quad = \mathbf{56°}$

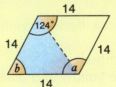

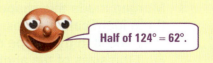

Half of 124° = 62°.

Shape, Space and Measures

Remember the angle properties you learnt in Chapter 10.

Remember the dashes on the sides mean the sides are equal.

1 Find the angles named with letters.
Use the properties of the shapes to help.
The first 3 are started.

a

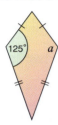

$a =$ ___
(a kite has one line of symmetry)

b

$b = 48°$ (_____)

c

$d = \dfrac{180°}{3}$

$=$ ___

(all angles in an equilateral triangle are 60°)

d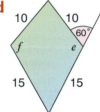

e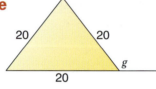

f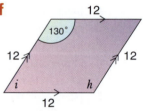

2 Use the properties of the shape to find the green shaded angle and the purple length.
Show your reasoning clearly.

a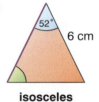

isosceles triangle

b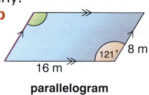

parallelogram

c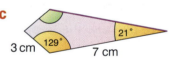

kite

d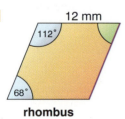

rhombus

e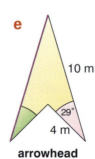

arrowhead

f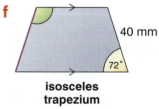

isosceles trapezium

*3 Morgan made this shape with three identical yellow and three identical blue tiles.
Each tile is in the shape of a rhombus.
Find the sizes of the angles in the yellow and blue tiles.

*4 Find k and j.
Use the properties of shapes to help.

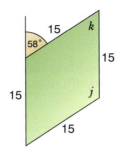

 Practical

You will need Logo.

Ask your teacher for the **Drawing Shapes with Logo** ICT worksheet.

---

*Investigation

Polygons and right angles
Try to draw a triangle with two right angles.

Try to draw a quadrilateral with two right angles.
Now try to draw one with three.
Is it possible to have four?

What is the greatest number of right angles it is possible to have in a five-sided polygon?

What if the polygon had six sides?
What if the polygon had seven sides?
What if ...

Investigate to find the maximum possible number of right angles in polygons with different numbers of sides.

Tessellations

A shape will **tessellate** if it can be used to fill a space.
There must be no overlapping and no gaps.
The shape may be translated, reflected or rotated to fill the space.

Examples

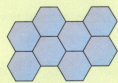

These hexagons tessellate.

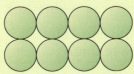

These circles do not tessellate.

Discussion

How would you describe this tiling pattern to someone over the phone? **Discuss**.
The pattern has been made with these two tiles.

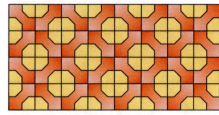

How could you construct these tiles? **Discuss**.

You could use a computer tiling software package to make this pattern.

Investigation

Tessellations

T

1 Choose one of the triangles shown.
Use some copies of this triangle.
On squared paper, tessellate this triangle to tile an area.
Colour your design, using at most three colours.

Do all triangles tessellate? **Investigate**.
Explain why or why not using angles on a straight line, alternate and corresponding angles.
You could use computer tiling software.

*2 Use computer tiling software to explore if other shapes will tessellate.

*3 Design a poster, using the tessellation of a shape.

Describing cubes and cuboids

Practical

You will need some Multilink or centimetre cubes and a partner.

Sit back to back with your partner.
Without letting your partner see, make a model using at least twelve cubes.

Example   Beth and Tony made these.

Take turns to tell your partner how to make the model.

Remember

1 We often construct 3-D shapes by first drawing a net.

There is more about nets on page 238.

2 We sometimes draw 3-D shapes on **isometric** (triangle dotty) paper.
   Make sure the paper is the right way round.

Right ✓   Wrong ✗

The dots make vertical lines   The dots make horizontal lines

When we make an isometric drawing, some edges are not shown.

Example

The dashed edges are not shown on an isometric drawing.

⭐ **Practical**

You will need a photograph or poster and a partner.

1 Look at a photograph or poster.
   Describe to your partner the 3-D shapes you can see in the picture or poster.
   Give as much detail as possible.
   Use words like face, edge, vertex, intersect, point, parallel, perpendicular.

2 **You will need** a copy of the nets below.
   Rosalee drew this net.

   She imagined how it would fold to make a cube.
   She then coloured
   a 1 set of edges that would be parallel red
   b 1 set of faces that would be parallel green
   c 1 set of edges that would meet at a point blue
   d 1 set of edges that would be perpendicular purple
   e 1 set of faces that would be perpendicular pink.

   Use a copy of this net.
   Do **a** to **e** on this net.

   Do the same for these two nets.

e Copy and complete the **top view** of the
model by shading the squares which are **red and blue**.

top view

f Imagine you turn the model **upside down**.
What will the new top view of the model look like?
Copy and complete the new **top view** of the model by
shading the squares which are **red and blue**.

new top view

⭐ Practical

You will need some Multilink or centimetre cubes and triangle dotty paper.

1 Build some 3-D shapes using cubes.
Draw the front elevation, side elevation and plan view of each.

2 Use cubes to build these shapes. Sketch your shapes.

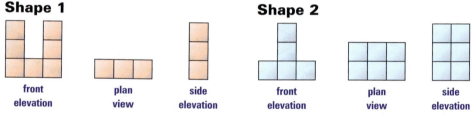

Shape 1

front
elevation   plan
view   side
elevation

Shape 2

front
elevation   plan
view   side
elevation

3 Which of these is it possible to build?
Is there more than one way to build some of them?

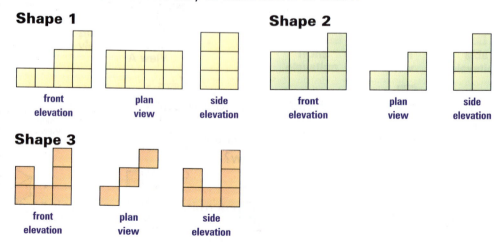

Shape 1

front
elevation   plan
view   side
elevation

Shape 2

front
elevation   plan
view   side
elevation

Shape 3

front
elevation   plan
view   side
elevation

4 *Work in pairs*.   Use cubes to build a 3-D shape.
Draw the front elevation, side elevation and plan view.
Ask your partner to build the shape from your drawings.

Construction

We can **construct the mid-point** and **perpendicular** bisector of a line segment.

This shows the **construction of the mid-point and perpendicular bisector of the line segment BC**.

A bisector cuts in half. Perpendicular means at right angles.

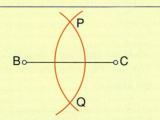

Open out the compasses to a little more than half the length of BC. With the point first on B and then on C, draw arcs to meet at P and Q.

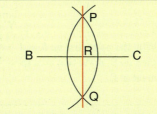

Draw the line through P and Q. R is the mid-point of BC. PQ is the perpendicular bisector of the line segment BC.

Keep the same length on your compasses.

Exercise 6

1 Use a copy of this.
Construct the perpendicular bisector of the line.

2 Use a copy of this.
  a Construct the perpendicular bisector of AB.
  b Measure the distance from A to where *your* line cuts AC.

3 Use a copy of this.
  a Construct the perpendicular bisector of AB.
  b Measure the length of your perpendicular bisector from where it cuts AB to where it cuts AD.

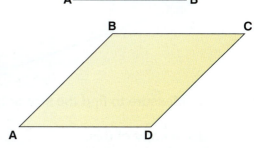

4 Use a copy of this.
Mr Smart wants to divide his living room in half.
  a Construct the perpendicular bisector of the wall AB.
  b Measure the distance from A to where the room divider meets AB.
  ＊c If 1 cm represents 1 m, how wide is each new room?

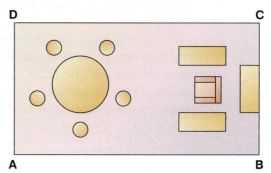

Shape, Space and Measures

We can **construct the bisector of an angle**.

This shows the construction of the bisector of the angle P.

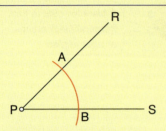

Open out the compasses to a length less than PR or PS. With the point on P, draw an arc as shown. Label the points A and B.

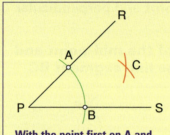

With the point first on A and then on B, draw two arcs to meet at C.

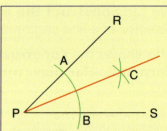

Draw the line from P through C. This line, PC, is the bisector of the angle P.

Exercise 7

T

1 Use a copy of this.
  a Construct the bisector of the angle ABC.

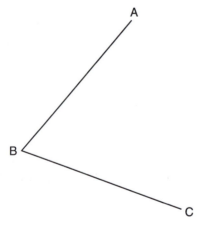

  b Measure to find the size of half of angle ABC.

T

2 Use a copy of this.
This diagram shows three fences in Peter's backyard.
He wants to put some paving stones in a straight line along the bisector of the angle between fences AB and BC.
  a Construct this angle bisector.
  b Measure the distance from C to the point where the angle bisector meets CD.

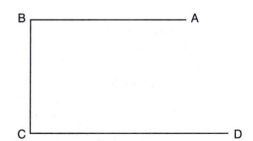

This shows the **construction of the perpendicular from A to the line segment BC**.

Keep the same length on your compasses.

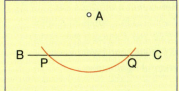

Open out the compasses.
With the point on A, draw an arc to cross BC at P and Q.

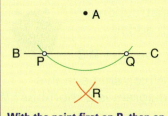

With the point first on P, then on Q, draw two arcs to meet at R.

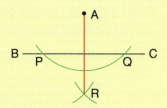

Join A and R. AR is the perpendicular from A to the line segment BC.

This shows the construction of the **perpendicular from a point P on a line segment BC**.

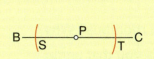

Open out the compasses to less than half the length of BC.
With the point on P, draw arcs, one on each side of P. Label where they cross BC as S and T.

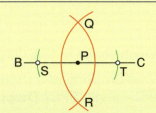

Open out the compasses a little more.
With the point first on S and then on T, draw arcs so they cut at Q and R.

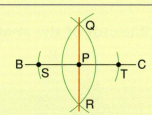

Draw the line through Q and R. QR is the perpendicular from P on the line segment BC.

Exercise 8

T

1 Use a copy of these.
  a Construct the perpendicular from P to the line segment MN.

  b Construct the perpendicular from P on the line segment MN.

• P

M——————————— N

M ——○———————— N
      P

T

2 Use a copy of this.
A, B and C show the position of three ships.
Ship A is due North of B.
Ship C is due East of B.
Ship D lies on the line perpendicular to AB at point P.
It also lies on the angle bisector of angle ABC.

Construct the lines to find where ship D lies.
How far is ship D from ship A?

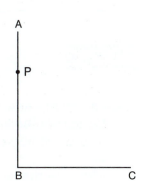

285

Shape, Space and Measures

*Discussion

The diagrams below are the final diagrams from each of the previous constructions. Some dashed lines have been added.

> Remember a rhombus has 4 equal sides, opposite angles equal and diagonals that bisect the angles and bisect each other at right angles.

Perpendicular bisector of the line BC

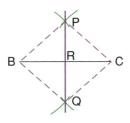

Which shape in this diagram is a rhombus? Why?

What property of a rhombus tells you that R is the mid-point of BC? **Discuss.**

What property of a rhombus tells you that PQ is the perpendicular bisector of BC? **Discuss.**

Bisector of the angle P

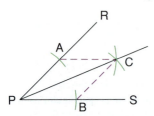

Which shape in this diagram is a rhombus? Why?

What property of a rhombus tells you that angle RPC = angle SPC? **Discuss.**

Line from A perpendicular to BC

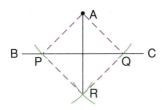

Which shape in this diagram is a rhombus? Why?

What property of a rhombus tells you that AR is perpendicular to BC? **Discuss.**

Perpendicular from a point P on a line segment BC

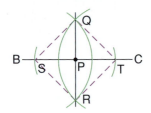

Which shape in this diagram is a rhombus? Why?

What property of a rhombus tells you that RQ is perpendicular to BC? **Discuss.**

Exercise 9

T

1 Use a copy of this diagram.
   a Construct the line through P that is at right angles to the line segment BC.
   b Label as X, the point where this line meets AB.
   c Measure the length of BX.

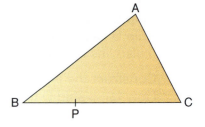

T **2** Three cycle tracks are shown: AD, DC and AC.
The council wants to put a new cycle track, perpendicular to AC, from B to DC.
Use a copy of the diagram.
Use your compasses to construct the line of the new cycle track.

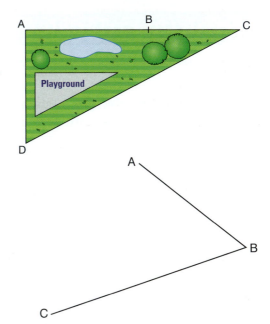

T **3** Use a copy of this diagram.
   a Bisect angle B
   b Draw the perpendicular bisector of the line segment BC.
         Label the mid-point as P.
   c Label as Q the point where the lines constructed in **a** and **b** meet.
   d Construct a perpendicular line from Q to the line segment AB.

4 Produce an accurate drawing of this diagram as follows.
   Draw a line DC, 6 cm long.
   Construct the square ABCD.
   Draw in the diagonals AC and BD.
   Bisect the angles between the diagonals and the sides.
   Shade as shown.

Puzzle

The 10 pieces of a circle puzzle may be constructed as shown.
1 On thin card, draw a circle, centre O and radius 6 cm.
2 Draw in the diameter AB.
3 Construct the diameter CD so that CD is perpendicular to AB.
4 Bisect OC. Label the mid-point as P.
   Bisect OD. Label the mid-point as Q.
5 Through Q, construct the line RS so that RS is parallel to AB.
6 Join AP, AQ, BP and BQ.
7 Carefully cut out the 10 pieces.

Rearrange the 10 puzzle pieces to make interesting shapes such as those shown below.

Constructing triangles and quadrilaterals

Remember

We can **construct a triangle** using a **ruler and protractor** if we are given two sides and the included angle or two angles and the side between them.

See page 238 for more.

A 31 mm 87° 44 mm C
B

Given
two sides and the angle between them (called SAS)

P
Q 67° 2·3 cm 73° R

Given
two angles and the side between them (called ASA)

If the three sides of a triangle are given we can construct the triangle using a ruler and compasses.

Example This shows how to construct the triangle XYZ. The diagram is not to scale.

Y ———— Z

Y Z

Y Z

X
Y Z

Draw a line 2.5 cm long.

Open the compasses out to 2.8 cm. With the point on Y, draw an arc.

Open the compasses out to 3.1 cm. With the point on Z, draw an arc to cross the first arc.

Complete the triangle.

X
2·8 cm 3·1 cm
Y 2·5 cm Z

Given
three sides (called SSS)

Be careful to keep the compasses open at the length you want.

To construct an accurate drawing you need to measure and draw carefully.

Discussion

How could you construct ABCD and PQRS?
Discuss.

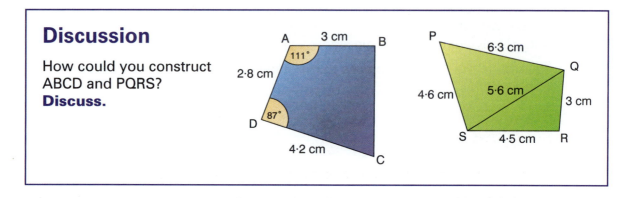

Exercise 10   **In this exercise, measure and draw very carefully.**

1 Use your ruler and compasses to draw triangles with sides of these lengths.
Measure the angle between the first two sides given.
   a 3 cm, 4 cm, 5 cm   **b** 5 cm, 5 cm, 4 cm
   c 8 cm, 4 cm, 6·6 cm   **d** 9 cm, 6 cm, 7 cm

2 Use your ruler and protractor or ruler and compasses to construct the quadrilaterals that are sketched.
Measure the dashed lengths and the red angle on *your* constructions.

   a 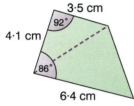   **b** 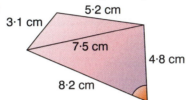   **c**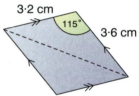

3 Linda drew this sketch of her sister's tractor.
Make an accurate drawing of it.
Note You will need a ruler, protractor and compasses.

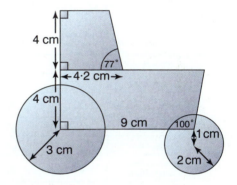

∗4 Use a ruler, compasses and set square to construct a net for each of these.

   a 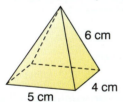   **b**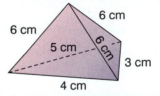

   c 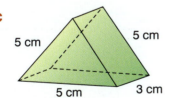   **d**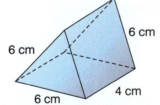

 Practical

You will need a dynamic geometry software package.

Ask your teacher for the **Constructions on a Line** and **Constructing a Triangle with Three Known Lengths** ICT worksheets.

Shape, Space and Measures

Summary of key points

 A We can **visualise and sketch 2-D shapes**.

 B **Properties of triangles**

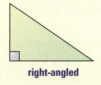
right-angled

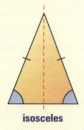
isosceles

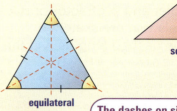
equilateral

scalene

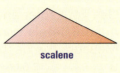
The dashes on sides show they are equal. Dotted lines are lines of symmetry

Properties of quadrilaterals

Quadrilateral	Diagonals equal	Diagonals cross at right angles	Diagonals bisect each other	Diagonals bisect the angles	Sides
Square	✓	✓	✓	✓	4 equal
Rectangle	✓	✗	✓	✗	4 equal
Parallelogram	✗	✗	✓	✗	opposite sides equal
Rhombus	✗	✓	✓	✓	4 equal
Kite	✗	✓	✗	✗	2 pairs of equal
Isosceles trapezium	✓	✗	✗	✗	1 pair of equal
Trapezium	✗	✗	✗	✗	no sides equal
Arrowhead	✗	✓ (outside arrowhead)	✗	✗	2 pairs of equal

A shape which has no reflex angles is **convex**.

A shape which has one or more reflex angles is **concave**.

 C A shape will **tessellate** if it can be translated, reflected or rotated to fill a space without gaps.

Example

 D We can use isometric paper to draw **3-D shapes**.

Example

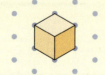

3-D shapes are often made by first drawing a **net**.

290

E The view from the top of a 3-D shape is the **plan view**.

The view from the front is the **front elevation**.

The view from the side is the **side elevation**.

Example

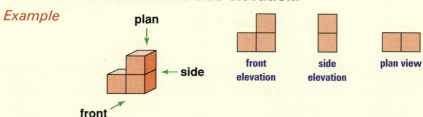

plan

← side

front

front
elevation

side
elevation

plan view

Constructions

See pages 283, 284, 285 and 288 for the following constructions:

mid-point and perpendicular bisector of a line segment

bisector of an angle

perpendicular from a point to a line segment

perpendicular from a point on a line segment

a triangle given 3 sides

Test yourself

1 Imagine two identical scalene triangles put together along sides of equal length. **A**
What shape is formed?
Is this the only possible shape? Explain your answer.

2 Name all the special triangles and special quadrilaterals that have **B**
  a 1 axis of symmetry   **b** rotation symmetry of order 4   **c** equal diagonals.

3 Which special quadrilateral could I be? **B**
Choose from the box.
  a My diagonals intersect at right angles.
    I have four equal sides and no right angles.
  b My diagonals bisect each other but not at right angles.
    My diagonals are not lines of symmetry.

> rhombus
> parallelogram
> square
> rectangle
> kite

4 Find the angles marked with letters. **B**
  a

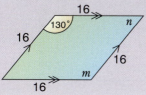

  b

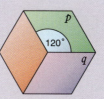

This shape is made with three
identical rhombuses.

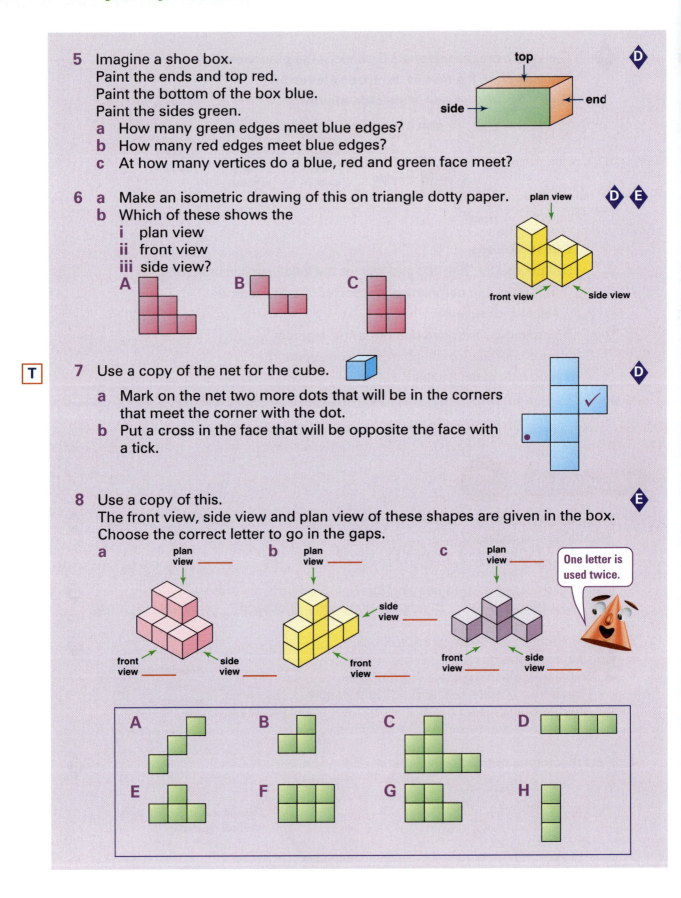

5   Imagine a shoe box.
    Paint the ends and top red.
    Paint the bottom of the box blue.
    Paint the sides green.
    a   How many green edges meet blue edges?
    b   How many red edges meet blue edges?
    c   At how many vertices do a blue, red and green face meet?

6   a   Make an isometric drawing of this on triangle dotty paper.
    b   Which of these shows the
        i   plan view
        ii front view
        iii side view?

    A   B   C

7   Use a copy of the net for the cube.

    a   Mark on the net two more dots that will be in the corners
        that meet the corner with the dot.
    b   Put a cross in the face that will be opposite the face with
        a tick.

8   Use a copy of this.
    The front view, side view and plan view of these shapes are given in the box.
    Choose the correct letter to go in the gaps.
    a   plan
        view _____
        front   side
        view _____   view _____
    b   plan
        view _____
                                 side
                                 view _____
        front
        view _____
    c   plan
        view _____

        One letter is
        used twice.

        front   side
        view _____   view _____

    A   B   C   D
    E   F   G   H

9 Use a copy of this diagram.
  a Using a pair of compasses construct the line through A that is at right angles to XY.
  b Name the point where this line meets YZ as B.
  c Bisect angle AYZ.
  d Name the point where this bisector meets AB as C.
  e Measure the length YC to the nearest millimetre.

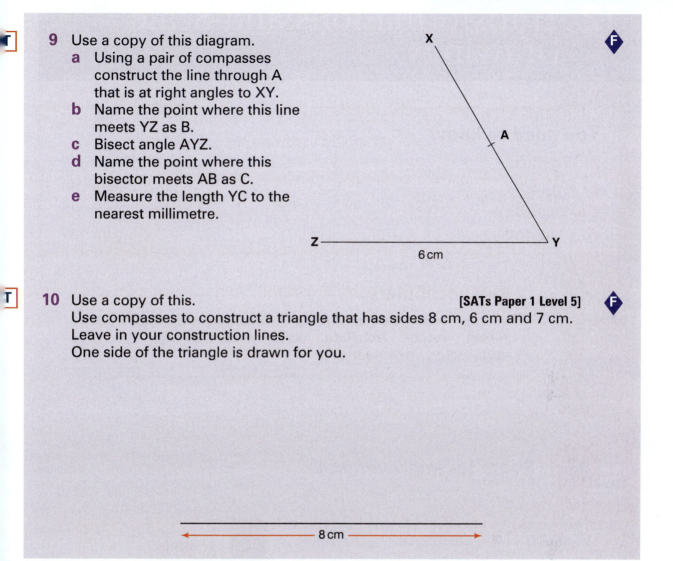

10 Use a copy of this.   **[SATs Paper 1 Level 5]**
Use compasses to construct a triangle that has sides 8 cm, 6 cm and 7 cm.
Leave in your construction lines.
One side of the triangle is drawn for you.

8 cm

12 Transformations and Scale Drawings

You need to know

Key vocabulary

centre of enlargement, enlarge, enlargement, map, plan, reflect, rotate, translate, scale, scale drawing, scale factor, symmetry

▶▶ Room for improvement

This is a picture of Cassandra's room.

This is the floor plan.

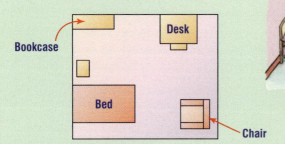

Bookcase

Desk

Bed

Chair

She decides to change her bedroom around.
She
     rotates the bookcase 90°
     translates the desk left
     rotates the bed and table 180°
     and then translates them right to the wall
     rotates the chair 180° and then translates it left to the wall.
Draw a possible floor plan for Cassandra's new arrangement.

Transformations

Remember

ABC has been reflected in the mirror line to give the image A′B′C′.
A is on the mirror line so it does not move.
B′ and C′ are the same perpendicular distance behind the mirror line as B and C are in front.

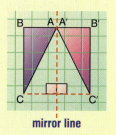

mirror line

We need to know the **centre of rotation** and the **angle of rotation** to **rotate** a shape.

Example   ABCD has been rotated 180° using (0, 0) as the centre of rotation.

When no direction is given for a rotation, we always rotate anticlockwise.

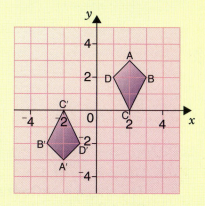

When we **translate** a shape we slide it without turning it.

Example   ABCDEF has been translated 3 units left and 5 units down.

The **inverse of a translation** is an equal move in the opposite directions.

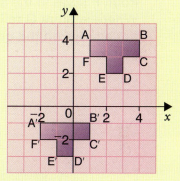

Worked Example

The points (⁻2, 4), (3, 2), (⁻1, 1) and (⁻3, 2) are the vertices of a quadrilateral.
Write down the coordinates of the vertices of the quadrilateral after
a reflection in the *x*-axis   **b** rotation 270° about the origin.

Answer

a

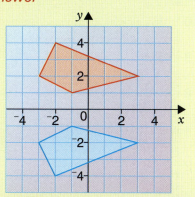

b

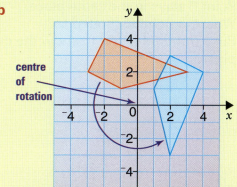

centre of rotation

The red shape has been reflected in the *x*-axis to give the blue shape. The coordinates of the image are (⁻2, ⁻4), (3, ⁻2), (⁻1, ⁻1) and (⁻3, ⁻2).

Use tracing paper and a pin to rotate the red shape 270° about (0, 0) to give the blue shape. The coordinates of the image are (4, 2), (2, ⁻3), (1, 1) and (2, 3).

Shape, Space and Measures

T

1 Use a copy of this.
Reflect each shape in the mirror line.

a Mirror line

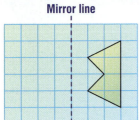

b Mirror line

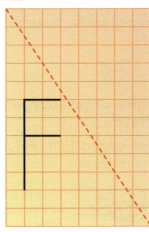

T

2 a Use a copy of this.
The shape is **reflected** in a mirror line.
Point P stays in the same place.
Where is point Q reflected to?
Put a cross on the grid to show the correct place.

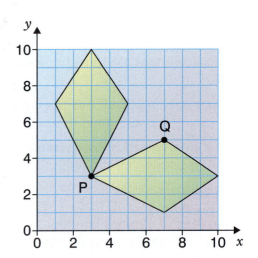

b Now the shape is **rotated**.
Point P stays in the same place.
Where is point Q rotated to?
Put a dot on the grid to show the correct place.

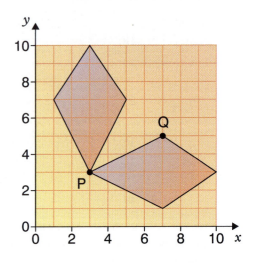

3 The rectangle PQRS is reflected in the y-axis.
What are the new coordinates of P, Q, R and S?

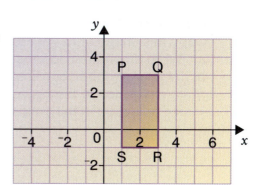

4 The rectangle WXYZ is reflected in the line $y = 1$.
What are the new coordinates of W, X, Y and Z?

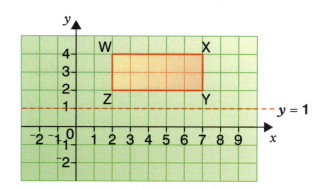

5 Use a copy of this.
Rotate each shape 270° about (0, 0).
Write down the new coordinates of A.

Remember to rotate anticlockwise.

a

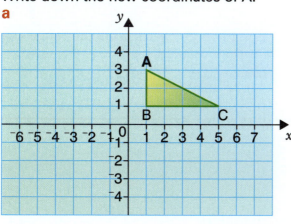

b

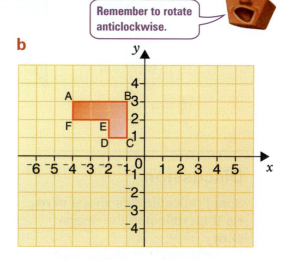

6 Three of the vertices of a square are (2, 3), (4, ⁻1), (0, ⁻3).
Plot these points on a grid.
  a Write down the coordinates of the fourth vertex.
  b Write down the coordinates of the four vertices:

Start with the original coordinates each time.

  i after translation 3 units to the right and 1 unit up
  ii after rotation 180° about the origin
  iii after reflection in the y-axis.

7 Draw a set of axes with *x*-values and *y*-values from ⁻8 to 8.
Copy this shape onto your axes.
a Write down the coordinates of A, B, C and D.
b Write down the coordinates of A′, B′, C′ and D′ after
  i reflection in the *x*-axis
  ii reflection in the *y*-axis.

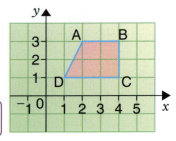

Start with the original coordinates each time.

T

8 Use a copy of this.   **[SATs Paper 2 Level 5]**
a You can **rotate** triangle **A** onto triangle **B**.
Put a cross on the **centre of rotation**.
You may use tracing paper to help you.
b You can **rotate** triangle **A** onto triangle **B**.
The rotation is **anticlockwise**.
What is the **angle** of rotation?
c **Reflect** triangle **A** in the mirror line.
You may use a mirror or tracing paper to help you.

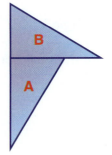

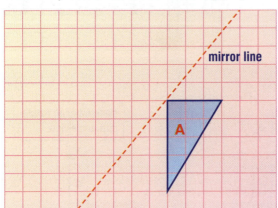

mirror line

Discussion

Sanjit reflected this shape in **m₁**.
He then reflected the image
in **m₂**.
He reflected that image in **m₃**
and so on.

What is the effect of repeated
reflections in parallel lines?
Discuss.

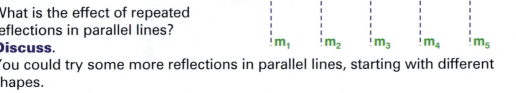

You could try some more reflections in parallel lines, starting with different shapes.

Exercise 2

T

1 Use a copy of this.
 a Reflect the shape in m_1.
 b Reflect the image you got in **a** in m_2.

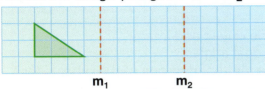

 c Repeat **a** and **b** for this diagram.

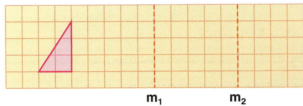

 d What *single* transformation is the same as reflection in two parallel lines?
 A translation **B** rotation **C** reflection

T

2 Rick is making patterns by reflecting a shape in two
 perpendicular lines.
 Use a copy of the diagrams below. For each
 1 reflect the shape in the x-axis
 2 reflect the image you got in **1** in the y-axis.
 What *single* transformation is the same as **1** and **2**?
 A reflection **B** rotation of 180° **C** rotation of 90°

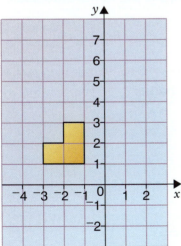
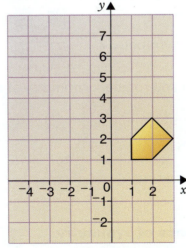

T

3 Use another copy of the diagrams in question **2** for each part of this question.
 a Translate each shape 2 units right and 4 units up and **then** translate the image
 1 unit left and 2 units down.
 What single translation is the same as these two?
 b Translate each shape 2 units left and 1 unit up and then translate the image 1
 unit left and 3 units down.
 What single translation is the same as these two?

Practical

You will need a copy of this triangle.

Rotate the triangle about the mid-point of one of its sides.
Name the shape formed by the object and image.
* Identify the equal sides and equal angles.
* What if you rotate it about the mid-point of one of the other sides?

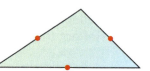

Symmetry

A shape has **reflection symmetry** or **line symmetry** if one or more lines can be found so that one half of the shape reflects to the other half.

Cassie drew this shape for the cover of her science folder.
It has three lines of symmetry.

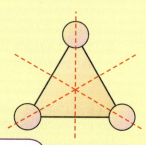

A shape has rotation symmetry if it fits onto itself **more than once** during a full turn.

> Shapes with rotation symmetry of *order* 1 do *not* have rotation symmetry.

The **order of rotation symmetry** is the number of times a shape fits onto itself in a 360° turn.
Ginny drew this shape.
It has order of rotation symmetry 4.

Worked Example
Describe the reflection and rotation symmetry of these shapes.

a 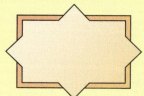   b

Answer
a Shape **a** has **two lines of symmetry**.

   It fits onto itself twice as it is turned through 360°.

   It has **rotation symmetry of order 2**.

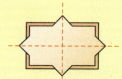

b Shape **b** has **no lines of symmetry**.

   It fits onto itself four times as it is turned through 360°

   It has **rotation symmetry of order 4**.

> We count the starting position when rotating or the ending position but not both.

Practical

A You will need a 3 × 3 pinboard or square dotty paper.

Marina made this shape on her 3 × 3 pinboard.
She wrote this.

> It has 7 sides.
> It is not regular.
> It has one line of symmetry and no rotation symmetry.

Make some other shapes on a 3 × 3 pinboard.
Decide if each of your shapes is regular.
Write down the symmetry properties of each.
*What is the biggest number of sides a shape on this pinboard can have?

B You will need a 5 × 5 pinboard or square dotty paper.

Helen divided her 5 × 5 pinboard into two congruent halves.
She wrote, 'The pattern has rotation symmetry or order 2'.

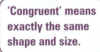

Find other ways to divide a 5 × 5 pinboard into congruent halves.
Describe the rotation symmetry of each.

Note You could see who can find the most ways.

> 'Congruent' means exactly the same shape and size.

There is more about symmetry and 2-D shapes on page 240.

Exercise 3

1 Describe the reflection and rotation symmetry of these.

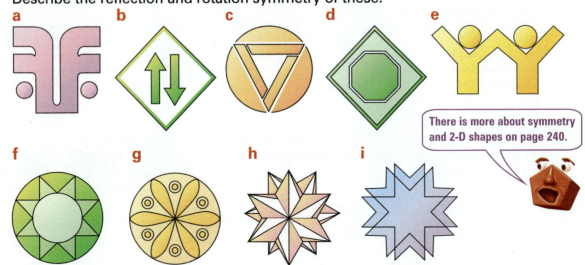

Shape, Space and Measures

2 I have a square grid and two rectangles.

[SATs Paper 1 Level 4]

grid   **two rectangles**

I make a pattern with the grid and two rectangles.
The pattern has **no** lines of symmetry.

T Use a copy of this.

a Put both rectangles on the grid to make a
pattern with **two** lines of symmetry.
You must **shade** the rectangles.

b Put both rectangles on the grid to make a
pattern with **only one** line of symmetry.
You must **shade** the rectangles.

c Put both rectangles on the grid to make a
pattern with **rotation** symmetry of **order 2**.
You must **shade** the rectangles.

3 Describe the reflection and rotation symmetries of these.

a
equilateral
triangle

b
regular
hexagon

c
square

d
isosceles
triangle

e
parallelogram

f
rhombus

g
rectangle

h
trapezium

i
isosceles
trapezium

j
kite

4 Show how you can put these shapes together to
make a single shape with reflection symmetry *and*
rotation symmetry.

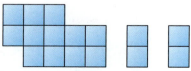

Use squared paper to help
with questions 4 and 5.

5 a Rochelle put these two shapes together, edge to edge.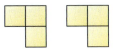
Her new shape had just 1 line of symmetry.
Show two ways Rochelle might have done this.

b Show a way of putting the two shapes together edge to edge to get a new shape with

i rotation symmetry of order 2 but no lines of symmetry (four ways)

ii 2 lines of symmetry *and* rotation symmetry of order 2.

*6 a Angus thought 'The base angles of an isosceles triangle are equal.'
How could he show this is true using the symmetry properties of the shape?

b Nick thought 'The opposite angles and opposite sides of a parallelogram are equal.'
How could he show this is true using the symmetry properties of the shape?

Enlargement

The green shape has been **enlarged** to the pink shape.
Each length on the pink shape is **2** times as long as it is on the green shape.

The **scale factor** of the enlargement is 2.

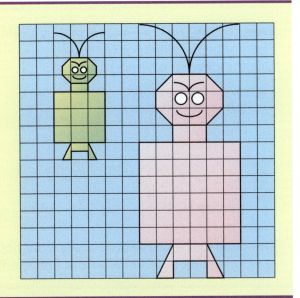

Exercise 4

1 Each shape on the right has been enlarged to the shape on the left.
Give the scale factor for each of these enlargements.

a  **b**  **c**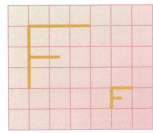

2   PQR has been enlarged to P'Q'R'.
    What is the scale factor of the enlargement?

a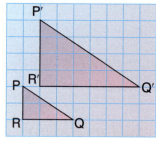

b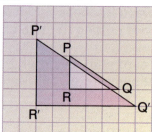

c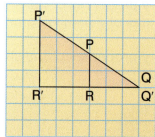

Triangle ABC has been enlarged by scale factor 2.
The image is called A'B'C'.
O is the **centre of enlargement**.

OA' = 2 times OA
OB' = 2 times OB
OC' = 2 times OC

centre of enlargement

$\dfrac{OA'}{OA} = 2$ and $\dfrac{OB'}{OB} = 2$ and $\dfrac{OC'}{OC} = 2$

2 is the scale factor.

The centre of enlargement is the only point that does *not* change its position after the enlargement.

Discussion

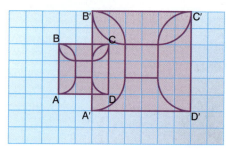   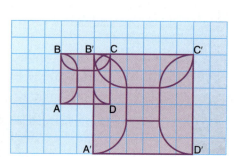

In both of these diagrams ABCD has been enlarged to A'B'C'D', by scale factor 2.

Why is A'B'C'D' in a different position in each diagram? **Discuss**.

To **draw an enlargement** we need to know the scale factor **and** the centre of enlargement.

Example   To enlarge ABCD by scale factor 2, centre of enlargement P,

1 draw a line from the centre of enlargement P, to some of the points on the smaller kite,
2 now make each line drawn in **1** twice as long (×2),
3 draw in the larger kite,
4 label the image points A′, B′, C′, D′, E′ and F′.

You can do this by measuring or by counting squares.

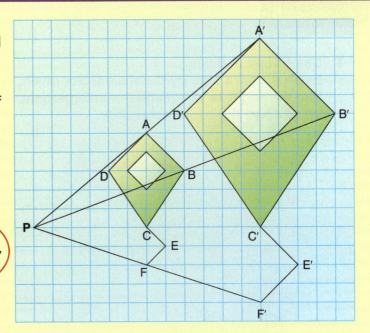

 Practical

You will need three A4 sheets of 1 cm square dotty paper and 3 copies of the table below.

1 Draw this 'alien' in the middle of a sheet of 1 cm square dotty paper.
2 Use the red dot as your centre of enlargement.
   Enlarge the alien by a scale factor of 2.
3 Use the blue dot as your centre of enlargement.
   Enlarge the alien by a scale factor of 3.
4 Use the blue dot again as your centre of enlargement.
   Enlarge the alien by a scale factor of 4.
5 Fill in your table for these enlargements.

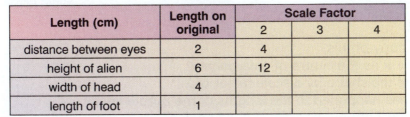

Length (cm)	Length on original	Scale Factor		
		2	3	4
distance between eyes	2	4		
height of alien	6	12		
width of head	4			
length of foot	1			

Rose noticed that for scale factor 2

height of alien on enlargement : height of alien on original = 12 : 6 = **2** : 1

Do you get the ratio 2 : 1 for all the other lengths?

What is this ratio when the scale factor is 3?
What about scale factor 4?

Do the angles in your shapes change when it is enlarged?

Shape, Space and Measures

Exercise 5

1 Use a copy of these.
Enlarge each shape by the scale factor given.
Use the origin as your centre of enlargement.
Write down the coordinates of A′.

a

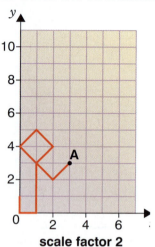

scale factor 2

b

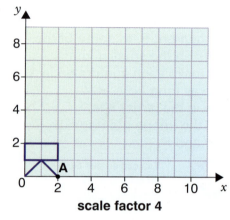

scale factor 4

c

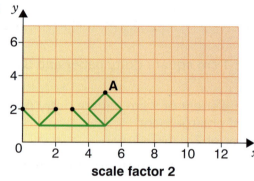

scale factor 2

d

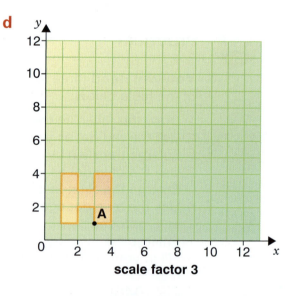

scale factor 3

2 a Use a copy of this.
Use the **×** as the centre of enlargement.
Enlarge the shape by a scale factor of 2.
What has happened to the lengths of the purple lines?
What has happened to the shaded angles?
Measure them if you need to.

b Repeat **a** for scale factor 3.
Try other centres of enlargement.

c Copy and finish this sentence.
Use these words to fill in the gaps: angles, lengths.

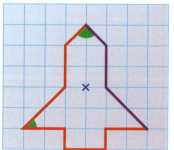

When a shape is enlarged the _____ stay the same size but the _____ change.

306

3 Use a copy of this.
This cuboid is made from **4** small cubes.

a Draw a cuboid which is **twice** as **high**, **twice** as **long** and **twice** as **wide**.

b Graham made this cuboid from **3** small cubes.

Mohinder wants to make a cuboid which is **twice** as **high**, **twice** as **long** and **twice** as **wide** as Graham's cuboid.

How many small cubes will Mohinder need altogether?

Scale drawing

A **scale drawing** is drawing of something in real life.
It has a scale so we can work out what each length in the drawing represents in real life.

A scale of 1 cm to 1 m means that 1 cm on the scale drawing represents 1 m in real life.

Worked Example

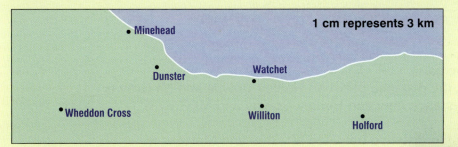

1 cm represents 3 km

Minehead
Dunster
Watchet
Wheddon Cross
Williton
Holford

a A helicopter flies from Minehead to Holford.
About how far, in kilometres, did the helicopter fly?
b The helicopter flies from Holford to Wheddon Cross.
Estimate this distance, in **miles**.
(Use 8 km = 5 miles.)

Shape, Space and Measures

Exercise 6   **Only use a calculator if you need to.**

1 What goes in the gaps?
Jafar made an accurate floor plan of his house.
He used a scale of **1 cm represents 400 cm**.
On his plan the games room is 2 cm wide.
  2 cm represents ____ in real life.
The games room is 3 cm long on the plan.
  3 cm represents ____ in real life.
The games room is ____ cm by ____ cm.

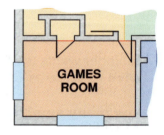

2 This map shows some farms in an area.
  1 cm represents 20 km.

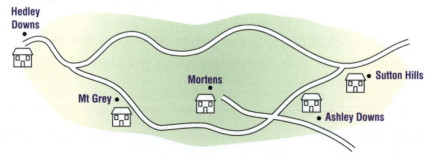

What goes in the gap?
a On the map, the distance between Hedley Downs and Mt Grey is ____ .
b The actual distance between Hedley Downs and Mt Grey is ____ × 20 km which is ____ .
c On the map, the distance between Mt Grey and Mortens is ____ .
d The actual distance between Mt Grey and Mortens is ____ .
e On the map, the distance between Mortens and Sutton Hills is ____ .
f The actual distance between Mortens and Sutton Hills is ____ .
g Find the actual distance between Mortens and Ashley Downs.
h A helicopter flew from Hedley Downs to Sutton Hills.
    About how far did if fly?

3 At the scene of an accident the police made a scale drawing.
They used the scale **1 mm represents 20 cm**.
On the drawing one of the skid marks was 84 mm long.
a 84 mm represents   84 × ____ cm
                              = ____ cm in real life
b How long, in metres were the skid marks in real life?

4 Peter made a scale drawing of his bedroom.
The scale he used was **1 cm represents 50 cm**.
On his drawing his room is 8 cm by 9 cm.
a What real width does 8 cm represent?
b What real length does 9 cm represent?

1 cm to 1 m can also be written as 1 : 100.

5 A scale of **1 cm represents 500 m** was used to draw a plan of a new road.
On the plan this road was 75 cm long.
a What was the actual length, in m, of this road?
b What was the actual length, in km, of this road?

Remember
1 km = 1000 m

6 Carrie made a scale drawing of a yacht.
She used the scale **1 cm represents 3 m**.
What is the real life length of these?
The length on the scale drawing is given.
a mast 4 cm
b length of yacht 3·5 cm
c width of yacht 1·2 cm
d width of sail at bottom 1·8 cm

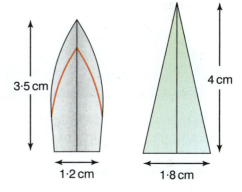

3·5 cm

4 cm

1·2 cm

1·8 cm

7 The scale of a map is **1 cm represents 25 km**.
What actual distance, in km, do these measurements, on the map, represent?
a 5 cm   *b* 5 mm   *c* 150 mm

We can **estimate heights or lengths on a scale drawing** by comparing them with heights and lengths we know.

Example   In the scale drawing the man is 1 cm high and the building is 3 cm high.

The building is three times as tall as the man.
The height of an average man is about 1·8 m.
The building is about 3 × 1·8 = 5·4 m.

The building is about **$5\frac{1}{2}$ m** tall.

Shape, Space and Measures

Exercise 7

1 Estimate the height of these.
  a the child
  b the garage
  c the tree
  d the light

Remember: an average man is about 1·8 m tall.

2 This picture shows an average adult woman.

An average woman is about 1·6 m tall.

Estimate the length of the fence.

3 This picture shows a whale.

The man is 1·8 m tall.

A man of average height is standing by it.
Estimate the length of the whale.

 ＊Practical

You will need a metre ruler and a tape measure.

Bryn wanted to know the height of the school flagpole.
He measured the length of the shadow of a one metre ruler.
The shadow was 64 cm long.
He measured the shadow of the flagpole.
It was 252 cm long.

He worked out the shadow of the flagpole was *about* four times longer than the shadow of the one metre ruler. $252 \div 64 \approx 4$
So the flagpole is about $4 \times 1 = 4$ m high.

Use Bryn's method to estimate the height of a tall tree, a lamp-post, a flagpole, ...

Making a scale drawing

Worked Example

A decorator wants to make a scale drawing of Kate's living room.
She made this sketch.
She used a scale of 1 cm represents 2 m.
Make the scale drawing.

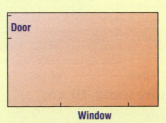

Answer

We must work out the lengths for the scale drawing.

      2 m is drawn as 1 cm.
So   1 m would be drawn as 0·5 cm.
      8 m is drawn as 8 × 0·5 cm = 4 cm.
      5 m is drawn as 5 × 0·5 = 2·5 cm.
      3·6 m is drawn as 3·6 × 0·5 cm = 1·8 cm.
      1·6 m is drawn as 1·6 × 0·5 cm = 0·8 cm.
      1·2 m is drawn as 1·2 × 0·5 cm = 0·6 cm.

The scale drawing is shown.

Practical

You will need a ruler, tape measure and scissors.

1 Kate made a scale drawing of her bedroom.
        She measured its length as 4 m.
        She measured its width as 3·5 m.
   Using a scale of 1 cm represents 50 cm she drew the length of her bedroom as 8 cm and the width as 7 cm.

   Make a scale drawing of a room in your school or house.

   Make separate scale drawings of the furniture. Cut them out.
   Design a layout for this room that is different from the present layout.

   Good scales to choose might be 1 cm represents 1 m or 2 cm represents 1 m.

*2 Make a scale drawing for something you have studied in another subject, for example an item of clothing.

   You will need to make a sketch first and estimate measurements.

1 Make an accurate scale drawing from each of these sketches.
Use the scale given in green.

a

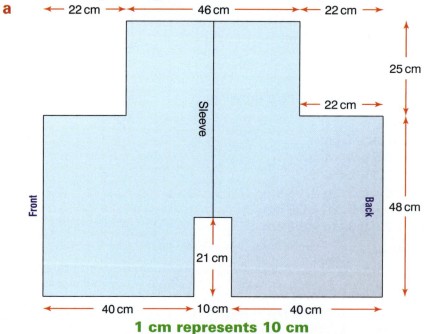

1 cm represents 10 cm

b

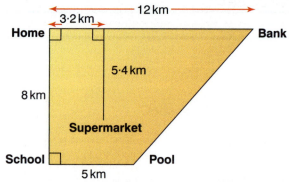

1 cm represents 2 km

c

1 cm represents 5 m

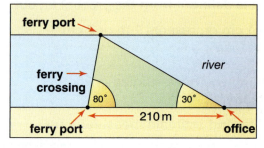

2 Use a copy of this.
Here is a plan of a ferry crossing.

[SATs Paper 2 Level 5]

Not drawn accurately

a Complete the accurate scale drawing of the ferry crossing below.

ferry → • ──────────────────────── • ← office
port ← ─────────── 210 m ─────────── →

b What is the length of the ferry crossing on **your** diagram?

c The scale is **1 cm** to **20 m**. Work out the length of the real ferry crossing.
Show your working, and **write the units with your answer**.

* # Finding the mid-point of a line

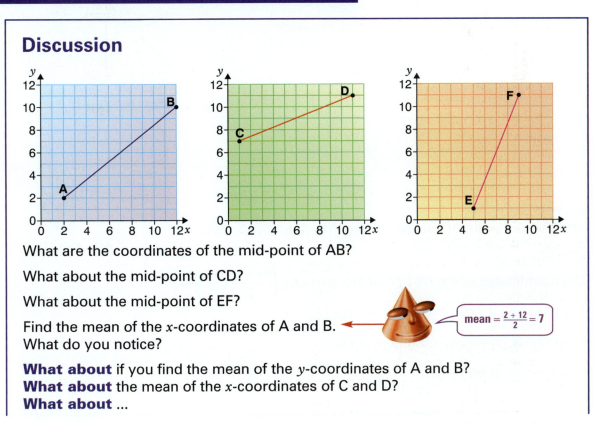

Discussion

What are the coordinates of the mid-point of AB?

What about the mid-point of CD?

What about the mid-point of EF?

Find the mean of the x-coordinates of A and B.
What do you notice?

mean $= \frac{2 + 12}{2} = 7$

What about if you find the mean of the y-coordinates of A and B?
What about the mean of the x-coordinates of C and D?
What about ...

313

Shape, Space and Measures

Does finding the mean of the x- and y-coordinates of A and B work for finding the mid-points of these? **Discuss**.

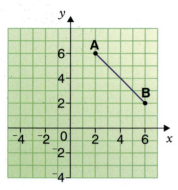 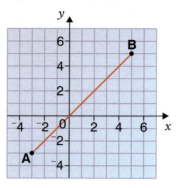 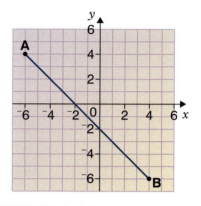

The **mid-point of a line segment joining A(x_1, y_1) to B(x_2, y_2)** is given by $\left(\dfrac{x_1 + x_2}{2}, \dfrac{y_1 + y_2}{2}\right)$.

mean of x-coordinates   mean of y-coordinates

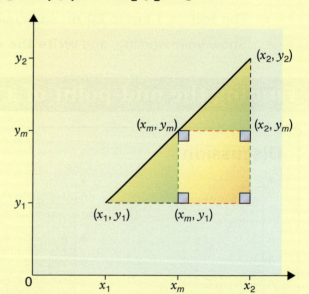

Worked Example

Find the mid-point of the line joining P(**3**, **4**) and Q(**5**, **10**).

Answer

The coordinates of the mid-point are given by $\left(\dfrac{x_1 + x_2}{2}, \dfrac{y_1 + y_2}{2}\right)$.

Call P the point (x_1, y_1) and Q the point (x_2, y_2).

Then $x_1 = $ **3**, $y_1 = $ **4**, $x_2 = $ **5** and $y_2 = $ **10**.

Coordinates of mid-point $= \left(\dfrac{\mathbf{3} + \mathbf{5}}{2}, \dfrac{\mathbf{4} + \mathbf{10}}{2}\right) = \left(\dfrac{8}{2}, \dfrac{14}{2}\right) = \mathbf{(4, 7)}$

Exercise 9

1 a Find the coordinates of the mid-points of these lines by looking at the diagrams.

i

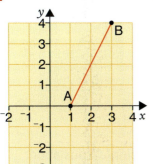

ii

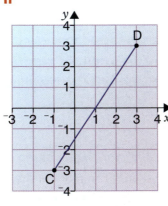

iii

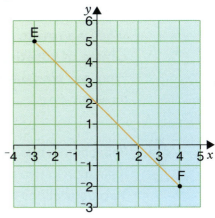

b Use a copy of this.
Check your answers to **a** by using the formula to find the mid-points.
Fill in the gaps.

i $\dfrac{x_1 + x_2}{2} = \dfrac{\underline{\quad} + \underline{\quad}}{2}$

$\qquad\qquad = \underline{\quad}$

$\dfrac{y_1 + y_2}{2} = \dfrac{\underline{\quad} + \underline{\quad}}{2}$

$\qquad\qquad = \underline{\quad}$

ii $\dfrac{x_1 + x_2}{2} = \underline{\quad}$

$\dfrac{y_1 + y_2}{2} = \underline{\quad}$

iii $\dfrac{x_1 + x_2}{2} = \underline{\quad}$

$\dfrac{y_1 + y_2}{2} = \underline{\quad}$

2 Plot the points A and B on a grid.

Find the coordinates of the mid-point of the line joining A and B using $\left(\dfrac{x_1 + x_2}{2}, \dfrac{y_1 + y_2}{2}\right)$.

Check on your grid that this gives the correct answer.

The first one is started.

a A(2, 6), B(4, 8)

The mid-point is $\left(\dfrac{x_1 + x_2}{2}, \dfrac{y_1 + y_2}{2}\right)$

$= \left(\dfrac{2 + 4}{2}, \dfrac{6 + 8}{2}\right)$

$= \underline{\quad}$

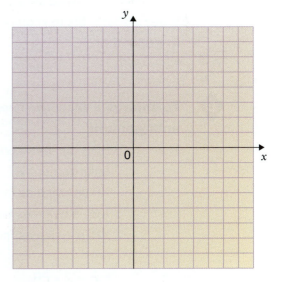

b A(0, 3), B(4, 7)
d A(3, 2), B(9, 8)
f A(1, 1), B(7, 8)

c A(0, 4), B(4, 8)
e A(1, 6), B(7, 1)
*g A(⁻3, ⁻6), B(⁻5, ⁻8)

Summary of key points

A **Transformations**

Example The red shape has been **reflected** in the *y*-axis, to give the green shape.

When we **rotate** a shape we need to know the **centre of rotation** and the **angle of rotation**. When we **translate** a shape we slide it without turning.

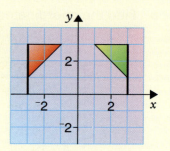

B Reflection in two parallel mirrors is the same as a single translation. Reflection in two perpendicular mirrors is the same as a single rotation. Two translations can be combined to give a single translation.

C A shape has **reflection symmetry** if one half of the shape can be reflected in a line to the other half. The line is a **line of symmetry**.

A shape has **rotation symmetry** if it fits onto itself **more than once** during a full turn.

This shape has 2 lines of symmetry and rotation symmetry of order 2.

The **order of rotation symmetry** is the number of times a shape fits exactly onto itself in a 360° turn.

If a shape has rotation symmetry of order 1, we say it does not have rotation symmetry.

D To **enlarge** a shape we need to know the **scale factor** and the **centre of enlargement**.

Example PQR has been enlarged by scale factor **2** and centre of enlargement X.
XP′ is twice as far as XP.
XQ′ is twice as far as XQ.
XR′ is twice as far as XR.
and
P′R′ = 2 × PR
P′Q′ = 2 × PQ
Q′R′ = 2 × QR

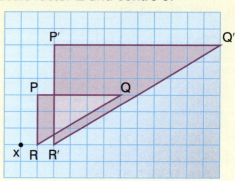

A **scale drawing** represents something in real life.

Example This is a scale drawing of a car.

Scale   1 mm represents 6 cm

Each millimetre on the drawing represents 6 cm in real life.

The car is 50 mm on the drawing.

In real life it is 50 × 6 = 300 cm

$$= 3\text{ m}$$

 We can **estimate heights or lengths** by comparing them with a height or length we know.

Example This ladder is 3 m or 300 cm high.

The ladder is about 3 times as high as the child.

The child is about 1 m or 100 cm tall.

 The **mid-point** of the line segment joining A(x_1, y_1) to B(x_1, y_2) is given by

$$\left(\frac{x_1 + x_2}{2}, \frac{y_1 + y_2}{2}\right).$$

Example The mid-point, M, of the line joining AB is

$$\left(\frac{3 + 13}{2}, \frac{4 + 10}{2}\right) = (8, 7).$$

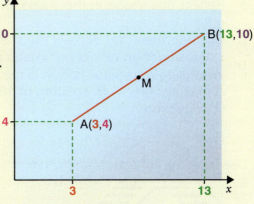

Shape, Space and Measures

T **1** Use a copy of this.
Reflect each shape in the mirror line.

a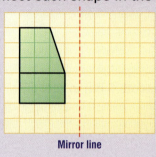

Mirror line

b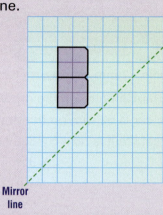

Mirror
line

T **2** Use three copies of this diagram.

Write down the coordinates of A, B and C
after these transformations.

a translation 4 units to the right
b reflection in the *x*-axis
c rotation 90° about (0, 0)

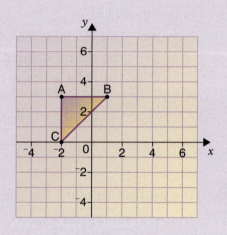

3 Use a copy of this.

a Reflect this shape in the *y*-axis and then
reflect the image in the *x*-axis.
b Use another copy. Reflect the shape in the
x-axis and then reflect the image in the *y*-axis.
c What *single* transformation is the same as **a**?
  A translation   **B** rotation of 180°
  C reflection

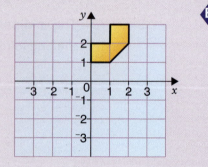

4 Describe the reflection and rotation symmetry of each of these.

a    **b**    *****c**

5 a The A on the left has been enlarged to the A on the right. What is the scale factor?

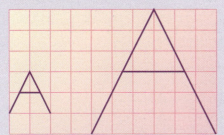

b ABC has been enlarged to A'B'C'. What is the scale factor?

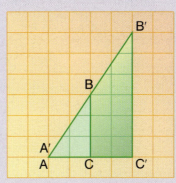

6 a Use a copy of this.
Enlarge this 'fish' by a scale factor of 2, centre of enlargement (0, 0).
Write down the coordinates of A', the image of A.

b Use another copy of the fish.
Enlarge it by a scale factor of 2, centre of enlargement (3, 2).
Write down the coordinates of A', the image of A.

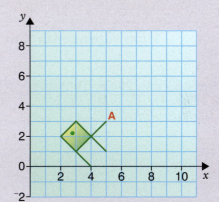

7 Jess made a scale drawing of the school grounds.
She used a scale **1 mm represents 2 m**.

a On her plan the distance between the gymnasium and the tennis courts is 15 mm.
What goes in the gaps?

15 mm represents   15 × ___ m

= ___ m in real life.

b On her plan the distance from the main gate to the main door is 32 mm. How far is this in real life?

8 Estimate the length of the house in this picture.
The man is about 1·8 m tall.

Shape, Space and Measures

9 Make an accurate scale drawing of this crane from this sketch. **E**

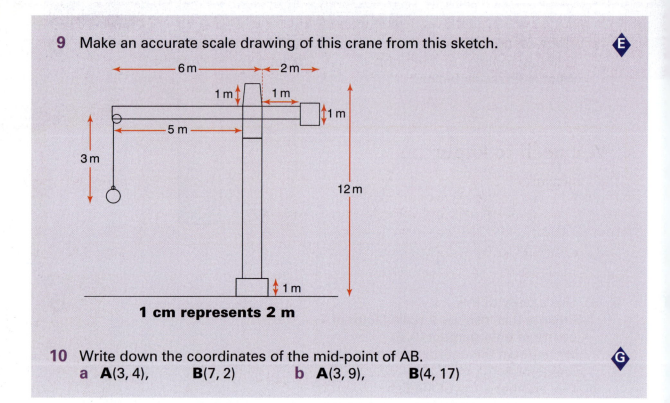

1 cm represents 2 m

10 Write down the coordinates of the mid-point of AB. **G**

   a **A**(3, 4),   **B**(7, 2)   **b** **A**(3, 9),   **B**(4, 17)

13 Measures, Perimeter, Area and Volume

You need to know

✓ measures   page 240
  – metric conversions
  – metric and imperial equivalents
  – reading scales

✓ perimeter, area and surface area   page 241

Key vocabulary

foot, hectare, yard, tonne, volume

Greener than green

Heather has a lawn mowing business.
She is asked to give a quote for mowing this rectangular park.

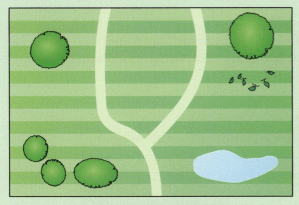

scale: 1 mm represents 1 m

She draws this scale drawing of the park.
If Heather charges £10 for every 1000 m² mowed, what

Shape, Space and Measures

Metric conversions

There is more about metric conversions on page 240.

Exercise 1

Find the answers to these mentally if possible.

1 Change
  a 700 mm to m
  b 37 mm to cm
  c 280 cm to m
  d 1450 cm to m
  e 800 mℓ to ℓ
  f 1564 g to kg
  g 360 mm to m
  h 36 mm to m
  i 86 cm to m
  j 50 mℓ to ℓ
  k 3 mm to m
  l 9 g to kg.

2 A pupil recorded how much rain fell on 5 different days.   [SATs Paper 1 Level 4]
  Results:

	Amount in cm
Monday	0·2
Tuesday	0·8
Wednesday	0·5
Thursday	0·25
Friday	0·05

  a On which day did most rain fall?
    On which day did the least rain fall?
  b How much **more** rain fell on Wednesday than on Thursday?
  c How much rain fell altogether on **Monday**, **Tuesday** and **Wednesday**?
    Now write your answer in millimetres.

3 How many minutes in
  a 4 hours
  b 5·5 hours
  c 3·25 hours
  d 6 hours 35 minutes?

4 How many hours and minutes in
  a 180 minutes
  b 95 minutes
  c 500 minutes
  d 426 minutes?

5 How many hours in
  a 4 days
  b $6\frac{3}{4}$ days
  c 1 week 3 days
  d 3 weeks 1 day 15 hours
  e 462 minutes?

6 **a** Sara and Adam are making fruit punch.
They add 7.6 *l* of lemonade to 800 m*l* of fruit
punch mix.
How many m*l* will this be altogether?

  b Sara and Adam divide the punch mixture into
4 bowls.
How many *litres* will be in each bowl?

7 **a** Chris and Dewi were stacking bricks.
Each brick was 8·5 cm high.
How high would a stack of 20 bricks be?
Give your answer in metres.

  b Chris and Dewi must put some bricks into a box.
The box is 1·35 m high.
How many layers of bricks will fit in the box?

8 **a** A book weighs 280 g.
How much will 85 of these weigh?
Give your answer in kilograms.

  b A poster weighs 17 g.
How much will 180 of these weigh?
Give your answer in kilograms.

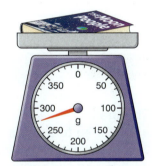

9 Pete wrote down what he and his friends drank.
    1·2 *l* cola   1·5 *l* juice   700 m*l* water
How many litres is this altogether?

10 Tyrone is a tour guide for 'Lakes Tours'.
Each tour takes 4 days.
Last year he guided 20 tours.
How many weeks and days did this take?

Other metric units

Mass
We use **tonnes** to measure very heavy objects.

Worked Example
Lump, the elephant, weighs 4520 kg.
How many tonnes is this?

> If you need to remind yourself how
> to multiply and divide by 10, 100
> and 1000 go to page 1.

Answer
4520 kg = (4520 ÷ 1000) tonnes
        = **4·52 tonnes**

Shape, Space and Measures

Area
Hectares are used to measure large areas.
1 hectare (ha) = 10 000 m^2

Examples   The area of a school's grounds might be about 5 hectares.
The area of a farm might be about 80 hectares.

Worked Example
A rectangular park is 150 m by 200 m.
Find the area of the park in hectares.

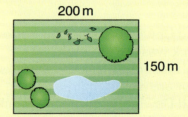
200 m
150 m

Answer
Area = length × width
    = 150 × 200
    = 30 000 m^2

30 000 m$^2 = \frac{30\ 000}{10\ 000}$ ha
    = **3 ha**

Exercise 2

   Only use a calculator if you need to.

1 Change
a   7000 kg to tonnes   b   5000 kg to tonnes   c   8700 kg to tonnes
d   5·4 tonnes to kg   e   8·7 tonnes to kg   f   9·83 tonnes to kg
g   4750 kg to tonnes   h   420 kg to tonnes.

2 Change
a   4 ha to m^2 b   3·6 ha to m^2 c   83 ha to m^2
d   5·72 ha to m^2 e   50 000 m^2 to ha   f   70 000 m^2 to ha
g   120 000 m^2 to ha h   5000 m^2 to ha.

3 How many m^2 in each of these parks?
Hyde Park 255 ha   Kensington Gardens 111 ha   Regents Park 197 ha
Kew Gardens 120 ha

4 A rectangular playing field is 200 m by 350 m.
How many hectares is this?

Remember
area = length × width

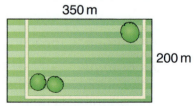
350 m
200 m

5 The mass of an empty lorry is 14·2 tonnes.
When the lorry has 16 containers on it its mass is 22·68 tonnes.
What is the mass of each container, in kg?

*6 Roydon School grounds are rectangular in shape.
One side is 150 m long.
The area of the grounds is 3 ha.
What is the length of the other side?

*7 What will the time and date be 8520 minutes after the start of 2005?

 *Puzzle

In three minutes time it will be three times as many minutes to 9 o'clock as it was past 8 o'clock nine minutes ago. What is the time now?

Metric and imperial equivalents

Length

These are rough **metric and imperial equivalents** for **length**.

1 mile ≈ 1·6 km	1 km ≈ 0·625 miles
1 yard ≈ 1 metre	(8 km ≈ 5 miles)
1 inch ≈ 2·5 cm	1 m ≈ 1 yard or 3 feet.

Remember these are estimates.

Worked Example
Dee walked 17 miles each week.
About how many km is this?

Answer
1 mile is about 1·6 km.
$17 \times 1·6 = 27·2$
Dee walked about **27 km** each week.

Exercise 3 **Only use a calculator if you need to.**

1 About how many miles to London?

London 16 km

2 Joel's car is about 12 feet long.
About how many metres is this?

3 Distances on a map were given in km.
Josefa wanted to convert them to miles.
About how many miles are there in the following?
  a 40 km   **b** 80 km   **c** 48 km   ***d** 36 km

Shape, Space and Measures

4 Argene is visiting England.
In her country, distances are measured in km.
Convert these distances to km.
  a 20 miles   **b** 200 miles   **c** 50 miles   ***d** 12 miles

5 Measurements for scenery for the school play were taken in inches and feet.
Amanda needed these measurements in cm.
Convert the following for her.
  a 6 inches   **b** 2 inches   **c** 18 inches
  d 3 feet   **e** 9 feet

6 Jenufa's family own a 15 foot caravan.
  a Find the approximate length of this caravan, in metres.
  b How many caravans like this could be stored end to end in a shed 20 m long?

*7 A book is 25 cm by 18 cm.
Find the length and width of this book, in inches.

Mass

These are rough **metric and imperial equivalents for mass**.

  1 pound (lb) is a bit less than $\frac{1}{2}$ kg.   1 kg ≈ 2·2 lb   1 oz ≈ 30 g

Worked Example
Alice bought a 3 kg bag of oranges.
a About how many pounds is this?
b She weighed one orange and it was $\frac{1}{4}$ lb.
About how many oranges were in the bag?

Answer
a 1 kg is about 2·2 lb.
    $3 \times 2\!\cdot\!2 = 6\!\cdot\!6$
There are about **6·6 lb** in the bag.
b 1 orange weighs about $\frac{1}{4}$ lb or 0·25 lb.

$\frac{6\cdot6}{0\cdot25} = 26\cdot4$ **Key**
There are about **25** oranges in the bag.

Exercise 4   **Only use a calculator if you need to.**

1 Menna bought this bag of potatoes.
She used 4 lb for a potato stew.
About how many pounds did she have left?

Potatoes
5 kg

2 Packages to send overseas were weighed.
   About how much do these weigh in kg?
   a 4·4 lb   **b** 13·2 lb   **c** 20 lb   *__d__ 20 oz

3 A recipe for raspberry jam uses 6 lb of sugar.
   About how many kg of sugar is used?

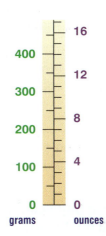

4 A scale measures in **grams** and **ounces**.
   Use the scale to answer these questions.
   a About how many ounces is **400 grams**?
   b About how many grams is **8 ounces**?
   c About how many ounces is **1 kilogram**?
      Explain your answer.

*__5__ At the supermarket, Ruski bought this fruit.
   a About how many pounds of fruit did she buy?
   b Apples are £0·89 per pound.
      How much would 4 kg of apples cost?
   c About how many ounces is 250 g?

MEMO

Apples	5 kg
Pears	250 g
Peaches	500 g
Bananas	250 g
Oranges	1 kg

Capacity

These are rough **metric and imperial equivalents for capacity**.

   1 pint ≈ 600 mℓ   1 litre ≈ 1·75 pints
   1 gallon ≈ 4·5 litres

Worked Example
A measuring jug can hold 2·5 litres.
About how much does this jug hold in pints?

Answer
1 litre is about 1·75 pints.
2·5 litres = 2·5 × 1·75 pints   **Key** [2·5] [×] [1·75] [=]
      = 4·375 pints
We could give the answer as **about $4\frac{1}{2}$ pints**.

Shape, Space and Measures

  Only use a calculator if you need to.

1 Travis bought a 3 litre bottle of juice.
  a About how many pints is this?
  b He poured 2 pints into a drink bottle.
     About how many litres are left?

2 About how many litres does a 2 gallon water container hold?

3 'Delicious Drinks' packages its juice in the following sizes.
  About how many pints is each?
  a 10 litres   **b** 5 litres   **c** 0·5 litre

4 A washing machine holds about 27 ℓ of water.
  About how many gallons of water does this washing machine hold?

5 *Highlife* paint can be bought in 2 ℓ and 5 ℓ tins.
  About how many more pints does the larger tin hold?

? Puzzle

This is part of a letter Marcia sent to her Swiss penfriend Yurg.
Yurg converted each of the measures given in this story to the unit in the brackets.
What answers should he get?

> Melanie took all of her pets to the school 'pet day'.
> She lives 30 miles (km) from her school so she left home very early.
> The goldfish tank had 4 gallons (litres) of water so it was heavy.
> The dog was on a 4 metre (feet) long lead.
> The cat was entered in the large cat section because it weighed 11 lb (kg).
> The cat won the 'longest tail' section.
> Its tail is 15 inches (cm) long.

Discussion

● Suggest an imperial or metric unit you could use to measure these. **Discuss**.

mass of a nail
distance from Mars to Jupiter
amount of petrol in a petrol tanker
time to grow a water cress plant
diameter of a 20p coin.

width of a pencil
mass of a car
time to travel to the moon
thickness of a hair

Suggest a measuring instrument and a way to measure each. **Discuss**.

- Melanie is making scones.
  Should she measure the flour to the nearest g,
  nearest 5 g, nearest 10 g or nearest 50 g? **Discuss**.

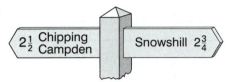

SCONES

375 g flour
3 tsp baking powder
1/4 tsp salt
1 cup milk
25 g butter

- The distance to Chipping Campden is given in
  miles.
  Do you think the distance on this road sign is
  correct to the nearest mile, nearest $\frac{1}{2}$ mile or
  nearest $\frac{1}{4}$ mile? **Discuss**.

$2\frac{1}{2}$ Chipping Campden   Snowshill $2\frac{3}{4}$

Practical

1 Design an experiment, such as measuring reaction time, to measure
small intervals of time.

2 The pendulum clock was invented in the 17th century.
Before this, time was sometimes measured using the clocks described
below.

The Candle Clock
Marks were made at equal distances down the side of a
candle.
How close do you think these marks might be if it took
5 minutes to burn the length of a candle between two
marks?

The Water Clock
A container, such as that shown, was filled with water.
The water dripped out of a hole at the bottom.
Marks were made on the inside of the container.
Equal intervals of time were read off by watching the level
drop from one mark to the next.
Do you think the marks should be equally spaced for this container?

continued ...

Sinking Bowl Clock
A small bowl, with a hole in the bottom, was put in a large container of water.
The time it took for the bowl to sink measured a particular interval of time.

Design and make a clock to measure an interval of time, maybe
2 minutes, 5 minutes or 15 minutes.
Make your clock as accurate as possible.

Discussion

Jonathon was asked to estimate the time it would take him to walk 5 miles.
It takes him about 20 minutes to walk the mile to school.

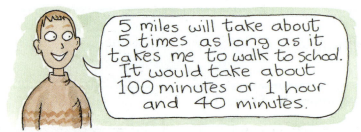

5 miles will take about 5 times as long as it takes me to walk to school. It would take about 100 minutes or 1 hour and 40 minutes.

Jonathon used what he already knew to help estimate the answer.

What other measurements do you already know that you could use when estimating? **Discuss**.

Use what you already know to help estimate these. **Discuss**.

the height of the tallest tree you can see   the capacity of a sink
the area of your school grounds   the area of the cover of this book
the mass of this book

Practical

You will need a tape measure and chalk.

1 Design an experiment to decide who is best at estimating 10 cm.
You *could* get everyone to cut 20 pieces of string to an estimated 10 cm.
Then measure each piece and write down the 'error'.
Find the mean error for each pupil.

2 'Invent' a person who is about your age.
Give your person a name.
Estimate sensible measurements for height, weight, arm length, foot length, waist measurement, ...
Make a poster showing your person.

Perimeter and area

Discussion

- **You will need** a sheet of paper and some scissors.

  Draw a triangle on the sheet of paper.
  Draw a rectangle around the triangle, as shown.
  Cut out the two pieces of the rectangle that are
  outside the triangle. (Pieces A and B.)
  Try to fit these two pieces exactly onto the triangle.
  Repeat this with other triangles.
  Do you think this statement is true? **Discuss**.

  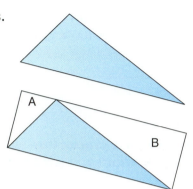

  > 'The area of a triangle is always half the area
  > of a rectangle.'

- A parallelogram can be made from a rectangle.

  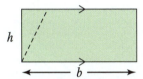

  Cut a triangle off one side of the rectangle.

  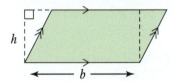

  Add it to the opposite side of the rectangle.

  Is the area of the parallelogram the same as the area of the rectangle?
  What is the formula for the area of the parallelogram? **Discuss**.
  Can all parallelograms be made this way? **Discuss**.

Remember

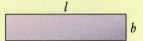

The **area of a rectangle** $= l \times b$

The **area of any triangle** is given by $A = \frac{1}{2}bh$ where b is the base of the triangle and h
is the perpendicular height

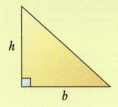
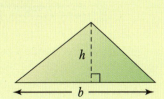
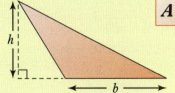

$$A = \tfrac{1}{2}bh$$

Shape, Space and Measures

The **area of a parallelogram** is given by $A = bh$ where b is the base of the parallelogram and h is the perpendicular height.

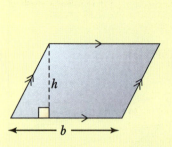 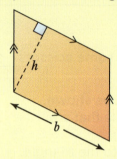 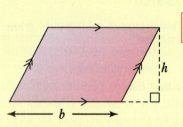

$$A = bh$$

The **area of a trapezium** is given by $A = \frac{1}{2}(a + b)h$ where a and b are the parallel sides of the trapezium and h is the perpendicular distance between them.

$$A = \frac{1}{2}(a + b)h$$

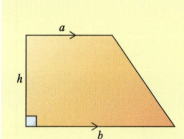 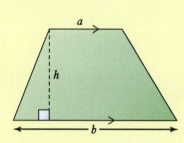 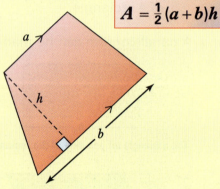

Examples

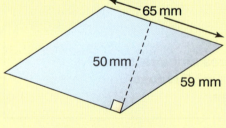 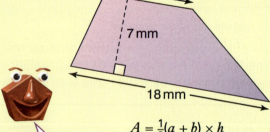

$A = bh$

$= 65 \times 50$

$= 3250 \text{ mm}^2$

Remember to put the units with the answers.

$A = \frac{1}{2}(a + b) \times h$

$= \frac{1}{2}(8 + 18) \times 7$

$= 91 \text{ mm}^2$

Worked Example

The ends of Lucy's roof are identical triangles.
She wants to know their area so she can paint them.
What is the area of *both* ends?

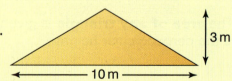

Answer

Area of one end $= \frac{1}{2}bh$

$= \frac{1}{2} \times 10 \times 3$

$= 5 \times 3$

$= 15 \text{ m}^2$

So both ends will be $2 \times 15 \text{ m}^2 = \textbf{30 m}^2$.

Exercise 6   **Only use a calculator if you need to.**

1   Alika has a box of square tiles.   [SATs Paper 2 Level 4]
       The tiles are three different sizes.

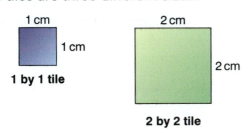

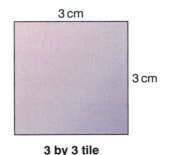

She also has a mat that is 6 cm by 6 cm.
36 of the 1 by 1 tiles will cover the mat.

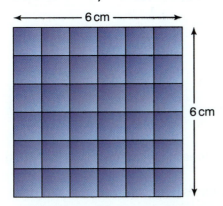

a   How many of the **2 by 2 tiles** will cover the mat?
b   How many of the **3 by 3 tiles** will cover the mat?
c   Alika glues three tiles on her mat like this:

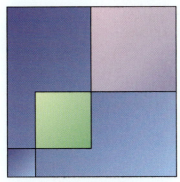

What goes in the gaps?
       She could cover the rest of the mat by using another **two** 3 by 3 tiles, and
       another ___ 1 by 1 tiles.

       She could cover the rest of the mat by using another **two** 2 by 2 tiles, and
       another ___ 1 by 1 tiles.

Shape, Space and Measures

2 Calculate the area of each of these triangles.

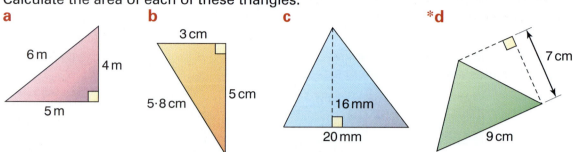

a

6 m 4 m 5 m

b

3 cm 5 cm 5·8 cm

c

16 mm 20 mm

*d

7 cm 9 cm

3 Calculate the perimeter of the triangles in question **2a** and **2b**.

4 Triangles of pastry are used to make savouries.
What area of pastry is used for each?

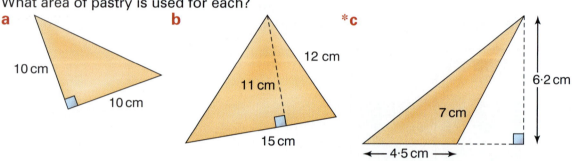

a

10 cm 10 cm

b

11 cm 12 cm 15 cm

*c

12 cm 6·2 cm 7 cm 4·5 cm

5 The 'Bermuda Triangle' is a part of the Atlantic Ocean in which more than 50 ships and 20 planes have mysteriously disappeared.
Find the area of the 'Bermuda Triangle' if the base is 1560 km and the height is 1320 km.

6 This diagram shows part of a tile pattern.

Each tile is a triangle like this.
a Find the area of one tile.
b What is the area of 12 tiles?

16 cm 34 cm 30 cm

7 Find the area of each of these parallelograms.

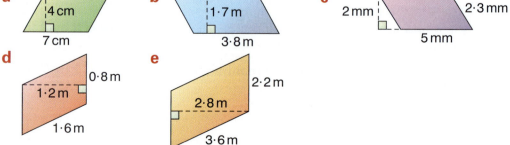

a

4 cm 7 cm

b

1·7 m 3·8 m

c

2 mm 2·3 mm 5 mm

d

0·8 m 1·2 m 1·6 m

e

2·2 m 2·8 m 3·6 m

8 Find the area of each of these trapeziums.
The first two are started.

a

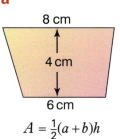

$A = \frac{1}{2}(a+b)h$

$= \frac{1}{2}(6+8) \times 4$

$= \underline{\qquad}$

b

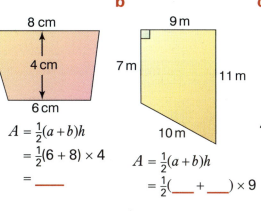

$A = \frac{1}{2}(a+b)h$

$= \frac{1}{2}(\underline{\quad} + \underline{\quad}) \times 9$

c

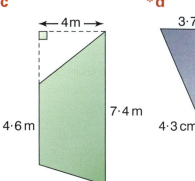

*d

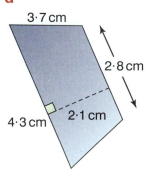

9 Use a copy of this.   [SATs Paper 2 Level 5]
In this question, all the grids are centimetre square grids.

a Draw a **rectangle** that has an **area** of **12 cm²**.

b Draw another rectangle that has an area of 12 cm².
This rectangle must have a **different perimeter** from the rectangle in part **a**.

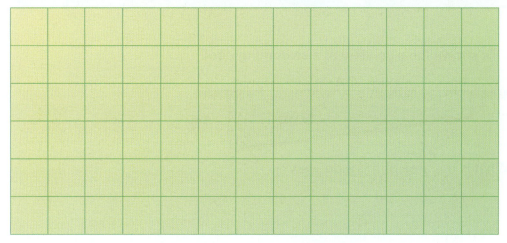

c Draw a **triangle** that has an **area** of **6 cm²**.

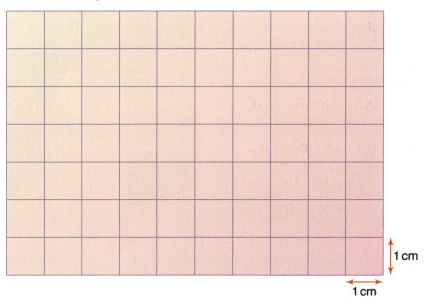

1 cm

1 cm

*10

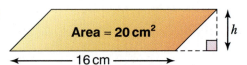

Area = 20 cm²   *h*

16 cm

The area of a parallelogram is 20 cm².
The base is 16 cm.
What is the height of this parallelogram?

Remember
Area = *bh*

*11 Even without measurements written on this diagram, we know that the areas of
the shaded figures are equal.
How do we know?

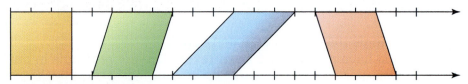

*12 This shows the side view of a swimming pool.

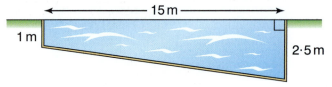

15 m

1 m

2·5 m

What is the area of this side view?

Investigation

Moving Vertices

You will need a pinboard or square dotty paper.

ABC is a triangle.
A has been translated 1 unit right to A_1.
How does this affect the area?

What if A is translated 2 units right to A_2?
What if A is translated 3 units right?
What if ...
Find the area of each triangle and explain what is happening.

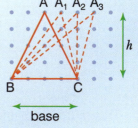

*Hexagons

Hetty made this shape with five regular hexagons of side n.
What is the perimeter of Hetty's shape?
Make a different shape with five regular hexagons.
What is the smallest perimeter you can get?
What about six regular hexagons of side n?
What about seven regular hexagons of side n?
What about eight, nine, ... regular hexagons of side n?

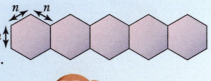

Put your results in a table.

Finding area and perimeter of shapes made from rectangles and triangles

Worked Example
Find the perimeter and area of the end of this building.
It is made up of a rectangle with a triangle on top.
The rectangle has base 8 m and height 4 m.
The triangle on top has a height of 3 m and base of 8 m.

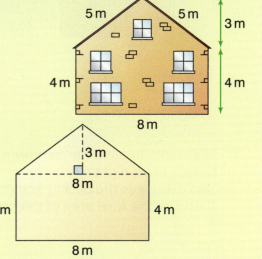

Answer
Perimeter $= 5 + 5 + 4 + 8 + 4$
$\qquad = \textbf{26 m}$
Area = area of rectangle + area of triangle
$\qquad = 8 \times 4 + \frac{1}{2} \times 8 \times 3$
$\qquad = 32 + 12$
$\qquad = \textbf{44 m}^2$

Always put units with your answers.

Shape, Space and Measures

1 This is the plan view of a cupboard.
Work out its perimeter.

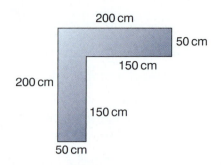

2 This is the floor plan of this room.
Work out the perimeter of the room.

First find the lengths of the red lines.

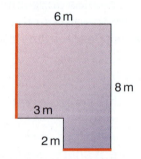

3 a Work out the perimeter of this room.
b Work out the area of the room.
Total Area = area of green shaded part
+ area of blue shaded part

The dashed line divides the shape into 2 rectangles.

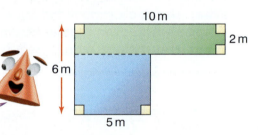

4 Work out the area of each of these rooms.

a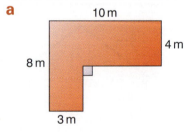

b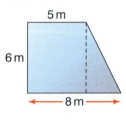

*c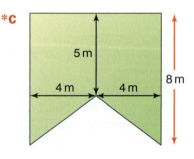

5 Ella wants to paint her bedroom ceiling dark blue
to look like a night sky.
She measures the ceiling and draws this plan.
Calculate the total area of the ceiling.

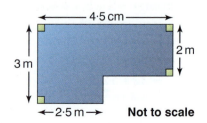

Not to scale

6 Kelsey wants to tile her bathroom.
She uses tiles that measure 10 cm by 10 cm.
She tiles this surface.

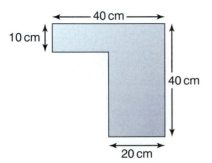

a How many tiles does she need?
b Calculate the area of this surface.
c Kelsey wants to put a wooden strip around the perimeter of this surface.
How many **metres** of wooden strip will she need?

7 A rectangular pool has a spa pool in one corner.
a What is the perimeter of the pool, **not** including the spa.
b What is the area of the pool and spa together?
c What is the area of the pool not including the spa?

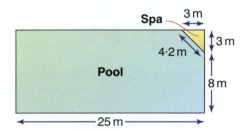

*8 The red shape is a road sign.
It needs painting.
Find the area of the sign.

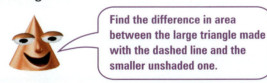

Find the difference in area between the large triangle made with the dashed line and the smaller unshaded one.

*9 Abel made picture frames.
Each was made from rectangles and triangles.

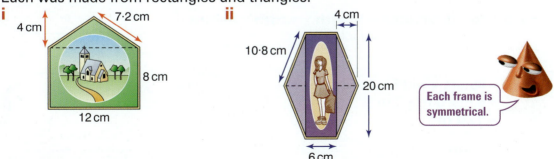

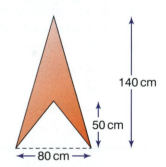

Each frame is symmetrical.

a Find the length of frame needed to make each picture frame.
b Find the area of each frame.
c Design a picture frame made from rectangles and triangles.
Put some measurements on.
Find the perimeter and area of your picture frame.

Surface area and volume

Remember
The surface area of a cuboid is given by this formula.

surface area = 2*lw* + 2*lh* + 2*wh*

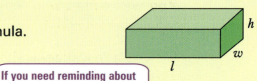

If you need reminding about surface area, go to page 242.

Discussion

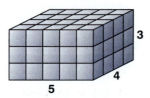

For this cuboid: length = 5 cubes
width = 4 cubes
height = 3 cubes

How many cubes are on the base of this cuboid?
Use this formula to find the volume.

Volume = number of cubes on base × number of layers

How many cubes are there altogether?

What is the volume of a cuboid *l* cubes long, *w* cubes wide and *h* layers high?
Discuss.

Volume is a measure of the amount of space something takes up.
We measure volume in mm^3, cm^3, m^3 or mℓ, ℓ, pints or gallons.

For **a cuboid, volume = length × width × height**
$= l \times w \times h$

Example
Volume = length × width × height
$= 8\,m \times 5\,m \times 3\,m$
$= \textbf{120 m}^3$

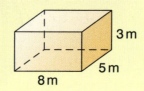

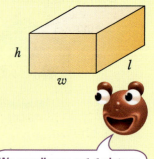

We usually use mℓ, ℓ, pints or gallons for fluids and mm^3, cm^3, m^3 for solid shapes.

Collecting data

Discrete data is sometimes grouped.
Groups must have **equal class intervals**.

Discrete data is sometimes called counting data.

Example This table shows the number of pupils at different schools.

Sometimes the last class interval is open.

Number of pupils	Tally	Frequency
0–99	II	2
100–199	I	1
200–299	IIII	4
300–399	IIII I	6
> 400	II	2

This is a tally chart or frequency table.

Planning a Survey

We follow these steps.

1 Decide what you want to find out.
2 Decide what data needs to be collected.
3 Decide where to collect the data from and how much to collect.
 ● You could use a **primary source**, for example
  a survey or a **sample** of people
  an experiment — observe, count or measure.
 ● You could use a **secondary source**, for example, reference books, CD-ROMs, websites, newspapers, ...

Once the survey has been planned, a **collection sheet** or a **questionnaire** often has to be designed.

Example Liz wrote down the colour of the chemicals in the science lab.

Colour	Tally	Frequency
Colourless	IIII IIII IIII	14
Blue	IIII I	6
Yellow	III	3

Frequency table

IIII is 5.

Link to science.

Example

How much time do you spend each week

(a) reading? 0 up to 2 hours ☐ 2 up to 4 hours ☐ 4 up to 6 hours ☐
    6 up to 8 hours ☐ > 8 hours ☐

(b) watching TV? 0 up to 3 hours ☐ 3 up to 6 hours ☐ 6 up to 9 hours ☐
    9 up to 12 hours ☐ > 12 hours ☐

Questionnaire

Practice Questions 15, 20

Displaying data

These diagrams show some ways of **displaying data**.

Pictogram

Number of drinks sold

Each symbol represents 10 drinks	
Hot chocolate	☕ ☕ ☕ ☕
Coffee	🍺 🍺 🍺 🍺
Juice	🥤 🥤
Coke	🍾 🍾

Note A whole symbol represents 10.
So half a symbol represents 5.

Bar chart

Sometimes bar charts are drawn sideways.

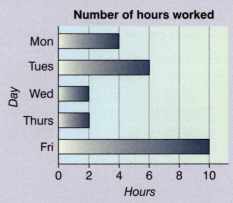

Bar-line graph

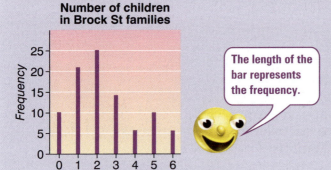

The length of the bar represents the frequency.

Bar chart

Sometimes two sets of data are shown on a bar chart. Always give a key.

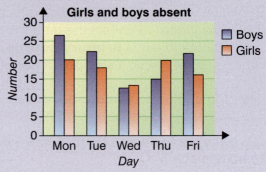

Pie chart

A pie chart shows the **proportion** in each category.

Always — give your graph a **title**
— **label** any axes
— have values at **equal intervals** on the axes.

Once you have displayed data on a graph, you can use the graph to help **interpret the data**.

Practice Questions 3, 4, 6b, 12, 17, 19, 21, 24

Mode, median, mean, range

Range = highest data value – the lowest data value

The **mode** is the most commonly occurring data value.
Sometimes a set of data has two modes.

Example   For   8, 9, 9, 3, 6, 9, 8, 8, 7, 5   the modes are 8 and 9.

If data is grouped, we find the **modal class**.
It is the class interval with the highest frequency.

Example   **Age of people in a village**

Age (years)	0–9	10–19	20–29	30–39	40–49	50–59	60–69	70–79	80+
Frequency	3140	2987	2864	3346	3162	2834	2172	1832	1436

The modal class is 30–39 because this has the highest frequency.

Mean $= \dfrac{\text{sum of data values}}{\text{number of data values}}$

Example   4, 8, 6, 2, 3, 9, 3, 5

$$\text{Mean} = \frac{4+8+6+2+3+9+3+5}{8}$$
$$= \frac{40}{8}$$
$$= 5$$

The **median** is the middle value when a set of data is arranged in order of size.
When there is an even number of values, the median is the mean of the two middle values.

Example   5, 6, 3, 9, 6, 4, 2, 1
In order these are 1, 2, 3, (4, 5,)6, 6, 9
Median $= \frac{4+5}{2}$
$= 4\cdot5$

We can **compare data** using the range and the mean or median.

Practice Questions 5, 6a, 7, 8, 9, 10, 11, 23, 25

Probability

We can describe the probability of an event happening using one of these words.

certain   likely   even chance   unlikely   impossible

Examples   It is **certain** that June follows May.
It is **likely** that you will eat breakfast tomorrow.
There is an **even chance** of getting a head when you toss a fair coin.
It is **unlikely** that you will see the Queen in person tomorrow.
It is **impossible** to draw a triangle with 4 sides.

Handling Data

Some events are **more likely** to happen that others.

Example   It is more likely you will spin a 2 than a 1 with this spinner.

Probability is a measure of the likelihood of an event happening.
We can show probabilities on a **probability scale**.

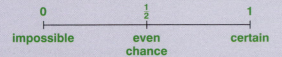

All probabilities lie from 0 to 1.

When a dice is rolled it could land on 1, 2, 3, 4, 5 or 6.
These are the possible **outcomes**.

Equally likely outcomes have an equal chance of happening.

Example   A counter is taken at random from the bag.
Getting red, getting blue or getting purple are all equally likely outcomes, because there is the same number of each colour counter.

For equally likely outcomes

   probability of an event = $\dfrac{\textbf{number of favourable outcomes}}{\textbf{number of possible outcomes}}$

Example   The letters in the word HORSES are put in a tub.
A letter is taken at random.
The probability of getting an H is $\frac{1}{6}$ (one in six).
The probability of getting an S is $\frac{2}{6}$ or $\frac{1}{3}$ (two in six).

Practice Questions 1, 2, 13, 14, 16, 18, 22

Practice Questions

1   Decide if each of these is **certain, very likely, likely, unlikely, very unlikely** or **impossible**.
   a   Someone will visit your house in the next week.
   b   A shopper at your local supermarket will buy sweets today.
   c   You will read 50 books this week.
   d   A polar bear will give you a kiss tomorrow.
   e   It will get dark tonight.

2   Put the events in question **1** in order from most likely to happen to least likely to happen.

3 This bar-line graph shows sales of football shirts at the Shamrock football club.

  30 of size 14 were sold.
  22 of size 16 were sold.

a Use a copy of this bar-line graph.
  Finish it.

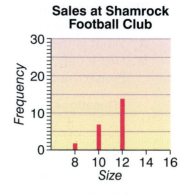

Sales at Shamrock Football Club

b How many size 12 were sold?

This bar-line graph shows sales of football shirts at the Avon football club.

One club was for senior players.
One was for junior players.

c Look at the two graphs.
  Which do you think was the junior club?
  How can you tell this from the graphs?

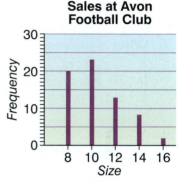

Sales at Avon Football Club

4 These pictograms show the items sold at a café one Saturday and Sunday.

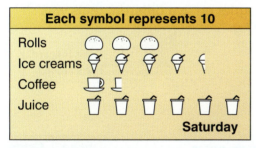

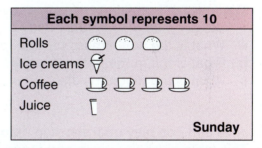

a How many coffees were sold on Sunday?
b How many more ice creams were sold on Saturday than Sunday?
c One of the days was hot and the other was cooler.
  Which day do you think was which?
  Explain.

5 Mr Salt drew this table to show the sizes of women's trainers he sold last week.
  a What is the modal shoe size?
  b What is the range of shoe sizes?
  c Why might Mr Salt want to know the modal shoe size sold?

Shoe size	Tally	Frequency
4	IIII III	8
5	IIII IIII I	11
6	IIII II	7
7	III	3
8	I	1

T

6 The table below shows the number of questions pupils got correct in a multi-choice test.
  a What is the modal class?
  b Use a copy of the grid.
  Finish the bar chart to show the data.

Questions correct	Tally	Frequency
0–9	ЖН II	7
10–19	ЖН ЖН III	13
20–29	ЖН ЖН ЖН II	17
30–39	ЖН II	7

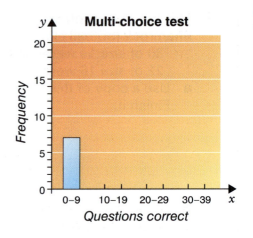

7 What is the range of each of these sets of data?
  a 5, 7, 8, 12, 16, 19, 24
  b 20, 32, 32, 38, 41, 52, 72
  c 2 g, 4 g, 3 g, 3 g, 2 g, 1 g, 2 g, 3 g, 4 g, 2 g

8 What is the median of each of the sets of data in question **7**?

9 What is the mean of each of the sets of data in question **7**?

10 The mass of iron filings picked up by a magnet on five different trials was
     85 g   96 g   110 g   97 g   97 g.
  a What is the mean mass picked up?
  b What is the range of the masses?
  c What is the median mass?

This is linked to science.

11 Julia and Zoe play goal shoot in netball.
This table gives the mean and range for their goals scored in 10 matches.
Who do you think is the better goal shoot?
Explain.

	mean	range
Julia	18	3
Zoe	10	8

12 This pie chart shows the homework time Rick spent on each subject last week.
  a Which subject did he spend the most time on?
  b Which subject did he spend the least time on?
  c Are these true? Write Yes or No.
    A Rick spent about 25% of his time on maths.
    B Rick spent less time on maths than an science.
    C Rick spent more time on science and art together than on maths.
    D Rick spent more than half the time on maths and geography together.

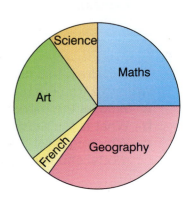

13 This spinner is spun.
Write 'certain', 'likely', 'even chance', 'unlikely', 'impossible'
for the likelihood of the spinner stopping on
a red   **b** green   **c** yellow
d 4   **e** red 4   **f** green 2.

14 These twelve letters are put in a box.
One letter is chosen without looking.
a Copy and finish this list of outcomes.
    A, D, ...
b Does each letter have an equal chance of being chosen?
c What is the probability of getting M?

15 For each of these questions, how would you collect the data?
    A questionnaire or data collection sheet
    B experiment
    C secondary source, such as website, book, newspaper, CD-ROM, ...
a What percentage of countries in the world speak English?
b Do adult males watch more TV than teenage boys?
c Do boys or girls know their times tables better?

16 This spinner is spun.
a Which colour is it least likely to stop on?
Give a reason for your answer.
b Copy this scale.

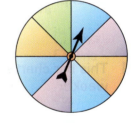

0        $\frac{1}{2}$ 1

Mark with an X the probability that the colour will be blue.
c Write down the probability that the colour will be red.

17 This graph shows the sales for 'Computers Today'.
a On which day did they sell £5000 of computers?
b What were the sales, in £, on Saturday?

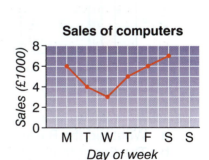

Sales of computers

18 Nick has these number cards.

He takes a card without looking.
What is the probability he will get
a [2]   **b** a red card   **c** an odd number

d a number smaller than 7?

You need to know

✓ collecting data page 347
  – discrete data
  – planning a survey

Key vocabulary

primary source, sample, secondary source, two-way table

⏩ The customer is always right

BMW

British Telecom

McDonalds

Big companies often carry out surveys to find out

– what customers want
– if customers are happy
– about other similar products.

What might companies such as BMW, British Telecom and McDonalds want to find out in a survey?

What data might they collect?

Discrete and continuous data

Discussion

1 Write down some possible values you could get if you counted the number of library books on a shelf in the library.
What if you were finding the number of visitors hospital patients had?

2 Write down some possible values you could get if you were measuring handspans, in cm.
What if you were measuring the temperature of water in an experiment?

The data in **1** and the data in **2** are different in some way.
Discuss what this difference might be.

Discrete data can only have certain values.
It is often found by counting.

If we collect data about the number of people at bus stops we might get 24, 12, 16, 3, 18, but **not** $26\frac{1}{2}$, $19\frac{1}{4}$, $3\frac{1}{3}$, ... because it is only possible to have whole numbers of people.

Examples – the number of goals scored by a soccer team this season.
– the number of trees in the gardens in your street

Continuous data is data which is measured.
It can have any values within a certain range.

Pete wanted to collect data on the height of the big cats at London Zoo.
He might get any height between 0·8 m and 1·2 m.
A height might be 0·94 m or 1·07 m.

Examples the masses of people in a weight lifting class
the fingernail length of netball players

Exercise 1

1 Which of the following are discrete data and which are continuous data?

  a the number of teeth people have
  b the height of newborn lambs
  c the number of windows in houses
  d the number of pupils in classes
  e the lengths of classrooms
  f the area of playgrounds at schools
  g the marks of pupils in a science test
  h the masses of eggs
  i the shoe sizes of pupils
  j the number of pupils who wear glasses
  k the cost of fruit
  l the number of rooms in houses
  m the amount of cola in bottles
  n the lengths of cats' tails

Grouping continuous data

Remember
Discrete data is sometimes grouped.

Example This frequency table shows the numbers of hours spent doing jobs each week.

Number of hours	Tally	Frequency
0–2	卌 II	7
3–5	卌 IIII	9
6–8	IIII	4
9–11	IIII	4
over 11	II	2

When we collect **continuous data** we group it into equal class intervals (equal-sized groups). We use a **frequency table**.

Time (t min)	Tally	Frequency
$10 \leqslant t < 12$	I	1
$12 \leqslant t < 14$	II	2
$14 \leqslant t < 16$	卌 卌 II	12
$16 \leqslant t < 18$	卌 卌 卌 I	16
$18 \leqslant t < 20$	IIII	4

equal class intervals

Example Libby drew this frequency table to show the times taken to run the cross-country course.

 These groupings mean the same thing.

$10 \leqslant t < 12$ and 10– both mean 'times greater than or equal to 10 minutes but less than 12 minutes'.

Time (t min)	Tally	Frequency
10–	I	1
12–	II	2
14–	卌 卌 II	12
16–	卌 卌 卌 I	16
18–20	IIII	4

If we want to know how many pupils had a time less than 16 minutes we must add the frequencies for the first three class intervals. $1 + 2 + 12 = 15$

Fifteen pupils had a time less than 16 minutes.

Discussion

- The class intervals for discrete data and the class intervals for continuous data are written differently. Why is this? **Discuss**.

- In the example above about Libby's cross country, in which class interval would a time of 18 minutes be put? **Discuss**.

What if the intervals had been
$10 < t \leqslant 12$, $12 < t \leqslant 14$, $14 < t \leqslant 16$, $16 < t \leqslant 18$, $18 < t \leqslant 20$?

Exercise 2

1 Derek wrote down the masses, to the nearest gram, of some apples.

155	172	186	162	168	196	153	160	201
185	175	168	180	194	186	163	170	210

a Use a copy of this table.
Fill it in.

b How many apples had a mass in the class interval $160 < m \leqslant 170$?

c How many apples had a mass less than or equal to 160 grams?

d How many apples had a mass of more than 180 g?

Mass of apple (g)	Tally	Frequency
$150 < m \leqslant 160$		
$160 < m \leqslant 170$		
$170 < m \leqslant 180$		
$180 < m \leqslant 190$		
$190 < m \leqslant 200$		
$200 < m \leqslant 210$		

2 The lengths, in metres, of the shot put throws of 20 people are listed.

8·2	16·0	5·3	6·8	9·5	15·7	12·0	17·4	16·3	15·3
3·9	18·6	7·9	8·5	10·5	8·0	11·6	12·1	13·2	14·8

a Use a copy of this frequency chart and fill it in.

b How many throws were greater than or equal to 4 metres but less than 8 metres?

c How many throws were less than 12 m?

Length of throw (x m)	Tally	Frequency
$0 \leqslant x < 4$		
$4 \leqslant x < 8$		
$8 \leqslant x < 12$		
$12 \leqslant x < 16$		
16+		

d How many throws were greater than or equal to 8 m?

3 The amount of ice cream, in milliletres, in 20 containers is given below.

1985,	1860,	2004,	2103,	2004,	1996,	1987,	2000,	2150,	2190
1994,	1832,	2100,	2096,	1900,	2056,	1934,	1864,	1999,	2001

a Use a copy of this frequency table.
Fill it in.

b How many containers had 1900 mℓ or more but less than 2000 mℓ?

c How many containers had 2000 mℓ or more?

Amount of ice cream (x mℓ)	Tally	Frequency
1800–		
1900–		
2000–		
2100–		

d How many containers had less than 2100 mℓ?

*4 This data gives the handspan, in mm, of some 13-year-old girls.

171	182	202	164	198	211	164	173	177	173
192	205	228	223	189	186	191	194	197	205
192	207	169	173	184	209	176	181	207	164
196	203	221	198	190	169	184	207	197	

Draw a frequency chart for this data.
Use the class intervals $160 < h \leqslant 170$, $170 < h \leqslant 180$, ...

Handling Data

Two-way tables

A **two-way table** can be used to display two sets of data.

Example   This table shows the number of males
and females who passed their driving
test on the first and second tries.

	1st try	2nd try
Male	27	13
Female	35	5

Worked Example

This table shows the number
of accidents before and after
traffic lights were put at an
intersection.

	Before lights	After lights
Minor	27	12
Serious	5	2
Resulting in death	2	0

a How many accidents in
total were there before
the lights were put up?
b How many minor accidents were there in total?
c Compare the accidents before and after the lights were installed.

A two-way table is often used to compare data.

Answer

a We add up the numbers in the 'before lights' column.
27 + 5 + 2 = **34**
There were 34 accidents before the lights were put up.
b We add up the numbers in the minor accidents row.
27 + 12 = **39**
There were 39 minor accidents.
c A possible answer is:
**There were fewer accidents after the lights were
installed and none of them resulted in death.**

Exercise 3

1 This table shows the hair colour of the boys and girls in
Deepak's class.
  a How many black-haired boys were in Deepak's class?
  b How many boys were in the class?
  c How many brown-haired pupils were in Deepak's
  class?

	Boy	Girl
Black	8	6
Blonde	2	3
Red	1	0
Brown	5	7

2 The children and teenagers in a book club
were asked what type of book they liked best.
The results are shown in this table.
  a How many teenagers in total were asked?
  b How many in the club liked mysteries best?
  c How many people were asked in total?
  d Compare the choices of children and teenagers.

	Child	Teenager
Mystery	11	5
Animal	6	1
Romance	0	12
Science fiction	5	4

3 This table shows the main meal choices of people at a café and the time taken, to the nearest minute, for the meal to arrive at the table.

Time taken (minutes)	Beef	Lamb	Chicken	Vegetarian	Pasta
1–5	0	0	2	5	1
6–10	4	3	1	2	6
11–15	3	1	4	0	3
15+	4	3	1	0	0

a How many of the beef meals took longer than 15 minutes?
b How many people ordered chicken?
c How many people's meals took 6–10 minutes to arrive?
d How many vegetarian meals took more than 10 minutes to arrive?
e If you were in a hurry, what would be the best meal to order?

 4

There are two local tourist attractions near Ronan's school.
     cathedral   castle
Visitors can get to each by car or bus.
Design a two-way table Ronan could use to collect information on how visitors get to each of the attractions.

Planning a survey

1 Deciding on the question

Before doing a survey we must **decide what question** we want answered.

Example   We could use a questionnaire to find out what people think of the local taxi service.

It is often helpful to think of **related questions** to ask.

Example   For the question 'What do you think of the local taxi service?', related questions might be:
           How long does it take to come to a call?
           How many taxis are available to hail?
           Are the taxis comfortable?
           Are the drivers polite?

Handling Data

Discussion

For each of the questions below, think of related questions that could be asked.

1 Do shoppers use different ways to get to a shopping centre in the middle of town than to get to an out-of-town shopping centre?
   Hint: Think about the time of day/week/year.

2 Are there different numbers of fish and other life forms in different parts of a stream:
   Hint: Think of time of year, depth of stream, how light it is.

3 At what time during a sports match are goals/points most likely to be scored?

4 Does the amount of TV school pupils watch differ from the amount adults watch?

5 What is the life expectancy in different countries?

2 Deciding what data to collect

After deciding the question or questions, we can decide what **data to collect**.

Discussion

For each of the questions in the above discussion, decide what data needs to be collected.

3 Sources of data

Remember
Data can be gathered from

primary sources

1 a questionnaire or survey of a sample of people

2 an experiment which may use technology, such as a data logger, graphical calculator or computer

secondary sources

reference books, websites, CD-ROMs, newspapers, historical records, **interrogating** a database ...

Discussion

● Elijah wanted to know which type of pie sold at the canteen was most popular. He could gather the data by doing a survey **or** he could use sales records for the canteen.

  Discuss the advantages and disadvantages of each method of collection. Think about which method will
  answer the question better
  take less time
  give the greatest information about the question and related questions.

● Where might you get data from to answer the questions in the discussion on page 362. **Discuss**.
  Is the source primary or secondary?

4 Sample size

When we give a questionnaire or survey it is often not practical to ask everyone possible. We choose a **sample** to ask.

Example   Fred is doing a survey on whether people think the local taxi service is good.
It is probably not possible to ask everyone who uses taxis in the local area.

A sample of people is a group that represents everyone.

A **sample** should be as large as it is sensible to make it.

Discussion

● Megan wanted to know which month Welsh people liked best.
  She asked 10 Welsh people.
  Is this sample big enough? **Discuss**.

  About how many should she ask so that her sample represents the Welsh population?

● Liam wanted to know what percentage of homes in Manchester were able to use the Internet.
  He sent a questionnaire to 50 000 homes in Manchester.
  Is this sample size sensible?
  What problems does it create? **Discuss**.

5 Collecting the data

To collect data, you need to design a **questionnaire or a data collection sheet**.
You need to decide

- what units to use for any measurements
- how accurate you want the data to be.

Example   Suzie is doing a survey on the masses of eggs.
She needs to decide whether to weigh the eggs to the nearest 1 g or
nearest 10 g.

Here are some **guidelines for writing a questionnaire**.

1 Allow for any possible answers.

Example

not at all ☐	*rather than*	up to 1 hour ☐
up to 1 hour ☐		1 up to 2 hours ☐
1 up to 2 hours ☐		2 up to 3 hours ☐
2 up to 3 hours ☐		
more than 3 hours ☐		

2 Give instructions on how you want the questions answered.

Example   Tick one of these boxes.

3 If your questions are asking for opinions, word the questions so that your opinion is
not evident.

Example

Tick one box to show what you thought of concert.
Excellent ☐   Very good ☐   Good ☐   Not very good ☐
Very poor ☐

rather than

Do you agree our concert was wonderful?   Yes ☐   No ☐

4 Keep the questionnaire as short as possible.

Example

Questionnaire on health
How many times have you been to the doctor in the last year?
Tick one box.

0 ☐   1 ☐   2 ☐   3 ☐   more than 3 ☐

How many days off work or school have you had in the last year?
Tick one box.

0 ☐   1–3 ☐   4–6 ☐   7–10 ☐   more than 10 ☐

Once you have written a questionnaire, it is a good idea to **trial it**.
Have a few people try it.
Then improve it if you need to.

Discussion

Discuss how to design a collection sheet or questionnaire to collect the data needed for the questions given in the discussion on page 362.

As part of your discussion, design a collection sheet for each.
Remember to decide what units to use for measurements and how accurate you want the measurements to be.

Exercise 4

1 Some pupils plan a survey to find the most common types of tree in a wood.
   [SATs Paper 2 Level 5]

Design 1	**Design 2**	**Design 3**
Instructions:	**Instructions:**	**Instructions:**
Write down the type of each tree that you see.	Use these codes to record the type of each tree that you see.	Use a tally chart to record the type of each tree that you see.
	Ash A	
For example:	Birch B	**For example:**
Elm, oak, oak, oak, sycamore, ash, ...	Elm E	
	Oak O	
	Sycamore S	
	For example:	
	E, O, O, O, S, A, ...	

Type of tree	Tally
Ash	I
Birch	
Elm	I
Oak	III
Sycamore	I
Other	

The pupils will only use one design.
a Choose a design they should **not** use.
   Explain why it is not a good design to use.
b Choose the design that is the best.
   Explain why it is the best.

T

2 Use a copy of this.
   Choose one of these topics.
   A What is the life expectancy of people in different countries?
   B What factors affect whether a teenager chooses to smoke?
   C How much money do pupils at your school spend after school?
   D Do male or females improve more if given practice and feedback at estimating a minute?
   E How many messages do boys and girls send on their mobile phones each week?

Plan your survey

Question:

Other related questions:

Possible results:

Data to be collected and where from:

Make up a **collection sheet** or questionnaire and attach it.

Sample size:

Summary of key points

A **Discrete** data can only have certain values.

Example The number of crisps in crisp packets.

Continuous data can have any values within a certain range.

Example The diameter of crisps at their widest part.

When we collect continuous data we usually group it into **equal class intervals** on a **frequency chart**.

Speed (s) mph	Tally	Frequency
10 –	‖‖ ‖‖ ‖‖‖	13
15 –	‖‖ ‖‖ ‖‖ ‖‖	17
20 –	‖‖ ‖‖	7
25 –	‖‖‖‖	4
30 – 35	‖‖	2

Or we can use
$10 \leqslant s < 15$, $15 \leqslant s < 20$ ...

 A **two-way table** displays two data sets in a table.

	Netball	Football	Hockey
Under 14	16	25	27
14 to 16	36	53	26
Over 16	29	34	28

Example   This two-way table shows the ages of students in school sports teams.

To plan a survey follow these steps.

1   Decide on the question you want answered and any related questions.
2   Decide what data needs to be collected.
3   Data can be gathered from
   – a **primary source** such as a questionnaire, survey or experiment
   – a **secondary source** such as reference books, websites, ...
4   Decide how to collect the data and, if appropriate, the **sample** size.
5   Design a questionnaire or data collection sheet.

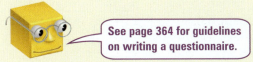

See page 364 for guidelines on writing a questionnaire.

You need to decide what units to use for any measurements.
You need to decide how accurate you want the data to be.

Test yourself

1   Which of these are discrete data and which are continuous data?
   a   the masses of dogs
   b   the number of fillings pupils have in their teeth
   c   the shirt size of men
   d   the length of material used for dresses

T

2 Rodney wrote down the time taken by 20 pupils to carry out a task.
These are the times in seconds.

33	39	38	34	31	28	35	34	27	34
32	21	32	26	30	35	28	36	25	23

a Use a copy of this frequency table. Fill it in.

time (t seconds)	Tally	Frequency
$20 < t \leqslant 25$		
$25 < t \leqslant 30$		
$30 < t \leqslant 35$		
$35 < t \leqslant 40$		

b How many pupils took more than 25 seconds but less than or equal to 30 seconds?

c How many pupils took less than or equal to 35 seconds?

d How many pupils took more than 30 seconds?

3 This table shows the instruments played by the pupils in a music group.

	Violin	Piano	Guitar
Male	1	3	6
Female	4	5	1

a How many males played the guitar?
b How many males were in the group?
c How many guitar players were in the group?
d Compare the instruments played by males and females in the group.

4 Choose one of these topics or choose one of your own. D
    How often do people eat take-aways?
    Are people happy with the local bus service?

Plan your survey
a Write down your question.
b Write down other related questions.
c Write down possible results.
d What data needs to be collected?
e Make up a collection sheet or questionnaire.
f What sample size would you use?

15 Analysing and Displaying Data

You need to know

✓ mode, median, mean, range   page 349

✓ displaying data   page 348

✓ comparing data   page 349

Key vocabulary

assumed mean, continuous, discrete, distribution, mean, median, mode, range, pie chart, stem-and-leaf diagram

Searching ...

Find each of these words in the square

DATA
MEAN
MEDIAN
MODE
RANGE
CLASS
TABLE
FREQUENCY
MODAL

N	P	O	M	E	D	N	M	O	D	A	L
A	D	M	M	E	D	I	R	N	R	C	S
N	A	E	E	U	R	D	A	A	A	L	S
G	T	A	M	D	E	O	N	C	N	A	T
R	P	N	E	Q	I	A	L	T	R	G	A
L	E	R	D	A	T	A	T	A	B	L	E
O	N	F	E	N	S	C	N	Y	E	D	B
M	A	A	N	S	M	G	M	I	A	N	L
W	E	M	E	A	E	D	O	M	O	C	E
H	M	T	A	F	M	O	D	S	A	L	S
M	E	F	R	E	Q	U	E	N	C	Y	F
A	N	C	L	A	S	S	A	L	C	L	A

Handling Data

Mode and range

Remember

Range = highest data value – lowest data value
It is a simple measure of the **spread** of the data.

The **mode** is the most commonly occurring data value.

If a frequency table has grouped data we find the **modal class**.

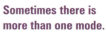

Sometimes there is more than one mode.

Time at the art gallery

Time (minutes)	5–	10–	15–	20–
Frequency	8	16	5	2

The modal class is 10– because this class interval has the highest frequency.

Discussion

Milly timed how long people spent at the art gallery.
The longest time spent there was 24 minutes 8 seconds.
The shortest time spent there was 5 minutes 6 seconds.
Milly rounded these to the nearest minute.
What is the range? **Discuss.**

Exercise 1

1 This chart gives the distances jumped from a standing start by the pupils of 8P.
  a What is the modal class for distance?
  b If the shortest distance jumped was 0·92 m and the longest was 1·73 m, what is the range?

Distance (d)	Frequency
0·8–	2
1·0–	10
1·2–	9
1·4–	5
1·6–	4

2 This table gives the fastest and slowest times for the cross country.
  a What was the range for this year?
  b What was the range for last year?
  c Why might it be useful to know these ranges?

This is linked to PE.

This year (Hours : minutes)		Last year (Hours : minutes)	
Fastest	1 : 20	Fastest	1 : 19
Slowest	2 : 10	Slowest	2 : 04

3 This table gives the number of minutes late 'Trans Tours' buses arrived.
  a What is the modal class?
  b If the shortest time was 1 minute 30 seconds and the longest time was 8 minutes 20 seconds, what is the range?

Minutes late	0–	2–	4–	6–	8–
Frequency	12	16	10	16	4

*4

	Jan	Feb	Mar	Apr	May	Jun	Jul	Aug	Sep	Oct	Nov	Dec
Toronto °C	⁻6·4	⁻5·6	⁻0·8	6·3	12·3	17·6	20·7	19·7	15·4	9·1	3·2	⁻3·3
Manchester °C	3·1	4·0	5·6	8·1	11·6	14·4	15·8	15·6	13·4	10·1	6·1	4·3

This table gives the average temperatures for Toronto and Manchester.
a What is the range for each place?
b Use the range to compare the temperatures.

> Link to geography.

*5 Use the Internet to find the average temperatures recorded last month in three different places in the world.
Compare the ranges.

Mean

Remember

$$\text{mean} = \frac{\text{sum of data values}}{\text{number of values}}$$

> Remember: To find the mean from a frequency table multiply the data values by the frequency then add to get the sum of the data values. Then divide by the sum of the frequencies.

When finding the mean of a large set of data, use a calculator or spreadsheet.

Example This table gives the scores of 285 entrants in a quiz competition.

Score	1	2	3	4	5	6	7	8	9	10
Number of people	8	16	23	26	31	87	41	32	17	4
Total of scores	1×8	2×16	3×23	4×26	5×31	6×87	7×41	8×32	9×17	10×4

Using a **calculator**, we can find the mean of the data.

$$\text{mean} = \frac{\text{sum of total of scores}}{\text{total number of people}}$$

total of (score × number of people)

$$= \frac{1\times8 + 2\times16 + 3\times23 + 4\times26 + 5\times31 + 6\times87 + 7\times41 + 8\times32 + 9\times17 + 10\times4}{8 + 16 + 23 + 26 + 31 + 87 + 41 + 32 + 17 + 4}$$

total number of people

$$= \frac{1626}{285}$$

$$= \mathbf{5\cdot71 \text{ (2 d.p.)}}$$

Exercise 2

Use a calculator to find the mean of these sets of data.
Give the answers to 2 d.p.
The first one has been started.

1 **Score rolled on a die**

Score	1	2	3	4	5	6
Number of throws	83	92	73	87	72	91
Total number of throws	1×83	2×92	3×73	4×87	5×72	6×91

> Do you think this is a fair die?

$$\text{mean} = \frac{1\times83 + 2\times92 + 3\times73 + \dots}{83 + 92 + 73 + \dots}$$

2 Visitors per patient at St George's Hospital

Visitors per patient	0	1	2	3	4	5	6	7	8
Frequency	6	25	14	9	23	14	17	3	2
Total visitors per patient	0×6	1×25	2×14	3×9	4×23	5×14	6×17	7×3	8×2

T **3** Kylie rolled a die with eight triangular faces.
It had the numbers 1 to 8 on it.
This table gives her results.
Fill in the empty boxes first.

Score	1	2	3	4	5	6	7	8
Number of throws	64	72	58	69	57	61	70	65
Total Score	1×64	2×72	3×58	4×69	5×57			

T **4** **Number of minutes taken by a person to solve a problem.**
Fill in the empty boxes first.

Number of minutes	0	1	2	3	4	5	6	7	8	9
Frequency	29	38	43	37	28	43	38	41	40	39
Total number of minutes	0×29	1×38	2×43	3×37	4×28	5×43				

Practical

We can also find the mean using a **spreadsheet**.

Example   This shows the spreadsheet for the data in the screen on the previous page.

	A	B	C	D	E	F	G	H	I	J	K	L
1	Score	1	2	3	4	5	6	7	8	9	10	
2	No. of people	8	16	23	26	31	87	41	32	17	4	=SUM(B2:K2)
3	Total	=B1*B2	=C1*C2	=D1*D2	=E1*E2	=F1*F2	=G1*G2	=H1*H2	=I1*I2	=J1*J2	=K1*K2	=SUM(B3:K3)/L2

- Ask your teacher for the **Finding Means with a Spreadsheet** ICT worksheet to learn how to use a spreadsheet to find the mean.
- Try finding the mean for some of the questions in **Exercise 2** using a spreadsheet.

Sometimes we can find the mean of some data using an **assumed mean**.
To find the mean using an assumed mean
  1 assume a possible value for the mean
  2 subtract this assumed mean from each data value
  3 find the mean of the differences you get in **2**
  4 add your answer to **3** to the assumed mean.

Guess the mean to get the assumed mean.

Worked Example

This data gives the scores of 5 bands at a festival.

22   28   25   26   24

Find the mean number of points scored.

You will probably always get some negative values in step 2.

Answer

1 Assume the mean is 26.

2 Subtract 26 from each data value.

22 − 26 = ⁻4   28 − 26 = 2   25 − 26 = ⁻1   26 − 26 = 0   24 − 26 = ⁻2

3 Find the mean of these differences.

$$\text{mean} = \frac{^-4 + 2 + ^-1 + 0 + ^-2}{5}$$

$$= \frac{^-5}{5}$$

$$= ^-1$$

4 Add ⁻1 to the assumed mean, 26.

26 + ⁻1 = 25

The mean is **25**

Exercise 3

1 Calculate the mean of these using the assumed mean method.
  a 4, 6, 10, 12   **b** 11, 10, 14, 16, 14
  c 16, 12, 20, 18, 23, 13   **d** 5, 8, 6, 13, 11, 4, 8, 9
  e 3·4, 3·6, 3·5, 3·9   **f** 4·6, 4·7, 4·9, 4·3, 4·5

2 Shannon wrote down how much she spent each day of her holidays.

	Mon	Tues	Wed	Thurs	Fri	Sat	Sun
Week 1	£2	£3	£3	£4	£5	£3	£8
Week 2	£3	£0	£2	£0	£3	£5	£1

Find
  a her mean daily spending in week 1 using an assumed mean
  b her mean daily spending in week 2 using an assumed mean
  c her mean daily spending for the two weeks.
     Could you find this by finding the mean of the answers to **a** and **b**?

*3 Adam wrote down these masses of salt used in an experiment.

2·6 g   3·2 g   2·4 g   3·6 g   2·7 g   3·5 g

Use the assumed mean method to find the mean.

Mean, mode, median, range

Remember

The **median** is the middle value when a set of data is put in order of size.

When there is an even number of values, the median is the mean of the two middle values.

See page 349 for more about the median.

Example   Mr Bradley's class had a hopping contest.
These are the distances they hopped, to the nearest 10 m.

The distances are in order.

60 m	60 m	80 m	90 m	100 m	110 m	120 m	140 m	150 m
160 m	170 m	170 m	190 m	190 m	200 m	205 m	205 m	

Mr Bradley said, 'There are 17 results.
The middle of these is the 9th result,
which is 150 m.
About half of the class hopped 150 m
or more'.

The **median, mean** and **mode** are all ways of summarising the data into a single number.

The median is useful for comparing with a middle value.

Examples   About half of Mr Bradley's class hopped more than 150 m.
Half of the class got more than 72% in a test.

The **mean** gives us an idea of what would happen if there were 'equal' shares.

The **mode** is useful for finding the 'most popular' or in summarising poll results.

Discussion

- Give some examples of when it might be useful to know each of these.
  Discuss.

      median   mean   mode

- What information about a data set does the range give us? **Discuss.**

*● Stan and Bob have both joined a golf club.
The club has a place for just one of them in its 9-hole competition team.
Stan and Bob have a play off.
This table shows the results.

Hole	1	2	3	4	5	6	7	8	9
Stan	3	4	4	3	4	3	4	4	21
Bob	4	5	5	4	5	4	5	5	4

Work out the mean, median, and mode for each player.

Is it better to use the mean, median or mode to decide who is the better player? **Discuss.**

The next exercise will give you practice at calculating the median, mean, mode and range.

Exercise 4   **Only use a calculator if you need to.**

1 Find the median, range and mean of these data values.
  a 5, 7, 7, 7, 9, 10, 16, 18, 20
  b 101, 236, 435, 167, 372, 462
  c 14·2, 27·6, 42·3, 96·1, 51·0, 47·7, 83·3, 32·6

> Remember to put the data in order for the median.

2 Fleur timed her classmates to see who could walk 100 m backwards the fastest.
This list gives their times (in seconds).

  25 26 28 28 29 29 29 29 30 31 31 32
  33 34 36 36 37 37 39 42 46 46 48 56

> How might the median, mean and range be useful?

  a Find the median time.
  b What goes in the gap?
     About half of Fleur's class walked 100 m backwards in less than ____ seconds.
  c Find the mean time to 2 d.p.
  d Find the range of the times.

3 Himesh tested to see which was his better kicking foot.
This table shows the number of goals scored out of each set of 5 kicks.
  a What is the **mean** number of goals scored out of 5 with
     i the left foot   **ii** the right foot?
  b What is the **modal** number of goals scored out of 5 with
     i the left foot   **ii** the right foot?
  c Which foot do you think is Himesh's better kicking foot? Explain.

Left foot		Right foot	
Goals out of 5	Frequency	Goals out of 5	Frequency
0	3	0	1
1	6	1	7
2	8	2	1
3	2	3	0
4	1	4	8
5	0	5	3

4 Broomsdale School bought 28 computers.
The costs are given by this table.
  a Find the total cost for the 28 computers.
  b Find the mean cost of each computer.

Item	Cost
28 computers @ £1425 each	____
Installation of 28 computers	£236
Total freight	£ 64
Insurance for 28 computers	£194
Total cost	____

5 a There are four people in Sita's family.
     Their shoe sizes are 4, 5, 7 and 10.
     What is the **median** shoe size in Sita's family?
  b There are three people in John's family.
     The range of their shoe sizes is 4.
     Two people in the family wear shoe size 6.
     John's shoe size is not 6 and it is not 10.
     What is John's shoe size?

[SATs Paper 1 Level 5]

6 James wrote down the prices of nine different CD players sold at 'Rock CD'.

£69   £87   £54   £99   £68   £72   £83   £91   £64

   a What is the median price?

   b 'Rock CD' had a sale and took £5 off the price of all CD players.
     What is the new median price?

  *c If the same amount, x, is subtracted from all the data
     values in a set of data, what happens to the median?

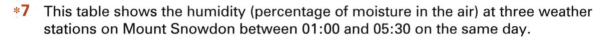

> What would happen to the mode, mean and range?

*7 This table shows the humidity (percentage of moisture in the air) at three weather stations on Mount Snowdon between 01:00 and 05:30 on the same day.

Time (GMT)	Humidity (%)		
	Summit 1085 m	Clogwyn 780 m	Llanberis 105 m
05:30	76·0	91·0	86·0
05:00	59·0	82·0	85·0
04:30	57·0	88·0	84·0
04:00	46·0	81·0	82·0
03:30	43·0	88·0	80·0
03:00	19·0	90·0	76·0
02:30	57·0	91·0	80·0
02:00	67·0	86·0	80·0
01:30	55·0	97·0	81·0
01:00	65·0	99·0	87·0

   a Find the mean, median and range of the humidity at each place.

   b Which place has the least consistent humidity readings?

*8 Olivia and Nishi played three rounds of a
computer game.
The mean of their scores was the same.
The range of Olivia's scores was twice the
range of Nishi's scores.
What are Nishi's two missing scores?

	Game 1	Game 2	Game 3
Nishi		100	
Olivia	40	100	160

Practical

Collect some number data on a topic that interests you.
Find the median of your data.
Find the mean of your data.

Ideas

Times taken to run your local marathon
Cost of a packet of crisps at various shops
Marks of Year 9 pupils in a test
Midday temperatures or hours of sunshine in your area
Prices of cars of a particular year, make and model

> You could use the Internet or a CD-ROM or database.

Puzzle

1 Pablo grew seven plants in an experiment.
The maximum height was 20 cm.
The range was 10 cm.
What might the heights of the seven plants be if the mean was 15 cm?

*2 In PE eight students were given marks out of ten for their folders.
Seven of the marks are 7, 4, 8, 8, 7, 3, 5
What is the eighth mark if
  a the mean is 6·25
  b the mode is 8
  c the median and the mode are the same
  d the mean is the same as the median?

Finding the median, range and mode from a stem-and-leaf diagram

Rob wrote down the temperatures one day for 21 cities in Europe.
He drew a **stem-and-leaf diagram** for the data.

21 °C is entered as	2	1
9 °C is entered as	0	9
33 °C is entered as	3	3

            ↑   ↑
          tens units

The diagram for all the data is

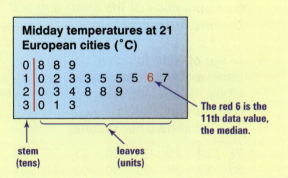

Midday temperatures at 21
European cities (°C)

```
0 | 8 8 9
1 | 0 2 3 3 5 5 5 6 7
2 | 0 3 4 8 8 9
3 | 0 1 3
```

The red 6 is the
11th data value,
the median.

stem   leaves
(tens)   (units)

To find the **median** count the data values.
There are 21.
The middle data value will be the 11th value.
To find the 11th value count from the top.
Start at the left of each row.

The **range** is   33 °C – 8 °C = 25 °C.
                     ↑   ↑
                  highest   lowest
                temperature temperature

The **mode** is 15 °C.
It occurs most often on the diagram.

Handling Data

1 This stem-and-leaf diagram shows the marks out of 60 for 33 students.
  a What is the highest data value?
  b What is the lowest data value?
  c What is the range?
  d What is the median data value?

Marks out of 60									
							stem = tens		
							leaves = units		
0	9								
1	2	4	7	9					
2	0	3	3	4	5	7	9		
3	0	0	4	5	6	6	7	7	7	9
4	0	0	5	6	7	8	8	9	
5	3	5	5						

> To find the median start at the lowest value and count each row from the left.

2

Hours of sunshine at weather stations									
						stem = hours			
						leaves = tenths			
0	5	8							
1	6	9	9						
2	1	1	4	7	7	9			
3	0	0	0	1	2	5	5	5	6	7	9
4	0	1	5	6	7	9	9		
5	0	1	5	5	6	8			
6	1	2	2	3	6	8	8	9	9	9	9
7	0	0	1	5	7	8			
8	0	0							

This stem-and-leaf diagram shows the marks out of 60 for 33 students.
  a What is the highest data value?
  b What is the lowest data value?
  c What is the range?
  d How many data values are there?
  e What is the median data value?

3 This stem-and-leaf diagram shows the weekly rainfall for 11 months of last year in a city in Europe.

Rainfall in cm																												
																				stem = centimetres								
																				leaves = tenths								
3	0	1	2	4																								
2	0	0	0	0	0	2	3	3	5	7	7	7	9																
1	0	0	1	1	1	1	2	2	2	3	3	3	4	4	4	4	5	5	6	6	7	8	8	9	9	
0	6	6	8	9	9																							

The rainfalls for the other weeks were 33 cm, 9 cm, 21 cm, 27 cm.
  a Use a copy of the stem-and-leaf diagram and add the extra rainfalls.
  b Find the median, mode and range of the weekly rainfall.

4 Libby wanted to compare the ages of the mothers and fathers of her classmates.
Libby asked each person in her class how old their mother was.
The list shows her results.

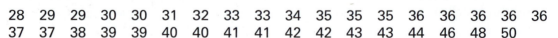

28	29	29	30	30	31	32	33	33	34	35	35	35	36	36	36	36	36
37	37	38	39	39	40	40	41	41	42	42	43	43	44	46	48	50	

  a Libby started a stem-and-leaf diagram to show her results. Use a copy of the diagram and finish it.
  b How many mothers were aged 30–39?
  c How many were aged 40–49?
  d Find the median, range and mode of the ages.

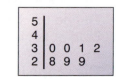

5				
4				
3	0	0	1	2
2	8	9	9	

Libby also asked each person in her class how old their father was.
This stem-and-leaf diagram shows her results.

  e Find the median, range and mode of the ages.
  f Compare the ages of the mothers and fathers.

5	1	1	2	3	4	7							
4	0	0	1	1	2	2	3	3	4	5	5	5	6	8
3	0	0	3	3	4	4	5	5	6	6	6	7	8
2	9	9											

Comparing data

Remember

To **compare data** we sometimes use the range and one or more of the mode, median or mean.

Example   This gives the mean, median and range for how many hours two brands of kettle lasted.

	Brand A	Brand B
Mean	632·1	629·6
Median	622·5	623
Range	701 − 585 = 116	913 − 321 = 592

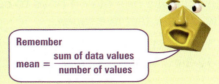

Remember

$$\text{mean} = \frac{\text{sum of data values}}{\text{number of values}}$$

The mean of brand A is slightly higher than the mean of brand B.
The median of brand A is slightly lower than that of brand B.
We could say that the brands are about the same.
However, if we look at the range, brand B has a much higher range than brand A.
This tells us that the number of hours brand B lasts is much more variable.

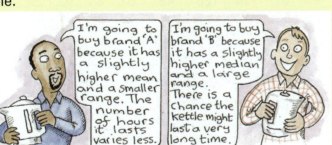

I'm going to buy brand 'A' because it has a slightly higher mean and a smaller range. The number of hours it lasts varies less.

I'm going to buy brand 'B' because it has a slightly higher median and a large range. There is a chance the kettle might last a very long time.

Exercise 6

1   Amy and Zoe both wanted to be chosen for a gymnastics team.
This table shows the mean and range of their scores in a trial.
One coach wanted Amy to be in the team.
Which of the explanations in the box could he use?
Another coach wanted Zoe in the team.
Which explanation could she use?

	Mean	Range
Amy	6·9	0·6
Zoe	7·1	1·5

A She has more consistent scores.
B She has more variable scores so could score higher.
C She has a higher mean.

2   This table gives the results of the last twelve public speaking contests for Lufta and Sophie.
Only one girl can represent the school at a local public speaking competition.
Which girl would you choose?
You can choose either as long as you use the results in the table to explain why.

	Mean	Median	Range
Lufta	17	16	12
Sophie	16	16	3

3 This gives the mean, median and range for how many hours two brands of light bulb lasted.

	Brand A	Brand B
Mean	161·25	176·35
Median	183·5	176
Range	205	23

   a Write some sentences comparing the brands using the range and one of the mean, median or mode.

   b Which brand would you buy?
      Explain your answer using the mean, median and range.
      It doesn't matter which one you choose as long as you give the reasons for your choice.

4 A coach must choose one of two boys to compete in the 100 m sprint. She looks at the results of their last five races.

    Ben   12·1 s   12·0 s   12·0 s   16·8 s   12·1 s
    Josh 12·3 s   12·4 s   12·4 s   12·5 s   12·4 s

   a Find the mean and range for each boy.

   b Which boy do you think the coach should choose?
      Explain why using your results from **a**.

*5 Hamish breeds mice.
This data gives the mass in grams of two breeds.

    Breed A (total 24)
    104   102   116   83   42   124   125   43   61   84   92   84
     51   41   119   120   53   62   84   104   112   79   51   42

    Breed B (total 23)
    87 76 84 87 88 91 73 62 58 73 94 82
    89 96 82 87 91 88 83 62 94 93 86

   a Find the mean, median and range for breed A.

   b Find the mean, median and range for breed B.

   c Write a brief report comparing the masses of the breeds.

   d Hamish wants to sell just one breed of mice. He has found that smaller mice are more popular.
      Which breed do you think he should choose to sell?
      Explain your answer using the range and one of the mean, median or mode.

*6 Adam was doing a survey on TV viewing patterns for adults and teenagers.
He showed his results on these stem-and-leaf diagrams.

Teenage hours watched/week
Stem = tens, leaves = units

```
4 | 0
3 | 0 1 3 5 7 7 8 9 9
2 | 0 3 4 4 5 6 7 8 9 9 9
1 | 0 2 3 4 5 5 8 9
0 | 0
```

Adult hours watched/week
Stem = tens, leaves = units

```
4 |
3 |
2 | 0 3 5 9
1 | 0 1 1 1 2 2 3 3 3 4 4 4 4 5 5 6 6 7 7 8 9 9
0 | 8 9 9 9
```

   a Find the mean, median, mode and range.

   b Compare and contrast teenage and adult viewing habits using your results from **a**.

Practical

Collect two sets of data you can compare.
Find the mean, median, mode and range of your data.
Use these to compare the data.

You could interrogate a database.

Suggestions
- Compare and contrast weather patterns in two different places.

    Example   Compare and contrast the weather in London and Bangkok using this data.

London						Average temperature						
	Jan	Feb	Mar	Apr	May	Jun	Jul	Aug	Sep	Oct	Nov	Dec
°C	4·9	4·6	7·1	9·0	12·6	15·6	18·4	17·8	15·2	12·0	7·7	6·1

						Average rainfall							
	Jan	Feb	Mar	Apr	May	Jun	Jul	Aug	Sep	Oct	Nov	Dec	**Year**
mm	61·5	36·2	49·8	42·5	45·0	45·8	45·7	44·2	42·7	72·6	45·1	59·3	590·4

Bangkok						Average temperature						
	Jan	Feb	Mar	Apr	May	Jun	Jul	Aug	Sep	Oct	Nov	Dec
°C	25·9	27·6	29·2	30·1	29·6	29·0	28·5	28·4	28·1	27·7	26·8	25·5

						Average rainfall							
	Jan	Feb	Mar	Apr	May	Jun	Jul	Aug	Sep	Oct	Nov	Dec	**Year**
mm	10·6	28·2	30·7	71·8	189·4	151·7	158·2	187·0	319·9	230·8	57·3	9·4	1445

- Compare and contrast reading habits of different age groups.
- Use the results of an experiment to find which type of battery lasts longer.

You could use the Internet to collect data on weather.

- Use the results of two top sports players at your school to decide who should represent the school if only *one* could be chosen.

You could ask your teacher for the **Interrogating a Database 1 and 2** ICT worksheets.

Bar charts

We often draw **graphs** to show the results of our data collection.

Examples

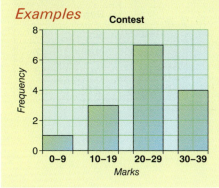

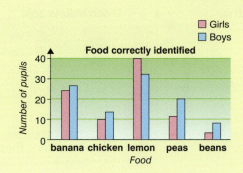

A picture can be worth a thousand words.

Handling Data

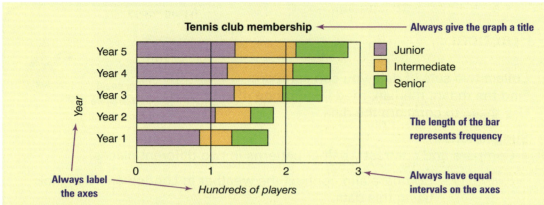

Tennis club membership ← Always give the graph a title

- Junior
- Intermediate
- Senior

Year (vertical axis)

The length of the bar represents frequency

Always label the axes

Hundreds of players

Always have equal intervals on the axes

Sometimes a **compound bar chart** like the one below is used to show information.

Example   **Sources of vitamin A in British diet**

| Meat | Dairy products, eggs and fats and spreads | Vegetables | Cereals | Other |

About half of the strip is meat. This means that about half of the vitamin A in the British diet comes from meat.
About a quarter of the strip is dairy products, eggs and fats and spreads. About a quarter of our vitamin A comes from these sources.

Exercise 7

T

1 This table gives the points scored by the people in a sports event.

Scores	5–9	10–14	15–19	20–24	25–29	30–34
Frequency	1	4	3	2	5	1

a Use a copy of the bar chart. Finish it.

b Scores of 25 or more won a prize. How many people won a prize?

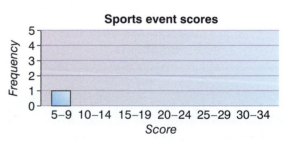

Sports event scores

T

2 The frequency table shows what colour disco tickets Eva and Charlie sold.
Use a copy of this bar chart.
Finish it.

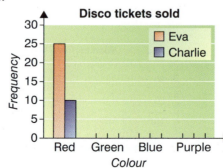

Disco tickets sold

	Frequency	
	Eva	Charlie
Red	25	10
Green	15	27
Blue	4	16
Purple	21	12

Colour (row label)

3 Jack, Jane and Grace were learning to type.
The table shows how many words per minute they could type at the end of each month.

		Month				
		Jan	Feb	Mar	Apr	May
Words per minute	**Jack**	8	12	18	24	28
	Grace	14	15	15	22	30
	Jane	10	10	14	18	24

Use a copy of this compound bar chart.
Finish it.

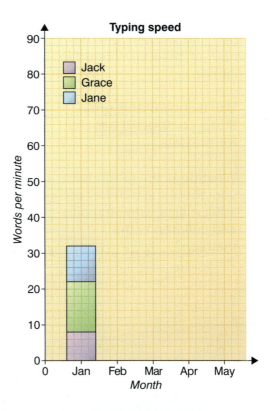

4 This table gives the amount of water used by Julia's family yesterday.

Activity	Toilet flush	Personal washing	Washing clothes	Washing up	Garden	Cooking	Drinking
Amount of water used (ℓ)	65	55	20	15	15	10	5

a Julia started this strip bar chart.
Use a copy of it.
Finish filling it in.

Toilet flush
0 10 20 30 40 50 60 70 80 90 100 110 120 130 140 150 160 170 180 190

b This strip bar chart shows the *average* daily water use by families of the same size as Julia's.

| Toilet flush | Personal washing | Washing clothes | Washing up | Garden | Cooking | Drinking |
0 10 20 30 40 50 60 70 80 90 100 110 120 130 140 150 160 170 180

Julia's family want to reduce the amount of water they use each day.
Use the graphs to decide an activity that could be targeted to reduce the amount of water used. Explain why you chose this.

Handling Data

T **5** This table shows the number of men and women and the number of children entered in a triathlon.

Year	2000	2001	2002	2003	2004
Men	125	150	180	195	240
Women	60	75	100	130	160
Children	10	10	15	25	30

a Use a copy of this grid. Plot a compound bar chart for the data.

b What does the graph tell you about the number of adult males, adult females and children in the triathlon?

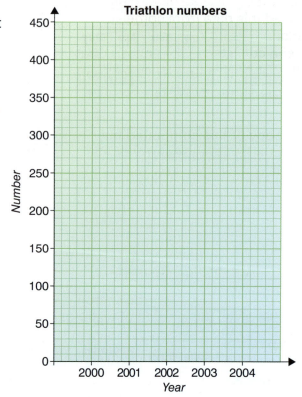

Triathlon numbers

Number (y-axis)

Year (x-axis)

Practical

Collect some data that can be displayed on a bar chart.
You could
- use the government statistics website
- do an experiment
- collect data using a collection sheet or questionnaire.

Ideas
Number of students away from school each day in each Year 8 and Year 9 class.
Time taken by two pupils to do homework on 5–10 nights.
Time taken for an experiment to stop bubbling when different amounts of metal are added to acid.

Frequency diagrams

Charlotte wanted to know if her thumb was longer than those of most other girls her age. She collected this data from some Year 9 girls.
The data she collected is continuous data. She graphed it on this frequency diagram.

Remember continuous data can take any value within a given range.

Class interval (ℓ in mm)	Frequency
30–	2
35–	2
40–	4
45–	9
50–	10
55–	2

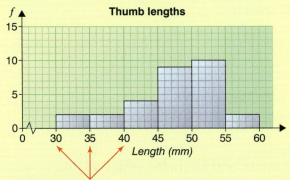

Thumb lengths

30– means all measurements 30 mm or more but less than 35 mm.

This is very important to remember.

For **continuous data**, we **label the division between the bars**.

Exercise 8

T

1 This gives the size of the last 25 strong earthquakes in the South Pacific. Use a copy of the frequency diagram and complete it.

Magnitude	Tally	Frequency
6·0–	ЖН II	7
6·2–	III	3
6·4–	ЖН II	7
6·6–	III	3
6·8–	III	3
7·0–7·2	I	1

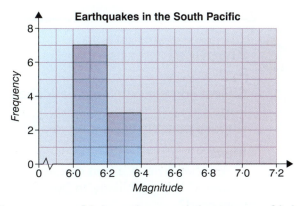

T

2 Rory wanted to know the most likely amount of injury time and the range of injury time in rugby.

a Use a copy of this grid. Draw a frequency diagram for the data.

Remember to give the graph a title and label the axes.

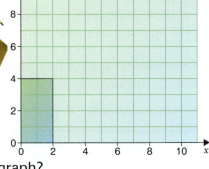

Injury time (t in minutes)	Frequency
0–	4
2–	5
4–	8
6–	4
8–	2

b Rory drew the graph for this data and said, 'My graph shows injury time is usually 5 minutes.' Is he correct?

c What can Rory tell from the graph?

T

3 Mrs Young wanted to improve fire drill.
She collected this data about the time taken to
evacuate the school.

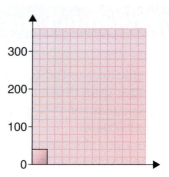

Time taken (min)	0–	1–	2–	3–	4–	5–	6–	7–8
Number of students	39	54	123	347	29	42	21	16

Use a copy of this grid.
a Draw a frequency diagram for this data.
b How many students evacuated the school in under
3 minutes?
c How many students took at least 5 minutes to
evacuate?
d Once all the students were out of the school, a
count was made.
How many students were counted?
e In which minute did most pupils evacuate the school?
f Comment on what this graph tells you about the fire drill at this school.

*4 Dr Menzies was studying blood pressure.
He took the blood pressure of 50 healthy males and 50 males who had flu.

Blood pressure	70–	80–	90–	100–	110–120
Frequency (healthy)	5	21	16	5	3
Frequency (flu)	1	17	22	5	5

a Draw separate frequency diagrams for the healthy men and the men who had
flu.
b Comment on what the graphs show.

Practical

Collect some continuous data about a topic or question that interests
you.
Choose suitable class intervals for your data.

Ideas
● At what time in a first division
match is a goal likely to be
scored?
● Collect data from an experiment
in one of your other subjects.
● Amount of time the pupils in your
class spend on the computer each
week.
● Times of the tracks on your five favourite CDs.
● Time spent on homework one week.

Draw a frequency diagram for your data.

> You could collect
> continuous data from a
> secondary source about a
> topic that interests you. You
> could use the Internet.

Drawing pie charts

A **pie chart** is a graph shown by a circle.
It shows how something is shared or divided.
The bigger the section of the circle the bigger the proportion it represents.

Example   This pie chart shows where students go on holiday.

Where students go on holiday.

Each section is called a **sector**.
The whole pie chart has an angle of 360° at the centre.
The angle at the centre of each sector is a fraction of 360°.

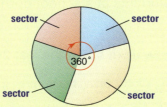

Example   A class of 30 students produced their own play.
Four produced the play, six were backstage, 15 acted in it and five made the costumes.
A pie chart is a good way to show this data.
To draw a pie chart follow these steps.

Step 1   Find the angle at the centre of each sector.
4 out of 30 or $\frac{4}{30}$ produced the play.

$$\frac{4}{30} \text{ of } 360° = \frac{4}{30} \times 360°$$
$$= 48°$$

$\frac{4}{30}$ is the proportion that produced the play.

48° is the angle at the centre of the 'producers' sector.
We can work out the other angles the same way.

backstage	$\frac{6}{30} \times 360° = 72°$	
actors	$\frac{15}{30} \times 360° = 180°$	
costumes	$\frac{5}{30} \times 360° = 60°$	

Check the angles add to 360°.

Step 2   Draw a pie chart with these angles at the centre of the sectors.

It is best to draw the smaller sectors first.

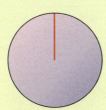

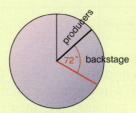
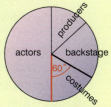

Draw a circle. Draw a radius on your circle. Start with a vertical line.

Use a protractor to measure an angle of 48° from the radius. Draw a sector and label it.

Use a protractor to measure an angle of 72° from the bold line. Draw a sector and label it.

Use a protractor to measure an angle of 60° from the bold line. Draw a sector and label it. Label the last sector.

Handling Data

Exercise 9

1 Which of these sets of data would be most suitable to display on a pie chart?
   a the temperature of salt water each minute as it is heated
   b the proportion of money a family spends on housing, food, clothing, heating and other
   c the distance jumped in the final of the long jump competition
   d the number of students leaving school at the end of each year.
   e the proportion of students at a school born in England, Scotland, Wales, Ireland, overseas

T

2 a This table shows how Jake spent the last 24 hours.

Activity	Sleeping	Playing sport	Doing homework	Eating	Friends
Hours	9	5	2	1	7

He wanted to draw a pie chart.
Make a copy of this and fill in the missing numbers to work out the angles of the sectors.

sleeping $\frac{9}{24} \times 360° = 135°$ playing sport $\frac{5}{24} \times 360° = \underline{\quad}°$

doing homework $\frac{2}{\underline{\quad}} \times 360° = \underline{\quad}°$ eating $\underline{\quad} \times 360° = \underline{\quad}°$

friends $\underline{\quad} \times \underline{\quad}° = \underline{\quad}°$

Use a copy of the circle.
Show Jake's information on a pie chart.

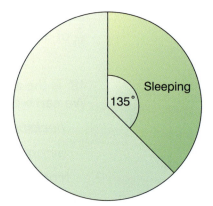

b This table shows how Jon spent the last 24 hours.

Activity	Sleeping	Playing sport	Doing homework	Eating	Friends
Hours	7	3	3	2	9
Pie chart angle	$\frac{7}{24} \times 360° = 105°$	$\frac{3}{24} \times 360° =$			

Use a copy of the circle.
Calculate and then fill in the pie chart angle.
Show Jon's information on a pie chart.
c Compare and contrast Jake's and Jon's last 24 hours.

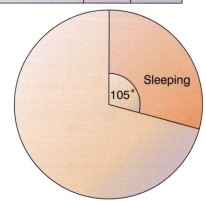

3 The owner of a shop that sold mobile phones noted the colours of the phones sold one month.
She wanted to draw a pie chart to show the results.

a Use a copy of the table below.
Calculate and then fill in the pie chart angles.

Mobile phone colour	Number of mobile phones	Pie chart angles
Black	75	$\frac{75}{120} \times 360° = 225°$
Yellow	5	
Grey	25	
Blue	15	
Total	**120**	**360°**

b Use a copy of the circle.
Show the information in the table on a pie chart.

c How might this information be useful to the shop owner?

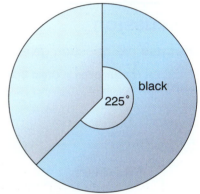

*4 This table shows the number of pupils by school type.

a Work out the angles for each sector.
Do the angles add to 360°? Why not?

b Draw a pie chart for this data.

Number of pupils by school type

Type of school	Thousands
State nursery	137
State primary	5 345
State secondary	3 886
Non-maintained schools	618
Special schools	114
All schools	**10 100**

Practical

You will need a spreadsheet package or statistical package.

Choose some data that could be displayed on a pie chart.
Enter the data into the spreadsheet or statistical package. Use the graph function to draw a pie chart.
You could enter the data from one of the questions in **Exercise 9**.

Interpreting graphs and tables

Robyn found these pie charts in a book.
They show the proportion of foot, car,
taxi and cycle traffic for two cities of similar
population.

She used the graphs to make some
comparisons.

There was about twice as much foot
traffic in city 1.
There was about half as much car traffic in city 1.

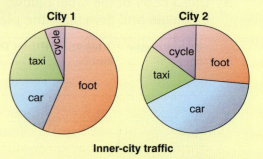

Inner-city traffic

Discussion

Discuss the question given under each graph or table.

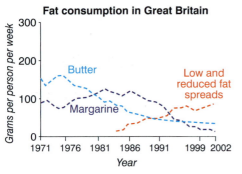

Fat consumption in Great Britain

How has fat consumption in Britain
changed over the last 30 years?

Reason why school pupils tried to give up smoking			
Current smokers			*England*
Why tried to give up	Boys	percentage Girls	Total
Worried about my health	47	55	52
Cost	28	35	32
To make me feel fitter	32	19	24
My family/friends persuaded me	12	10	11
Smoking made me smell or look nasty	7	10	9
Did not like/enjoy it	4	9	7
Other	12	13	13
	137	210	347

Percentages total more than 100 because some pupils gave more than one answer

Compare and contrast the reasons given by boys to
those given by girls for trying to give up smoking.

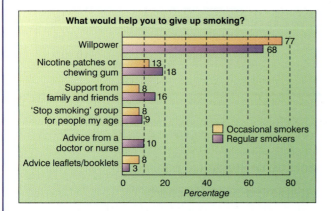

What would help you to give up smoking?

You are asked to develop a programme to help
pupils stop smoking. Funds are limited. Use the
graph to help choose two things your programme
would focus on. Explain why you chose these,
using the graph to support your explanation.

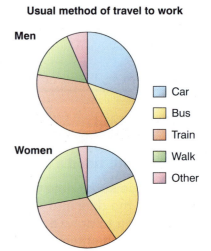

Usual method of travel to work

What are the differences between
the way men and women usually
travel to work?

Exercise 10

1 This graph shows the amount of pocket money Julia's friends get.

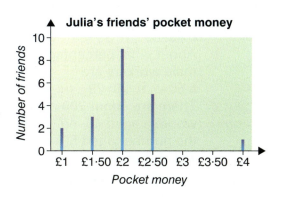

Julia's friends' pocket money

  a What is the modal amount of pocket money?

  b Julia gets £1·50 pocket money.
She asked her mother to increase it to £3.
She told her mother that most of her friends got from £2 to £4.
Is Julia's mother being misled?
Give a reason.

2 The diagrams show the number of hours of sunshine in two different months.

[SATs Paper 1 Level 5]

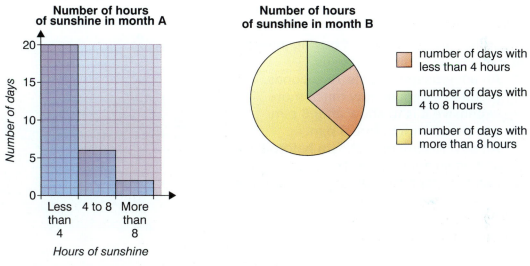

  a How many days are there in month A?

  b How many days are there in month B?

  c Which month had more hours of sunshine?
Explain how you know.

3 Camilla likes the weather to be still and hot when she goes on holiday.
These line graphs give the average wind speed and average temperature for the place where Camilla would like to holiday.
Use the graphs to choose which month you think Camilla should go on holiday.
Explain your reasons.

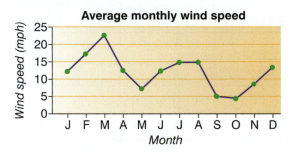

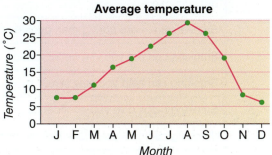

4 This compound bar chart shows the ages of workers at a factory.

a About what percentage of female workers are
  i 40–49 **ii** 30–39 **c** 50+?

b About what percentage of male workers are
  i 50+ **ii** 20–29 **iii** 40–49?

c There are about 200 male workers aged 40–49. Estimate the number aged 50+.

d There is about the same total number of male workers as female workers.
  Which of these is true?
  A Generally, the male workers are younger than the female workers.
  B Generally, the female workers are younger than the male workers.
    Explain how you used the chart to decide.

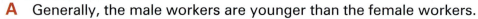

Age at factory

Age in years
- 50+
- 40–49
- 30–39
- 20–29

Male *Female*

5 This chart gives the cost of buying advertising at the cinema at different times.

a An advertisement lasts 75 seconds. Use the graph to estimate how much more expensive it is to show after 6 p.m. compared with before 6 p.m.

b An advertisement is shown before 6 p.m. and again after 6 p.m. The total cost is £950. Use the graph to estimate how long the advertisement lasted.

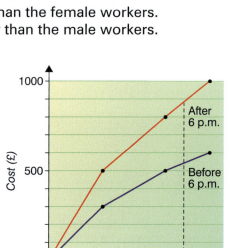

75 seconds

*6 This table shows the percentage of smokers and non smokers aged 11–17 who agreed with the statements given.
Comment on the relationship between people who smoke and what they believe.

Statements	Non smokers	Smokers
Smoking can help calm you down	24%	75%
Smokers tend to be more rebellious than people who don't smoke	32%	26%
Smokers are more boring than people who don't smoke	26%	7%
Smoking can put you in a better mood	12%	50%
Smoking can help you stay slim	12%	24%
Smoking helps give you confidence	7%	24%
Smoking can help you make friends more easily	8%	14%
Smoking makes you look more grown up	7%	13%

*7

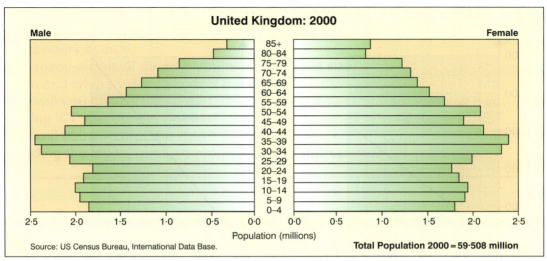

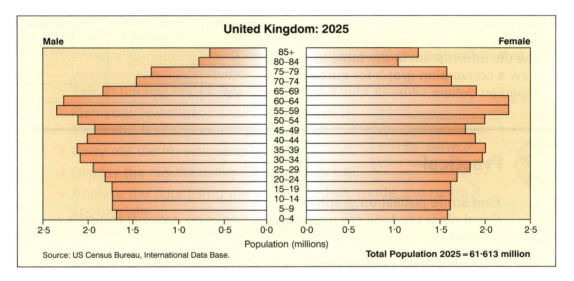

These diagrams show a **population pyramid** for the United Kingdom for 2000 and a predicted population pyramid for 2025.

a About how many people were 70 or older in 2000?

b About what percentage of the population was 70 or older in 2000?

c About how many people are predicted to be 70 or older by 2025?

d About what percentage of the population is predicted to be 70 or older by 2025?

> This is linked to percentages.

Suggestions
- How do communities of animals/plants in two habitats differ?
- How do methods of transport to and from two different schools or shopping centres or ... differ? What factors affect this?
- How does the weather at two different places differ?
- What factors affect the growth of plants?
- Simulate the cooling rates of penguins that huddle and those that don't.
- Compare and contrast TV viewing or radio listening habits of different age groups.

Summary of key points

range = highest data value – lowest data value

The **mode** is the **most commonly occurring** data value.

If a frequency table has grouped data we find the **modal class**.

Example

Age	0–	10–	20–	30–	40–
Frequency	3	7	18	22	19

The modal class is 30– because this class interval has the highest frequency.

$$\text{mean} = \frac{\text{sum of data values}}{\text{number of values}}$$

When finding the mean of a large set of data, we use a calculator or spreadsheet.

To find the **mean from a frequency table** multiply the data values by the frequency then add to get the sum of data values. Then divide by the number of data values.

Example

Score	2	4	5	7	8	9
Number	2	1	7	8	6	3
Total of scores	2×2	4×1	5×7	7×8	8×6	9×3

$$\text{mean} = \frac{\text{sum of total of scores}}{\text{total number of people}}$$

$$\text{mean} = \frac{2 \times 2 + 4 \times 1 + 5 \times 7 + 7 \times 8 + 8 \times 6 + 9 \times 3}{2 + 1 + 7 + 8 + 6 + 3}$$

$$= \frac{174}{27}$$

$$= 6 \cdot 4 \ (1 \ \text{d.p.})$$

We can find the mean using an **assumed mean**.

For an example, see page 372.

The mean, median and mode are all ways of summarising data into a single number.

The **median** is useful for comparing with a middle value.

The **mean** tells us what would happen if there were equal shares.

The **mode** is useful for finding the 'most popular.'

 We can find the median, range and mode from a **stem-and-leaf graph**.

Example If the time taken for lunch was 15 minutes we enter this as

Time taken for lunch
stem = hours, leaves = tenths

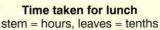

This diagram shows all the data on time taken for lunch.

There are 25 data values.

The median is the 13th value, 0·8 hours.

The range is 4·0 − 0·2 = 3·8 hours.

The mode is 0·5 hours.

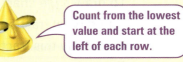
Count from the lowest value and start at the left of each row.

 To **compare data** we usually use the range and one or more of the median, mode or mean.

Example Robert found the mean, median, mode and range of the hours of sunshine per week in two places.

Place A mean 53 range 24 **Place B** mean 54 range 8

The two places have about the same mean sunshine hours but in Place A the number of sunshine hours varies much more than in Place B.

 Bar charts can be used for grouped discrete data. Compound bar charts can be used to show three sets of data.

Examples

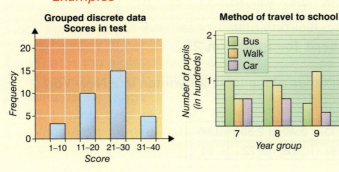

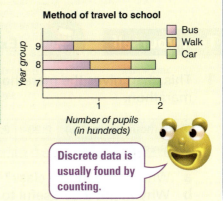

Discrete data is usually found by counting.

8 These are the results for two problem-solving teams.

	Toby's team	Gemma's team
Mean score	6·13	6·13
Median score	6	6
Mode	4	6
Range	6	3

Which team would you choose to enter in the competition?
Explain why using the range and one of the mean, median or mode.

T **9** Use a copy of this.
The table shows the handspans of
the pupils in Simon's class.
a Draw a frequency diagram.
b How many pupils had a
handspan of 24 cm or more?

Handspan in cm	20–	22–	24–	26–
Frequency	4	12	16	2

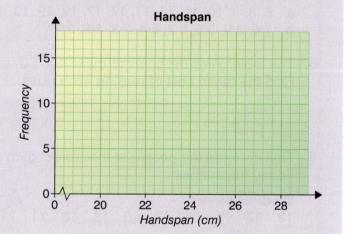

T

10 This table gives the number of times each number on a dice was tossed
by three pupils doing an experiment.

Number	1	2	3	4	5	6
Jessica	50	46	58	42	51	53
Nandoor	41	57	43	58	51	50
Rosie	58	48	49	51	48	46

Use a copy of the grid.
Finish the compound bar chart to
show this data.

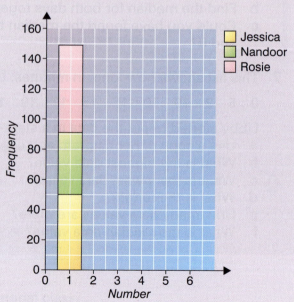

T

11 Reena recorded the ages of 36 Beatles fans.

Age	<20	20–	30–	40–	≥50
Frequency	3	5	9	5	14

She wanted to draw a pie chart.
Make a copy of this and fill in the missing
numbers to work out the angles of the
sectors.

<20 $\frac{3}{36} \times 360° = 30°$ 40–   __ × __ = __°

20– $\frac{5}{\underline{}} \times 360° = \underline{}°$ ≥50   __ × __ = __°

30–   __ × 360° = __°

Use a copy of the circle.
Show Reena's information on a pie chart.

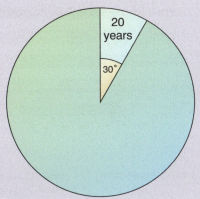

H

12 A newspaper predicts what the ages of secondary
school teachers will be in six years' time.
They print this chart.

[SATs Paper 2 Level 5]   **I**

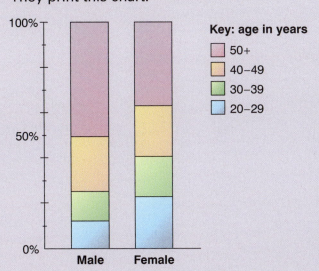

Key: age in years
- 50+
- 40–49
- 30–39
- 20–29

a The chart shows 24% of male teachers will be aged 40 to 49.
About what percentage of female teachers will be aged 40 to 49?
b About what percentage of female teachers will be aged 50+?
c The newspaper predicts there will be about 20 000 male teachers aged 40 to 49.
Estimate the number of male teachers that will be aged 50+.

d Assume the total number of male teachers will be about the same as the total number of female teachers.

Use the chart to decide which statement is correct.

Generally, male teachers will tend to be younger than female teachers.

Generally, female teachers will tend to be younger than male teachers.

Explain how you used the chart to decide.

13 This graph shows the amount spent by a school on computing equipment, paper and video/camera equipment

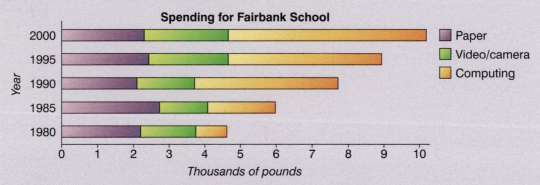

Spending for Fairbank School

a Has spending increased or decreased over the last 20 years on
  i paper   **ii** video/camera   **iii** computing?
b Predict what spending will be like next year and in ten years' time.

16 Probability

You need to know
✓ probability   page 349

········· **Key vocabulary** ···

calculated probability,   event,   experimental probability,
$p(n)$ probability of event n,   random,   relative frequency,
sample space,   theoretical probability,   unpredictable

Spiralling out of control

We can draw a spiral in two directions.

anticlockwise   **clockwise**

Ask about 50 people to draw a spiral.
Collect the results on a table like this one.

Is it equally likely that the next person
you ask to draw a spiral will draw a
clockwise one or an anticlockwise one?

	Tally	Frequency
Clockwise		
Anticlockwise		

Language of probability

Discussion

● The Taylor family are going on holiday to Madrid.

They travel by train to the airport.
They fly to Madrid.
They hire a car and drive to the Hotel.

Think of some things that are **likely** to happen to this family as they do this.
Think of some things that are **unlikely** to happen.
Discuss.

● Mandy has five cards.
She picks one card without looking.
Is this a **random** event? **Discuss**.
Is it possible to predict with certainty the outcome of a random event?
Discuss.

Sasha has two boxes of chocolates.
If she takes a chocolate without looking (at **random**) the outcome is **unpredictable**.
She might get a white chocolate or a dark chocolate.

Box A

Sasha wants a dark chocolate.
It is **more likely** she will get a dark chocolate from Box A than Box B.
$\frac{12}{18} > \frac{9}{18}$

There is a greater **proportion** of dark chocolates in Box A.

Box B

Some events are **equally likely** to happen.

Example   If a dice is tossed, the outcomes 1, 2, 3, 4, 5 and 6 are all **equally likely**.

Exercise 1

'Equally likely' means each has the same chance of happening.

1 Which of these events will have equally likely outcomes?
   a rolling a dice   **b** taking a card at random from these cards
   c spinning this spinner

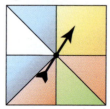

2

Colour	Number
Red	14
Green	2
Yellow	7
Blue	10

This table shows the colours of hats sold at a fête.
I walk around the fête.
Which colour hat am I most likely to see?
Why?

3 Jimmy has some gold and silver medals.
They are all the same size.
  a Jimmy put 3 gold and 2 silver medals in a bag.

Jess is going to take one without looking.
She says 'There are two colours so it is just as likely that I will get a gold medal as a silver medal.'
Explain why Jess is wrong.
  b How many gold medals should Jimmy take out of the bag to make it just as likely that Jess will get a gold medal as a silver medal?
  c Danny has a different bag with 10 medals in it.
It is more likely that Danny will take a silver medal than a gold medal from his bag.
How many silver medals might there be in Danny's bag?

4 Brookside school is selling 'scratch and win' cards.
You choose **one** square to scratch and if the square says 'win' you win £5.
On which of these cards are you most likely to get a 'win'? Explain.

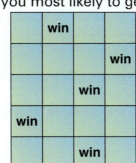

A

win		
	win	
		win
			win

B

	win		
			win
		win	
win			
		win	

C

	win	win
	win	win

∗5 Two family packs of muesli bars contain different numbers of chocolate-coated and plain bars.
  Pack 1 has 8 chocolate-coated out of 20 bars.
  Pack 2 has 6 chocolate-coated out of 10 bars.
Emily only likes chocolate-coated bars.
She can pick a bar at random from either pack.
Which pack should she pick from. Why?

*6   The diagram shows where pupils in Years 7, 8 and 9 went on their last holiday.

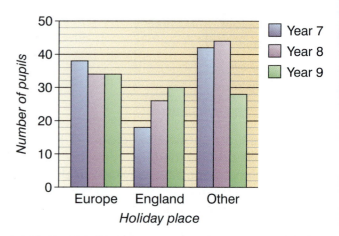

a   A pupil from Year 7 is chosen at random.
   Are they **most likely** to have holidayed in Europe, England or other?

b   Repeat **a** for a pupil from Year 8.

c   Repeat **a** for a pupil from Year 9.

Calculating probability by listing outcomes

Remember

If the outcomes of an event are **equally likely**,

Probability of an event = $\dfrac{\text{number of favourable outcomes}}{\text{number of possible outcomes}}$

Discussion

Elizabeth was tossing a coin twice.
She said, 'There are three possible outcomes, so there is a 1 in 3 chance I will get two heads.'
Craig said, 'There are four possible outcomes, so there is a 1 in 4 chance you will get two heads.'
Who was right? **Discuss**.

Tracy is vegetarian.
At 'Pizza palace' she can choose bean or spicy cheese.
She has pizza on Tuesday and Thursday.
All the possible outcomes is called the **sample space**.
This can be given as a list or table.

| **Tuesday** | bean | bean | spicy cheese | spicy cheese |
| **Thursday** | bean | spicy cheese | bean | spicy cheese |

or

Tuesday	Thursday
bean	bean
bean	spicy cheese
spicy cheese	bean
spicy cheese	spicy cheese

We can use the sample space to help calculate probability.

Worked Example
A dice and a coin are tossed together.
Find the probability of getting a head and a number greater than 4.

Answer
All the possible outcomes are:
H1   H2   H3   H4   H5   H6   T1   T2   T3   T4   T5   T6
There are 12 possible outcomes in the sample space.
There are two favourable outcomes: H5 and H6.
P (H and number greater than 4) = $\frac{2}{12}$ **or** $\frac{1}{6}$

Exercise 2

1 This spinner is spun.
   a List all the possible outcomes for colour.
   b Use the list to find the probability of spinning green.

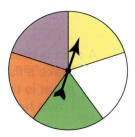

2 Kit had these 12 letters in a bag.
   He took one without looking.
   a List the possible outcomes.
   b What is the probability of getting a vowel.

The vowels are
A, E, I, O, U.

3 In a contest there were three people, Jan, Laura and Caryl.
   There was a prize for first and second.
   a Copy and finish this table for who could come first and who could come
   second.

1st	Jan	Jan				
2nd	Laura	Caryl				

   b What is the probability that Jan gets second?

4 What are the possible outcomes when
   a Josie and Charlotte choose a flavour of ice cream from vanilla, chocolate and
   raspberry

Josie	vanilla	vanilla					
Charlotte	vanilla	chocolate					

   a Copy and finish this table for their choices.
   b What is the probability that a vanilla ice cream is **not** chosen?

407

More calculating probability

Example   There were 20 countries at a sports event.
4 were African, 8 were European, 5 were Asian and 3 were other.
One country was chosen to lead the opening parade.
The probability that this was a European country

$= \frac{8}{20}$ ← number of European countries
                ← number of countries altogether

$= \frac{2}{5}$ **or 40% or 0·4**

> Probability can be given as a fraction, decimal or percentage.

The probability of an event happening can be given on a **probability scale**.

Worked Example
Sophie spins this spinner twice.
She multiplies the scores as shown.
Draw an arrow on the probability scale to show how likely Sophie's total will be
a an odd number
b not an odd number
c a number less than 6.

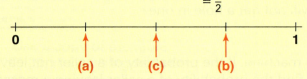

×	1	2	3	4
1	1	2	3	4
2	2	4	6	8
3	3	6	9	12
4	4	8	12	16

Answer
a probability (odd number) $= \frac{4}{16}$ ← There are 4 odd numbers on the table.
                                     ← Total number of outcomes

$= \frac{1}{4}$

b probability (not an odd number) $= 1 - \frac{1}{4}$

$= \frac{3}{4}$

c probability (number less than 6) $= \frac{8}{16}$ ← There are 8 numbers less than 6 on the table.

$= \frac{1}{2}$

```
+-------+-------+-------+-------+
0       ↑       ↑       ↑       1
       (a)     (c)     (b)
```

Exercise 4

1 Jody rolls a dice.
What is the probability of Jody getting
a 4   **b** an even number   **c** a number greater than 2?

T   **2** Use a copy of this.
Show, with an arrow, the probabilities for question **1a**, **b** and **c**.

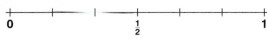

3 Sam had counters with the prime numbers less than 20 in a bag.
He draws one at random.

   a What is the probability of Sam drawing
      i 7   **ii** 2 or 3   **iii** a number less than 10
      iv a number greater than 5?
   b On a copy of the probability scale draw arrows to show the
      probabilities in **a**.

4 Martha uses this spinner for a game.
Use a copy of a probability scale.
Draw an arrow to show how likely Martha is to get these when she
spins the spinner.

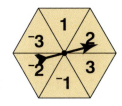

   a 2   **b** a number less than zero   **c** *not* a number less than 2

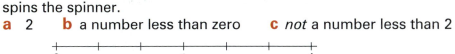

5 There were 200 tickets sold to the school musical.
They are numbered 1 to 200.
One number is drawn at random for a prize.

   a Brendon has number 182. What is the probability he will win?
   b The Tucker family have tickets with numbers 25, 26, 27 and 28.
      The Billens family have tickets with numbers 120, 121, 122 and 123.
      Which family has the better chance of winning the prize? Why?
  *c Marcia buys several tickets. She works out she has a 5% chance of winning.
      How many tickets has she bought?
  *d Three people have lost their tickets and cannot claim the prize if their number
      is drawn. What is the chance that nobody wins?

6 In a box of chalk, 10 pieces are white, 5 are yellow and 3 are red.
Emma chooses one piece at random.
Find the probability that this stick is
   a yellow   **b** not yellow.

7 a Jo has these 4 coins.   **[SATs Paper 2 Level 5]**

      Jo is going to take one of these coins at random.
      Each coin is equally likely to be the one she takes.
      Show that the **probability** that it will be a **10p** coin is $\frac{1}{2}$.
   b Colin has **4 coins** that total 33p.
      He is going to take one of his coins at random.
      What is the probability that it will be a 10p coin?
      You must show your working.

8 There are 12 bikes in a bike stand. Six are red, 4 are green and 2 are blue.
Calculate the probability that the next bike to be taken from this stand will be
   a red   **b** not red   **c** green   **d** not green   **e** red or green
   f neither red nor green.

9 Of 50 chocolates in a box, 25 contain a nut, 15 are pure
chocolate and the rest have a creamy centre.
One chocolate is chosen at random from this box.
Find the probability that this chocolate
   a contains a nut   **b** does not contain a nut   **c** is creamy centred
   d is not pure chocolate.

10 Kamal made an eight-sided dice from an octahedron.
It had the numbers 1 to 8 on it.

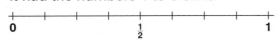

On a probability scale like this, draw an arrow to show the probability that when
he rolls the dice he will get
   a 3   **b** a number less than 7   **c** not a prime number.

11 Mark and Kate each buy a family pack of crisps.   **[SATs Paper 1 Level 5]**
Each family pack contains **ten bags** of crisps.
The table shows how many bags of each flavour are in each family pack.

flavour	number of bags
plain	5
vinegar	2
chicken	2
cheese	1

  a Mark is going to take a bag of crisps at random from his family pack.
     What goes in the gaps?
     The probability that the flavour will be _____ is $\frac{1}{2}$.
     The probability that the flavour will be cheese is _____.
  b Kate ate **two bags** of **plain** crisps from her family pack of 10 bags.
     Now she is going to take a bag at random from the bags that are left.
     What is the probability that the flavour will be **cheese**?
  c A shop sells **12 bags** of crisps in a large pack.
     I am going to take a bag at random from the large pack.
     The table shows the probability of
     getting each flavour.
     Use the probabilities to work out **how
     many bags** of each flavour are in this
     large pack.

flavour	probability	number of bags
plain	$\frac{7}{12}$	
vinegar	$\frac{1}{4}$	
chicken	$\frac{1}{6}$	
cheese	0	

*12 A school has a new canteen. A special person will be   [SATs Paper 2 Level 5]
chosen to perform the opening ceremony.
The names of all the pupils, all the teachers and all the canteen staff are put into a box.
One name is taken out at random.
A pupil says:

> There are only three choices.
> It could be a pupil, a teacher or one of the canteen staff.
> The probability of it being a pupil is $\frac{1}{3}$.

The pupil is **wrong**. Explain why.

*13 Yoghurt is sold in packs of 12.
Mel is going to take one without looking.
The scale shows the probability of
getting each flavour.

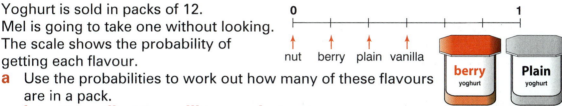

a Use the probabilities to work out how many of these flavours
are in a pack.
   i vanilla   **ii** plain   **iii** nut   **iv** berry
b What is the probability of getting a flavour *other than*
   i nut   **ii** berry   **iii** plain or berry?

*14

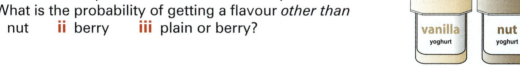

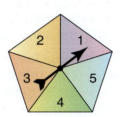

In a game Todd spins an arrow. The arrow stops on one of sixteen
equal sectors of a circle. Each sector of the circle is coloured. The
probability scale shows how likely it is for the arrow to stop on any
one colour.
a How many sectors are
   i coloured red   **ii** coloured blue   **iii** coloured yellow?
b What is the probability that if the spinner is spun it will not stop on red or blue?

Discussion

- Here is a spinner with five equal sections.
  A class is divided into two groups, A and B.
  They spin the pointer lots of times.
  If it stops on an odd number, group A gets a point.
  If it stops on an even number, group B gets a point.
  Is this a fair game? **Discuss**.
  What if group A gets 2 points when it stops on odd and group B gets 3 points
  when it stops on even?

- A **biased** dice is one that is not equally likely to land on each of the numbers
  on its faces.
  What is a biased spinner? Draw one. **Discuss**.

Estimating probability from experiments

Class 9D weighed 50 bags of crisps.
17 were below their given mass.

We can use the results to **estimate the probability**
that a bag of crisps, chosen at random, will be below
its given mass.
An estimate of the probability is $\frac{17}{50}$.

 ## Practical

You will need two fair dice.

Roll the two dice 50 times.
Add the numbers and record the total in a frequency table.

Total	2	3	4	5	6	7	8	9	10	11	12
Tally											
Frequency											

Compare your results with another group.
Are they different? Why?
Use the results of your experiment to estimate the probability of getting
each total the next time you roll two dice.

Predict what might happen if you repeat the experiment.

Combine your results with some other groups.

Use the combined results of your experiment to estimate the probability
of getting each total the next time you roll two dice.

Discussion

- If an experiment is repeated, will the outcomes be
  exactly the same each time? **Discuss**.

- Billie rolled a biased dice 50 times.
  He used the results of his experiment to estimate the
  probability of getting each number the next time he
  rolled the dice.

  Melanie rolled a biased dice 200 times.
  She used the results of her experiment to estimate the probability of getting
  each number the next time she rolled the dice.
  Who is likely to get the better estimate of the probabilities?
  Why? **Discuss**.

Exercise 5

1   Hayley has woken at 7 a.m. on 33 out of the last 40 days.
Estimate the probability that she will wake at 7 a.m. tomorrow.

2   100 out of the last 500 cars that passed the school gate were white.
Estimate the probability that the next car to go past will be white.

3   It has been cloudy on 150 out of the last 200 days.
Estimate the probability it will be cloudy tomorrow.

4   Ben has won 17 out of the last 20 tennis games he has played with Rob.
Estimate the probability that he will win the next one.

5   29 of the last 50 people to buy a burger have bought a cheeseburger.
Estimate the probability that the next person to buy a burger will buy a cheeseburger.

6   It has rained on Mr Wilson's birthday 17 out of 35 times.
Estimate the probability it will rain on his next birthday.

*7   These were the flavours of the last 50 ice
creams sold at Delicious Delights.
   a   Estimate the probability that the next ice
cream sold will be
      i   chocolate   **ii**   vanilla.
   b   If the table had given the results for the last
200 ice creams sold, would the estimated
probabilities be more accurate? Explain

Chocolate	27
Berry	8
Vanilla	1
Banana Choc Chip	14

*8   This table shows the results of a survey of the arm
lengths of 100 fifteen-year-old boys.
Another fifteen-year-old boy's arm is measured.
Use the information in the table to estimate the
probability that his arm
   a   has a length of 62 cm or more but less than 64 cm
   b   has a length of 58 cm or more but less than
60 cm.

Length of arm	Frequency
58–	5
60–	26
62–	54
64–	15
Total	100

Comparing calculated probability with experimental probability

Practical

A You will need a dice.

Roll a dice 60 times.
Record the results in a tally chart.

Draw a bar chart to show your results.

> Compare your experimental probabilities with the calculated probabilities.

Are the results what you would expect?
If you repeated this experiment would you expect to get the same results?
What results would you expect if you tossed the dice 600 times?

✳Repeat but roll two dice and add the numbers.

T **B You will need** a computer software package that simulates rolling two dice.

Ask your teacher for the **Throwing Dice 1 and 2** ICT worksheets.

a Simulate rolling two dice and adding the numbers 50 times.

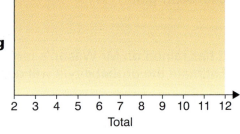

Draw a frequency diagram of the results.

b Calculate the theoretical probability of each total.
Draw a frequency diagram of these probabilities.
Compare the two diagrams.

c Simulate rolling two dice and adding the numbers 200 times.
Draw a frequency diagram of the results.
Compare this frequency diagram with the one you drew in **b**.

d Repeat **c** for 500 trials.

e What do you notice happens as you increase the number of trials?

C Jake thought that a buttered biscuit was more likely to land buttered-side down when dropped.

What is the theoretical probability of this?
Devise an experiment, using a card with the dots representing the buttered side, to test Jake's prediction.

Carry out the experiment.
Compare the theoretical and experimental probabilities.

Investigation

Are you a mind reader?

You will need five cards with a design on each, for example:

1 Shuffle the cards and place them face down.

2 Ask someone to take one of the cards and picture the symbol in his or her mind.

3 See if you can 'guess' what the symbol is.

4 Repeat this with at least 20 other people.
   Record your results in a table.

	Tally	Frequency
Symbol correct		
Symbol incorrect		

5 From these results, estimate the probability that the next time you try to guess the symbol you will be correct.

6 What is the theoretical probability of you guessing the correct symbol?

Summary of key points

 When an event is **random**, the outcome is **unpredictable**.
Some outcomes are **more likely** than others.

Example This spinner is more likely to stop on blue than on red when spun.

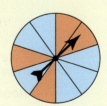

Some events have **equally likely** outcomes.

Example This spinner is equally likely to stop on red, blue, green, yellow or purple.

B To **calculate a probability** we often give all the possible **outcomes** using a list or table.

The set of all the possible outcomes is called the **sample space**.

Example These two spinners are spun. The possible outcomes could be shown in a table.

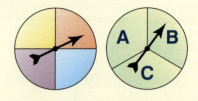

red	red	red	blue	blue	blue	purple	purple	purple	yellow	yellow	yellow
A	B	C	A	B	C	A	B	C	A	B	C

The probability of getting red A $= \frac{1}{12}$ ← ways of getting red A
← number of possible outcomes

 C If the probability of an event occurring is p, then the **probability** of it **not** occurring is $1 - p$.

Example If the probability of having to stop at an intersection is $\frac{1}{4}$, then the probability of not having to stop is $\frac{3}{4}$.

Probability of an event can be shown on a **probability scale**.

Example
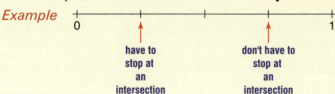

have to
stop at
an
intersection

don't have to
stop at
an
intersection

 D We sometimes use an experiment to **estimate a probability**.

Example A factory tested five hundred phones.
Twelve had faults.
We can use this to estimate that the probability that a phone chosen at random will be faulty is $\frac{12}{500} = \frac{3}{125}$.

Note

1 When an experiment is repeated, there will usually be different outcomes.

2 The more times an experiment is repeated, the better the estimate of probability will be.

 E When we **compare experimental probability** with **calculated probability**, the more trials in the experiment, the closer the experimental probability is to the calculated probability.

Test yourself

1 A prize is given if a counter, when dropped onto a board, lands on a purple square.
On which board are you most likely to win?

A 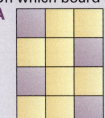 B 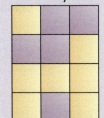 C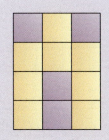

counter

2 Gary takes a marble at random from each of these bags.

  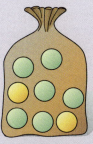

Bag A   Bag B   Bag C

From which bag is he most likely to get a yellow marble?

3 Three counters are put into a bag.
One is green, one is red and the other is blue.
A counter is drawn at random and its colour noted.
The counter is then put back in the bag.
Another counter is then drawn at random and its colour noted.
Copy the table and complete it to show the sample space.

1st counter	green	green	...
2nd counter	green	red	...

The sample space shows all the possible outcomes.

Find the probability that
a the first counter is green and the second one is red
b two red counters are picked
c two counters of the same colour are picked
d just one blue counter is picked
e neither counter is green.

4 Jimmy had these number cards.
He took one without looking.
a List all the possible outcomes.
b Use the list to find the probability of taking .

T **5** This table shows the colours of 100 tickets sold in a raffle.
One ticket is drawn without looking.
Use a copy of the probability scale.
Draw an arrow on the scale to show the probability of drawing a
   a green ticket   **b** blue ticket   **c** yellow ticket.

green	60
pink	20
blue	5
yellow	15
Total	100

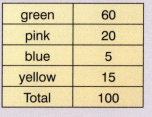

6 The probability of a pack of crisps not being underweight is 0·9.
What is the probability of a pack being underweight?

T **7** Emily spun these two spinners and added the numbers together.
   a Use a copy of this table.
      Fill it in to show all the totals.

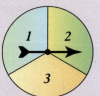
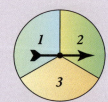

+	1	2	3
1			
2			
3			

   b How many possible outcomes are there altogether?
   c Find the probability of getting these.
      i a total of 3   **ii** a total less than 7   **iii** a total of 4 or more

8 A large dice with the numbers 1, 2, 3, 4, 5, 6, 7, 8, 9, 10, 11 and 12 on it is rolled.
What is the probability of the dice landing on these?
   a an even number
   b a number less than 5
   c a number greater than 10
   d not a prime number
   e not an even nor an odd number

9 Jo dropped a drawing pin 50 times.
It landed on its head 5 times.
   a Estimate the probability that the next time the drawing pin is dropped it will land on its head.
   b He dropped it another 50 times.
      It landed on its head 3 times.
      Use all 100 results to estimate the probability of it landing on its head.
   c Is the probability from 50 drops of the drawing pin more or less accurate than that from 100 drops of the drawing pin?

on head   on side

Test Yourself Answers

Chapter 1 page 32

1 a 1000   **b** 1 000 000   **c** 10
2 a 10^2 **b** 10^5 **c** 10^6 **d** 10^9
3 One point four nine times ten to the power of eight
4 3504
5 a 360   **b** 21 000   **c** 5·7   **d** 3·6   **e** 120   **f** 0·475   **g** 2·6   **h** 0·01024
6 a 1400   **b** 21 000   **c** 40   **d** 20   **e** 0·05
7 a 5·7   **b** 4730   **c** 0·083   **d** 68
8 a >   **b** <   **c** >
9 a 4·07, 4·27, 4·7, 4·72   **b** 0·007 kg, 0·06 kg, 0·689 kg, 0·798 kg, 0·869 kg
10 a x is greater than 1·0 kg and less than or equal to 1·5 kg
　　b blue
11 a i 623 000   **ii** 600 000   **b i** 832 000   **ii** 800 000
　　c i 7 084 000   **ii** 7 100 000   **d i** 986 000   **ii** 1 000 000
12 Three of: 315, 316, 317, 318, 319, 320, 321, 322, 323, 324
13 a 7500   **b** 8499
14 a 54·39   **b** 72·35   **c** 4·03   **d** 16·3   **e** 4·6   **f** 16   **g** 105·00   **h** 10·00
15 a 1·8   **b** 6·7 (1 d.p.)   **c** 0·9 (1 d.p.)   **d** 6·5
16 a 47·8 g (1 d.p.)   **b** 0·09 kg (2 d.p.)

Chapter 2 page 57

1 a Dead Sea   **b** Jerusalem   **c** 770 m   **d** 290 m   **e** 430 m
2 ⁻10 °C, ⁻7 °C, ⁻1 °C, 1 °C, 3 °C, 4 °C
3 a ⁻4   **b** ⁻23
4

×	2	⁻3	5	⁻5
2	4	⁻6	10	⁻10
⁻1	⁻2	3	⁻5	5
⁻4	⁻8	12	⁻20	20
3	6	⁻9	15	⁻15

5 a 3 × ⁻5 = ⁻15   **b** ⁻6 × ⁻4 = 24   **c** ⁻5 × ⁻7 = 35
　　⁻5 × 3 = ⁻15   　　⁻4 × ⁻6 = 24   　　⁻7 × ⁻5 = 35
　　⁻15 ÷ 3 = ⁻5   　　24 ÷ ⁻4 = ⁻6   　　35 ÷ ⁻5 = ⁻7
　　⁻15 ÷ ⁻5 = 3   　　24 ÷ ⁻6 = ⁻4   　　35 ÷ ⁻7 = ⁻5

6 a

+	⁻29	79	⁻82
46	17	125	⁻36
⁻47	⁻76	32	⁻129
⁻38	⁻67	41	⁻120

b

×	⁻5·5	42	⁻4·5
⁻34	187	⁻1428	153
8·6	⁻47·3	361·2	⁻38·7
⁻3·8	20·9	⁻159·6	17·1

7 a 1, 7   **b** 1, 2, 3, 6
8 11, 13, 17, 19, 23, 29, 31, 37
9 a

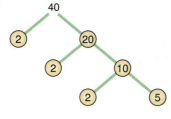

b

2	48
2	24
2	12
2	6
	3

10 a i 2 and 5   **ii** 2 and 3
   b i $40 = 2 \times 2 \times 2 \times 5$ or $40 = 2^3 \times 5$ **ii** $48 = 2 \times 2 \times 2 \times 2 \times 3$ or $48 = 2^4 \times 3$
11 a Factors of 18 are 1, 2, 3, 6, 9, 18. Factors of 42 are 1, 2, 3, 6, 7, 14, 21, 42
   b 6
   c 4, 8, 12, 16, 20, 24
   d 6, 12, 18, 24, 30, 36
   e 12
12 a

   b HCF = 4   LCM = 240

13 a 8 **b** 5 **c** 9 **d** 16 **e** 3 **f** 36
14 a 5 **b** 4 **c** 49 **d** 3
15 a 46·24 **b** 841 **c** 5·7 **d** 10·4
16 28
17 a 8 **b** 3 **c** 64 **d** 5 **e** 1000 **f** 2 **g** 3 **h** 64
18 a 1728 **b** 4913 **c** 3·8 **d** 10·2

Chapter 3 page 88

1 a 12 **b** 7 **c** 0·1
2 139
3 a 980 **b** 250 **c** 700 **d** 407
4 a 30 **b** 48 **c** 105 **d** 3500 **e** 2 **f** 4080 **g** 207
5

6 a 4·9 **b** 0·24 **c** 130 **d** 116 **e** 121 **f** 50
7 a Five point nine, five point seven   **b** A possible answer is 123.   **c** 84°
8 a B **b** C **c** A
9 A possible answer is: $25 \times 20 = 500$ or $20 \times 20 = 400$
10 a 61·26 **b** 10·17 **c** 178·37
11 £29·75
12 a 19 890 **b** 22 176 **c** 41·6 **d** 174·8 **e** 307·8 **f** 36·48
13 a £20·52 **b** £9·48
14 a 10 **b** 4
15 a 13·8 **b** 14·25
16 a 36·7 **b** 15·3
17 36 slices
18 a A **b** A or C
19 a 250 **b** 170 **c** 7100
20 a 9·8 **b** 4·356 **c** 1·1 (1 d.p.) **d** 0·8 (1 d.p.)

Chapter 4 page 116

1 a $\frac{1}{6}$ **b** $\frac{1}{2}$ **c** $\frac{1}{4}$ **d** $\frac{3}{5}$ **e** $\frac{3}{4}$
2 a $\frac{1}{6}$ **b** $\frac{1}{3}$
3 a 25 **b** $\frac{20}{25}$ or $\frac{4}{5}$
4 a $\frac{2}{5}$ **b** $\frac{1}{4}$ **c** $\frac{43}{100}$ **d** $\frac{39}{1000}$ **e** $1\frac{27}{100}$
5 a 0·9 **b** 0·19 **c** 0·03 **d** 0·35 **e** 0·12
6 a 90% **b** 19% **c** 3% **d** 35% **e** 12%
7 a $0·\dot{3}$ **b** $0·\dot{3}\dot{6}$

8

Decimal	Percentage	Fraction
0·5	50%	$\frac{1}{2}$
0·3	30%	$\frac{3}{10}$
0·35	35%	$\frac{7}{20}$
0·03	3%	$\frac{3}{100}$

9 a $\frac{3}{5}$ **b** $\frac{5}{12}$ **c** $\frac{7}{9}$ **d** $\frac{4}{6}$ or $\frac{2}{3}$

10 a 8 **b** $\frac{15}{12}$ or $1\frac{3}{12}$ or $1\frac{1}{4}$

11 a 3 **b** $\frac{4}{12}$ or $\frac{1}{3}$

12 a 5, 12 **b** $\frac{17}{20}$

13 a £24 **b** 28 m **c** 14·7 m

14 38·5 kg

15 a 68 m **b** 234 mℓ

16 a English **b** Maths **c** 3 out of 10 boys liked Science, that is 30%.
Only $\frac{3}{20}$ girls liked Science and that is 15%.

17 a 8 m **b** 11 kg **c** 12 m **d** 14 km

18 100 cm

19 a $\frac{6}{5}$ or $1\frac{1}{5}$ **b** $\frac{15}{8}$ or $1\frac{7}{8}$ **c** $\frac{35}{6}$ or $5\frac{5}{6}$

20 a $6 \div \frac{1}{8}$, 48 **b** $5 \div \frac{1}{3}$, 15

21 84

22 6

Chapter 5 page 133

1 a 1 : 2 **b** 1 : 3 **c** 1 : 4 **d** 2 : 3 **e** 2 : 5 **f** 3 : 6 : 5

2 3 : 5

3 1 : 3

4 a 3 : 2 **b** 2 : 3 **c** $\frac{2}{5}$ **d** 60%

5 a 1 : 3 **b** 2 : 3 **c** 1 carton of orange juice

6 Steph gets £500 and Becky gets £300.

7 a 200 g and 250 g **b** 105 g and 70 g

8 a 50 g **b** 150 g **c** 550 g

9 a 6 oz **b** 2 oz **c** 4 oz **d** $1\frac{1}{2}$ oz **e** 5 oz

10 8

11 18

12 a 600 mℓ **b** 50 mℓ **c** No. 1 part out of 5 parts is screenwash. This is 20%.

Chapter 6 page 159

1 a $y = 3x + 7$ function, $2x + 4 = 7$ equation, $v = u + at$ formula

 b $P = nRT$ formula, $2p + 7 = {}^-3$ equation, $y = \frac{x+3}{7}$ function

2 a $p + 400 = 900$ **b** $p + 100 = 400$

3 a $y = 15$ **b** $x = 3$

4 a $k = 2$ **b** $m = 2\frac{1}{2}$ or 2·5

5 a $n = 5$ **b** $x = 24$ **c** $p = 4·5$ **d** $b = 5$ **e** $h = 32$

6 $15n + 25 = 175$, $n = 10$ days

7 a $x = 0$ **b** $y = 7$ **c** $d = 1$ **d** $y = 5$ **e** $p = 5$

8 a $w + w + 3 + w - 2$ **b** $3w + 1$
 c $3w + 1 = 25$ **d** $w = 8$
 e $x - 20° + 2x + x = 180°$
    $4x - 20° = 180°$
    $4x = 200°$
    $x = 50°$

9 a $n = 2$ **b** $x = 9$ **c** $a = 11$ **d** $n = 15$ **e** $b = 4·5$
 f $a = 3·5$ **g** $a = 4$ **h** $n = 4$ **i** $x = 8$ **j** $n = 12$

Test Yourself Answers

10 a

Miles	5	10	15	20	25
Kilometres	8	16	24	32	40

    b 8 : 5. The ratio is constant, the same for every pair.

       16 : 10 = 8 : 5   24 : 15 = 8 : 5   32 : 20 = 8 : 5   40 : 25 = 8 : 5

    c Yes. The number of kilometres increases constantly as the number of miles increases constantly.

    d The points lie in a straight line or the ratio *kilometres : miles* is constant for all sets of values.

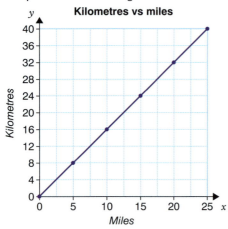

    e i 72 km   **ii** 168 km

Chapter 7 page 182

1 a $4p$ **b** $3(x + 2)$ **c** $4(b - 3)$ **d** xy **e** m^2 **f** $\frac{3p}{2}$ **g** $\frac{x + y}{a}$ **h** $5c^2$

2 a $8a + 11b$ **b** $2x + 3$ **c** $3m + 12$ **d** $7q + 3$ **e** $n - 1$ **f** $8a^2$ **g** $2p^2$ **h** $12y^2 - 2y$

3 a $20b$ **b** $21x$ **c** $24n$ **d** m^3 **e** $12x^2$

4 a a **b** $4m$ **c** 6 **d** $\frac{3i}{2}$ **e** c **f** 1 **g** $\frac{3n^2}{2}$

5 a $5y + 10$ **b** $7x - 28$ **c** $12p + 8q$ **d** $12a - 15b$ **e** $6x - 2y$ **f** $15m$

6 a m^3 **b** $11m$ **c** $3m$ **d** $\frac{m}{3}$ **e** $4 \times m - 3$ **f** $2m$ **g** $2m - 2$ **h** $3m + 3$

7 a 9 **b** 13 **c** 12 **d** 13 **e** 14

8 a 55 cm^2 **b** 85 cm^2 **c** 40 cm^2 **d** 52 cm^2

9 a 26 mℓ **b** 36 mℓ **c** 21 mℓ

10 a 7 **b** 240

11 a $n - 2$ **b** $2n$ **c** No. $3(n - 1) = 3n - 3$ not $3n - 1$ **d** $6n - 4$

12 a $14a$ **b** $10x + 5y$ **c** $10n + 10$ **d** $10x + 4$

13 $C = 18d + 20$

Chapter 8 page 207

1 20, 18, 16, 14, 12, 10, 8

2 a 5, 15, 25, 35, 45 **b** 1, 3, 9, 27, 81 **c** 1, 2, 3, 5, 8

3 a i Up **ii** Down **iii** Up **b i** No **ii** No **iii** Yes

4 a 51, 60, 69 **b** 44, 40, 36 **c** 20, 10, 5

5 a 3 km, 4 km, 5 km, 6 km, 7 km, 8 km **b** 3 km, 7 km, 12 km, 18 km, 25 km, 33 km

6 a 3, 6, 9, 12, 15 **b** 5, 6, 7, 8, 9 **c** 1, 3, 5, 7, 9

7 a 24, 32, 40 **b** 48 **c** B **d** Each time a star is added, 8 more dots are added. So the number of dots is 8 times the number of stars.

8 a $T(n) = 2n + 3$ **b** $T(n) = 3n - 1$

9

Input	Output
2	6
10	38
4	14
0	⁻2

10 a
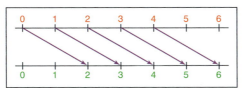

b
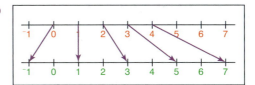

11 a $y = x + 2$ or $x \rightarrow x + 2$ **b** $y = 2x - 1$ or $x \rightarrow 2x - 1$

12 a Add 4 **b** Multiply by 2

13 a Add 1 **b** Multiply by 12

14 a 10, 5, 14 **b** 1, 13, 7

15 $x \rightarrow x - 4$ or $y = x - 4$ **b** $x \rightarrow 5x$ or $y = 5x$

Chapter 9 page 231

1 a

x	6	4	2	0	⁻1
y	2	0	⁻2	⁻4	⁻5

b
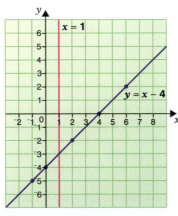

2 a

x	2	1	0	⁻1
y	2	0	⁻2	⁻4

b (2, 2), (1, 0), (0, ⁻2), (⁻1, ⁻4)

c
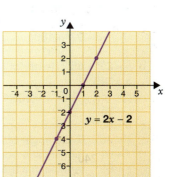

d Yes

e No. If you substitute $x = ⁻2$ into $y = 2x - 2$, you get $y = 2 \times ⁻2 - 2 = ⁻6$, not ⁻4.

3 a, **b** and **c**

4 a B **b** D **c** A **d** B **e** A

5 a C **b** B

6 a

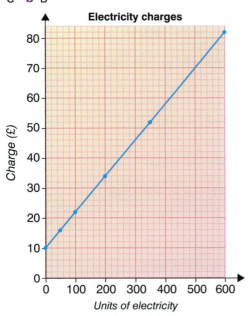

b £70   **c** 250

7 a About 5·5 gallons **b i** About 9 ℓ **ii** About 270 ℓ

Test Yourself Answers

8

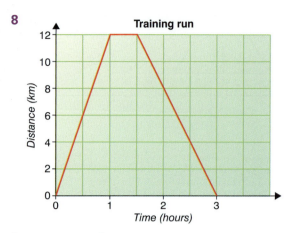

Training run

9 a Saturday **b** Thursday, Friday, Sunday
10 Tap A l_2, Tap B l_1

Chapter 10 page 264

1 a 26° **b** 146° **c** 102°
2 a $a = 78°$ (alternate angles on parallel lines are equal)
    $b = 78°$ (vertically opposite angles are equal)
   b $c = 82°$ (alternate angles on parallel lines are equal)
    $d = 64°$ (corresponding angles on parallel lines are equal)
   c $a = 45°$ (alternate angles on parallel lines are equal)
    $b = 135°$ (angles on a straight line add to 180°)
    $c = 45°$ (corresponding angles on parallel lines are equal)
3 a Supplementary angles   **b** Complementary angles
4 a $a + 61° + 36° = 180°$ (angles in a triangle add to 180°)
         $a + 97° = 180°$
           $a = 83°$
   b $b = 80° + 65°$ (exterior angles of triangle = sum of interior opposite angles)
    $b = 145°$
    $c = 180° - 145°$ (angles on a straight line add to 180°)
    $c = 35°$
   c $d = 88°$ (vertically opposite angles are equal)
    $e = 46°$ (base angle of isosceles triangle and angles in a triangle add to 180°)
   d $f = 50°$ (angles at a point add to 360°)
    $g = 65°$ (angles in a triangle add to 180°; base angles isosceles triangle)
   e $h = 70°$ (angles at a point add to 360°)
         $i + h + 32° = 180°$ (angles in a triangle add to 180°)
    $i + 70° + 32° = 180°$
        $i + 102° = 180°$
            $i = 78°$
           $j = h + i$ (exterior angle of a triangle = sum of interior opposite angles)
             $= 70° + 78°$
             $= 148°$
   f $k + 55° = 130°$ (vertically opposite angles are equal)
              $k = 75°$
               $l = 55°$ (alternate angles on parallel lines are equal)
        $k + l + m = 180°$ (angles in a triangle add to 180°)
   $75° + 55° + m = 180°$
       $130° + m = 180°$
            $m = 50°$
   g $n = 30°$ (angles on a straight line add to 180°)
        $q + n + 85° = 180°$ (angles in a triangle add to 180°)
   $q + 30° + 85° = 180°$
       $q + 115° = 180°$
            $q = 65°$
            $p = 30°$ (alternate angles on parallel lines are equal)

5 a $x + 102° + 90° + 73° = 360°$
$x + 265° = 360°$
$x = 95°$

b $x + 110° + 95° + 62° = 360°$
$x + 267° = 360°$
$x = 360° - 267°$
$= 93°$

c $60° + 60° + 35° + x + 85° + 60° = 360°$
$x + 300° = 360°$
$x = 60°$

6 $x + 80° + x + 20° = 180°$ (angles on a straight line add to 180°)
$2x + 100° = 180°$
$2x = 80°$
$x = 40°$

Chapter 11 page 291

1 An arrowhead (delta), kite, parallelogram, rectangle, triangle
2 a Isosceles triangle, kite, arrowhead, isosceles trapezium
   b Square
   c Square, rectangle, isosceles trapezium
3 a Rhombus   **b** Parallelogram or rectangle
4 a $m = 130°, n = 50°$ **b** $p = 120°, q = 60°$
5 a 2 **b** 2 **c** 4
6 a 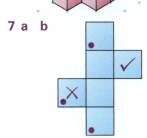   **b i** Plan view B   **ii** Front view C   **iii** Side view A

7 a b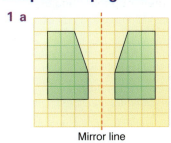

8 a plan view – F, front view – G, side view – B
   b plan view – D, front view – C, side view – H
   c plan view – A, front view – E, side view – E
9 e 32 mm
10 Pupil's accurate drawing

Chapter 12 page 318

1 a 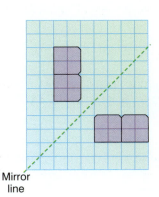   **b**

Test Yourself Answers

2 a A′ (⁻2, 3), B′ (5, 3), C′ (2, 0)
  b A′ (⁻3, ⁻2), B′ (⁻3, 1), C′ (0, ⁻2)
  c A′ (⁻2, ⁻3), B′ (1, ⁻3), C′ (⁻2, 0)
3 c B
4 a 1 line of symmetry, no rotation symmetry
  b 3 lines of symmetry, rotation symmetry of order 3
  c No lines of symmetry, rotation symmetry of order 2
5 a 3 **b** 2
6

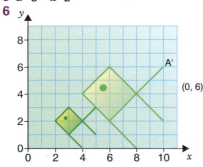 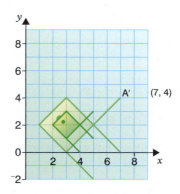

7 a 2; 30 m **b** 64 m
8 About 10·8 m.
9 On the drawing 12 m should be 6 cm, 2 m should be 1 cm, 6 m should be 3 cm, 3 m should be 1·5 cm, 1 m should be 0·5 cm.
10 a (5, 3) **b** $(3\frac{1}{2}, 13)$

Chapter 13 page 344

1 a 3·2 **b** 2·8 **c** 3·6 **d** 0·68 **e** 0·76 **f** 0·047 **g** 6·85 **h** 7000 **i** 60 **j** 0·058 **k** 10
  l 50 000 **m** 4·2 **n** 4·5
2 a 8 hours 20 minutes **b** 3 hours 48 minutes **c** 2·2 years **d** 105 years 2 months
  e 8 years 11 weeks **f** 2 days 2 hours
3 a 480 000 m² **b** 48 ha
4 1800 mℓ
5 250 kg
6 a 4 km, 2·4 ℓ, 2·7 kg, 90 g, 3 m, 10 cm
  b 10 miles, 8·8 lb, 2 yards or about 6 feet, 7 pints
7 a A, C and D
  b 32 cm
8 a 36 cm² **b** 1·25 m² **c** 80 cm²
9 a 28 cm² **b** 130 m² **c** 6·72 m²
10 a 24 cm **b** 53·2 m **c** 11 m
11 a 40 cm **b** 16 cm
12 a 78 cm² **b** 14 m²
13 a 27 cm³ **b** 162 cm³
14 a 54 cm² **b** 198 cm²
15 72

Chapter 14 page 367

1 a continuous **b** discrete **c** discrete **d** continuous
2 a

time (t seconds)	Tally	Frequency
20 < t ≤ 25	III	3
25 < t ≤ 30	ЖҺ	5
30 < t ≤ 35	ЖҺ IIII	9
35 < t ≤ 40	III	3

   b 5 **c** 17 **d** 12

3 a 6 **b** 10 **c** 7 **d** More females than males played the violin and piano. More males than females played the guitar.

4 Possible answers are:
  A b What factors affect how often people eat take-aways, hours of work, travel, how close the nearest take-away shop is, what choice of take-aways are available, ...
  c People eat take-aways, on average, once a week.
  d For each person, the data needed is how often, on average, he or she eats take-aways, how far away the nearest take-away is, how many hours they work, ...
  e How often, on average, do you eat take-aways?

  everyday ☐   every 2 or 3 days ☐   every 4 or 5 days ☐   once a week ☐
  once a fortnight ☐   once a month ☐   less than once a month ☐

  How far away is your local take-away?

  < 0·5 km ☐   0·5 up to 1 km ☐   1 up to 2 km ☐   2 up to 5 km ☐   > 5 km ☐

  How many hours do you work away from home each week?

  < 20 ☐   20–40 ☐   > 40 ☐

  Do you think the take-away choice is good?

  yes ☐   no ☐

  f A good sample size would be 20–50 people.
  g Primary
  B b What does 'happy with' mean? High frequency, comfort, price, number of stops, ...
  c Most people think the local bus service needs to be cheaper and run more frequently at the weekend.
  d Need data on what people think of frequency of service, time taken, comfort, number of bus stops, price, variation in service, ...
  e Do you think the local bus service
    i runs frequently enough in the

  Morning weekdays | YES / NO | , Afternoon weekdays | YES / NO | , Evening weekdays | YES / NO |

  Saturday | YES / NO | , Sunday | YES / NO |

    ii should be cheaper   YES / NO

    iii is comfortable   YES / NO

    iv needs more stops on its route   YES / NO
  f About 50 people

Chapter 15 page 398

1 a 3:00-
  b One possible answer is 'so that the organisers know when to have lots of timekeepers on'.
2 5·51 (2 d.p.)
3 12 cm
4 a median = 17, range = 14, mean = 15
  b median = 3·3, range = 4·6, mean = 3·7 (1 d.p.)
5 £3·02
6 a Wednesday median = 19·5, Friday median = 21
  b Overall median = 21   **c** No
7 a

Time spent on homework (min)
0 | 0 5 9
1 | 0 0 5 5 7 9 9
2 | 0 5 5 5 7
3 | 0 2

  b 32   **c** 0   **d** 32   **e** 17   **f** 19

8 The means and medians are the same. If you choose Toby's team you could say that their range was bigger so there would be a chance they could score highly. If you choose Gemma's team you could say their scores are more consistent so you could rely on them more than on Toby's team. Gemma's team has a higher mode so this could mean they score highly more often.

Test Yourself Answers

9 a
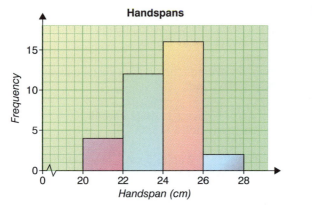

b 18

10
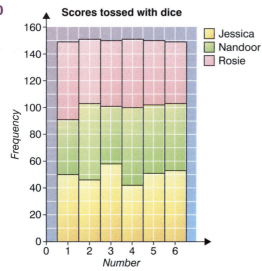

11 20- years 50°, 30- years 90°, 40- years 50°, 7, 50- years 140°

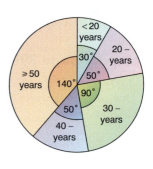

12 a About 21% **b** About 38% **c** About 48 000
 d Generally female teachers will tend to be younger than male teachers.
  There is a greater proportion of female teachers aged 20 to 29 than male teachers.
13 a i Overall the spending on paper did not change much.
 ii The amount spent on video and camera equipment has increased a bit.
 iii The amount spent on computing equipment has increased a lot.

Chapter 16 page 419

1 Board B
2 Bag B because this has the highest proportion of yellow marbles.
3

1st counter	green	green	green	red	red	red	blue	blue	blue
2nd counter	green	red	blue	green	red	blue	green	red	blue

 a $\frac{1}{9}$ **b** $\frac{1}{9}$ **c** $\frac{3}{9}$ or $\frac{1}{3}$ **d** $\frac{4}{9}$ **e** $\frac{4}{9}$
4 a 2, 5, 6, 7, 9 **b** $\frac{1}{5}$

5

```
0                    ½              1
├──┼──┼──┼──┼──┼──┼──┼──┼──┤
   ↑  ↑ ↑              ↑
   b  c                a
     pink
```

6 0·1

7 a

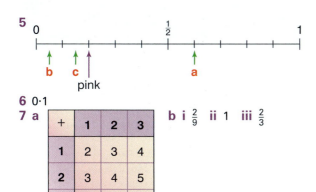

+	1	2	3
1	2	3	4
2	3	4	5
3	4	5	6

b i $\frac{2}{9}$ **ii** 1   **iii** $\frac{2}{3}$

8 a $\frac{1}{2}$ **b** $\frac{1}{3}$ **c** $\frac{1}{6}$ **d** $\frac{7}{12}$ **e** 0

9 a $\frac{5}{50}$ or $\frac{1}{10}$ **b** $\frac{8}{100}$ or $\frac{2}{25}$

c The probability estimated in **b** will be more accurate because the number of trials (times he dropped the drawing pin) is greater.

Index

Index